Science Images and Popular Images of the Sciences

Routledge Studies in Science, Technology and Society

Science Images and Popular Images of the Sciences

Edited by
Bernd Hüppauf and
Peter Weingart

Routledge
Taylor & Francis Group
New York   London

Most essays of this collection are revised papers presented at the conference *Images of the Sciences and Scientists in Visual Media* held at Deutsches Haus at New York University in November 2003. We gratefully acknowledge financial support for the publication by Deutscher Akademischer Austauschdienst (New York), the Institute of Science and Technology Studies (IWT), Bielefeld University, and the Ernst von Siemens Kunststiftung, Munich.

Routledge
Taylor & Francis Group
270 Madison Avenue
New York, NY 10016

Routledge
Taylor & Francis Group
2 Park Square
Milton Park, Abingdon
Oxon OX14 4RN

© 2008 by Bernd Hüppauf and Peter Weingart
Routledge is an imprint of Taylor & Francis Group, an Informa business

10 9 8 7 6 5 4 3 2 1

International Standard Book Number-13: 978-0-415-38381-3 (Hardcover)

Library of Congress Cataloging-in-Publication Data

Science images and popular images of the sciences / edited by Bernd Hüppauf and
  Peter Weingart.
     p. cm. -- (Routledge studies in science, technology and society ; 8)
  Includes bibliographical references and index.
  ISBN 978-0-415-38381-3 (hardback : alk. paper)
     1. Science--Study and teaching--Audio-visual aids. 2. Science fiction
films--History and criticism. 3. Motion pictures in science. 4. Visual communication.
I. Hüppauf, Bernd-Rüdiger. II. Weingart, Peter.

  Q190.S37 2007
  303.48'3--dc22   2007011628

Visit the Taylor & Francis Web site at
http://www.taylorandfrancis.com

and the Routledge Web site at
http://www.routledge.com

Contents

Figures and Tables

TABLES

Part I
Popularizing Science Images
Introduction

1 Images in and of Science

Bernd Hüppauf and Peter Weingart

TRANSFORMING SCIENCE IMAGES INTO POPULAR IMAGES OF SCIENCE

Images in science, or *science images*, as we prefer to call them, have been very influential in the history of the modern sciences. Yet, until recently, very little was known about them. The conditions of their emergence, genesis, and continued production, their fashions, impact on research, philosophy, and the general perception of nature had attracted surprisingly little interest (surveys of recent literature are Daum 1998; Kretschmann 2003; Oels 2005). This indifference was often associated with contempt for the visual in relation to abstract discourse. In recent years, this has changed. There is a new uncertainty about the relation between images and knowledge brought about by new theories of the sciences as well as a new type of images. The computer generated digital images have led to an intensive debate about the relationship between images and the sciences. The end of an epoch of illustrations in the sciences is undoubtedly approaching. Iconoclashes were always not only destructive but also linked to moves toward the construction of new images and techniques of representation (for a detailed and instructive illustration of the ambivalence of iconoclasm see Latour and Weibel 2002). More often than not, the tearing apart of images was implicated in a discarding of systems of belief and knowledge and was followed by enthroning new ideals, theories, and deities. The sciences were no exception and the current changes of *science images* seem closely related to a change of the *image of science*.

At the current turning point, the status of images is being radically redefined and the relationship between images and science has become an object of intensive research. Neglect has given way to a new assessment of the importance of the visual for the intrinsic processes of the production of knowledge and, as a result, of the position of the sciences in the public mind. Not only the history and sociology of science and science studies have taken an interest in these issues but art history and cultural studies have also discovered images in the sciences as an object of study. The proposition of a new discipline named *image theory* concerned with the "science of images" (Mitchell in this volume) was initiated, it can be assumed, by new developments in the production of science images that inspired attempts to make them comprehensible in terms of a general theory of the visible. The question of what we do when we *see* an image is being reconstituted as a result of an extension of the field of pictures by including the new electronically generated images. The pioneering work by Aby Warburg and art historians associated with his multi-disciplinary approach to images as icons had prepared a field of research that is only loosely connected with art history and all but neglected by the sciences. The extension of a theory of pictures as art makes it possible to attribute to images in the sciences a status of their own, independent of their artistic value, and develop an iconology directed at symbols by resemblance such as pictures, images, sketches, models, diagrams, graphs, and also linguistic images such as metaphors and allegories. Intensive research has made the common assumption untenable that images in the sciences are produced and communicated by and among scientists for ancillary purposes such as mere illustration or demonstrations for teaching. Visualization has been demonstrated to be an integral element of the reasoning in the sciences: while it is deeply involved in the abstract process of theory building, it cannot be understood in complete isolation from conventions of seeing developed over centuries in art history. This has implications for the image of the sciences that go far beyond the confines of the sciences.

Images produced in the sciences have never been limited to the closed communication circles of scientists but also appealed to a much wider group of recipients often for their aesthetic qualities. Little is known, however, about the impact of science images on the public's construction of general images of the sciences, attitudes toward science, and, indeed, the scientists' own concepts of the sciences. Essays of this collection support the hypothesis that a public image of science and scientists is based, among other aspects, on moving science images produced for a scientific community to a different context of reception. Our case studies try to substantiate the claim that science images, moved from a scientific context to the public sphere and distributed through various media, including popular and scientific periodicals, films, television, the visual arts, and the Internet can be interpreted as an important segment of designing public discourse on the sciences and ultimately a public image of science as such. The circulation of this transformation process has never been well understood and if the new science images are equally opaque and contingent as they are unavoidable and indispensable, we need to ask ourselves what consequences they will have on our concept of science and the science mediated relationship to reality. Turning a sentence by Mondzain upside-down one could say that traditional science illustrations laid claim to being images of the truth but that following the invention of digital images the truth no longer has/is an image (Mondzain 2005, 2006). This cannot be read but as a plea for developing a new image literacy capable of coping with a new situation brought about by the contingency of digital images. Essays of this volume are concerned with the composition and function of images from a variety of spheres of science of both their nineteenth century tradition and their definition en route to the unstable situation of the present.

This perspective requires not only the investigation of images as an instrument in scientific discourse but also recourse to a reconstitution of theories of knowledge production by including the material conditions of their production, such as working conditions in laboratories, the technology of instruments, processes of dissemination and reception by and for specialists and non-specialists alike. Recent theoretical approaches are based on the assumption that scientific research cannot be understood as a rational process of generating abstract theories only but is inevitably shaped by a variety of material factors. Empirical evidence is gathering that they also have deep effects on the negotiation of truth claims.[1] This collection of essays is an extension of these approaches by addressing the role of their contribution to the construction of public images of science. It is concerned with the changing relationships that link images produced in and for the sciences with the images of the sciences communicated in the broader public. This shift of perspective is suggested by the fact that images of science as an institution and scientific knowledge as its product reveal a surprising continuity over long periods of time and at the same time they are subjected to often abrupt and radical epistemic changes in the sciences. There is evidence for the assumption that the general image of science is based on archetypical stereotypes consistent with traditions that often reach back to pre-modern beliefs and, equally, on the rules and requirements of the modern popular media; this image must be understood as a surprisingly stable combination of persistent stereotypes and changing patterns. It is worth asking what impact changing science images have on the construction of popular images of the sciences. This is a particularly pertinent question in a period of the emergence of entirely new techniques of producing images. Conflicting descriptions are resulting and are a reflection of different conceptions of both science and images. Are the electronically generated images that only faintly resemble traditional pictures symptomatic of a change in the sciences?

Pictures are by their very nature a more accessible medium of communication than (specialists') language, more open to interpretation than the written word. Science images, therefore, can have a considerable impact on broader audiences and can be turned into powerful tools of persuasion. Hence, the intrinsic function of images in the science as a means of knowledge production needs to be complemented by looking at their function as media in public discourse. Their impact on a broader public has led Nikolow and Bluma to call for a combination of the history of visualization with the history of popularization of science; that is, an extension of the history of visualizations to include the history of public perceptions of the sciences. (Nikolow and Bluma 2002: 204). Thus, we suggest including another category, namely, images *of* science and scientists created outside science and communicated in public discourse. This shift of perspective is suggested by the fact that science as an institution and scientific knowledge as its product are embedded in society by way of a range of media. The genre of science images is one of these media that needs to be kept separate from the images that "society makes of science." The latter are often based on the former, as case studies of this volume demonstrate, yet they are created outside the framework of science. They explore and exploit the mirror images of science or scientists in the collective imagination. We therefore distinguish between two different kinds of images, based upon their respective functions and their target audiences:

- Science images produced *in* the sciences as visual elements in scientific research and processes and directed at the scientific community (e.g., technical illustrations that remain accessible to an expert community only);
- Science images produced *in* the sciences but directed at a broader public (e.g., colored images of the ozone hole prepared for wide distribution); and
- Images *of* science produced by scientists (e.g., pictures of laboratories or representative research instruments);
- Images *of* science produced and communicated by the public media (e.g., representations of famous scientists in movies, literature, art and comics).[2]

It is possible to a limited degree to identify and delineate these different types of images by their origins. But it is impossible to gauge their reception because it is not limited to the group of intended recipients.

The communication of these images is reciprocal. Images produced in the sciences for the expert community in their large majority reach this limited audience only. Yet some find their way into newspapers, art magazines or TV shows. This may be due to their news value or their aesthetic appeal. Images produced by scientists but designed to reach a non-expert public will be based upon the scientist's assessment of the public's knowledge of and interest in the sciences.[3] Whether they will have the intended effect is far from certain. Likewise, the impact of images of science developed by novelists, cartoonists, or script writers for books or films is even more uncertain. Can it be surmized that they reflect ingrained, common stereotypes about science? In what way do they also need to be surprising or shocking to gain attention from an audience of growing surfeit? These images find their way back to scientific discourse proper because it is not isolated from popular images; for example, as a perceived image of scientists as public heroes or, alternatively, a bad image of science that needs to be improved. In short, science images disseminated widely in various media open a view on the communicative interchanges between scientific communities and the public. They are also indicative of power relations that dominate these interchanges. Thus, the unknown effects that these different types of images have within and outside the group of intended recipients make them a significant object of investigation concerned with the construction of images of science believed in and supported by scientists and non-scientists. At a time when images are gaining increasing importance as a medium of communication and shifting power relations between the community of experts and the public, this perspective promises to be more revealing than analyses of verbal communication.

SCIENCE IMAGES: TRUE TO ARTISTIC CONVENTIONS OR TRUE TO NATURE, AND THE NINETEENTH CENTURY DEFINITION OF OBJECTIVITY

While images in the sciences are distinct in many respects, their theory needs to be extended and included in a general category of images. This view contradicts theories of the past. Science images were not considered part of the history of art and therefore interpreted through their correspondence to the ideal of *objectivity* in the sciences that, in spite of changing definitions of objectivity, has dominated modern society and its sciences. It has, from Leonardo, Newton, and Descartes on, undergone fundamental changes that were implicated in redefinitions of truth and its image.

From early on, the properties of the science images were in correspondence with both the self-definition of science and the images projected to non-scientists reflected in the arts and literature of the time. Joan Blaeu's *Atlas Major* of 1665, to take one example, was a work of science true to the nature represented in its pictures and at the same time it could be perceived as a distinct work of art and aesthetic imagination. The combination of these two distinct but complementary definitions of scientific pictures was symptomatic of the image that the sciences projected of themselves to the small group of those educated to have an image of the sciences at all. Albertus Seba's *Thesaurus* combined naturalistic images and pictures of fantastic animals often based on stories from the new world. They illustrate the liberties of scientific illustrators to make use of the imagination as long as it could be justified by popular narrative.[4]

In the course of changing epistemic practices, definitions of scientific objectivity and expectations directed at science, a new ideal of images produced in the sciences and for scientific purposes emerged. It was based on a separation of aesthetic values from the information value. The science image itself, as well as its function in the production of scientific knowledge and in public presentations of the sciences, were increasingly based on a clear definition of the sciences on the one side and pictures as a supporting tool on the other. Scientific knowledge was understood "to be encapsulated in theories which…are interpreted by logical empiricists as axiomatic systems.... Thinking—and not visualizing—is held to be conducive to this activity (the sciences)" (Baigrie 1996: xvii). It was taken for granted that the sciences followed their specific rules of abstract reasoning, whereas images were attributed to the ancillary function of applying artistic means in order to translate the sciences in a code comprehensible to many (Topper 1996). How was this *translation* done? The eye of the science illustrator, an experienced illustrator writes, "captures" his object, observes it with great care, and then attempts to reproduce it on a two-dimensional sheet as accurately as possible (Papp 1968: 1; compare Holländer's well informed introduction to his anthology: Holländer 2000: 9–15). The statement emphasizes that science images are nothing but pictures, which serve the purpose of illustrating knowledge gained elsewhere and with completely independent techniques. It is significant that this expert identifies images produced for scientific texts with illustrations of objects constructed through scientific methods. There is no doubt as to the nature of the object illustrated. It is nature defined by the rules of the sciences and prepared through scientific research, primarily the presumed objective logic of the experiment. This was common practice before the introduction of photography as a means for illustrating science texts and continued well into the twentieth century.[5]

While images had been part of scientific practices contributing to a culture of the visual on an equal basis since the beginning of methodical modern research, they were subjected to fundamental change in the nineteenth century. Science images were now interpreted as illustrations in the sense of helpful but dispensable additions to texts and as visual representations of objects of the sciences with no noticeable impact on the processes of research and knowledge production. Questions of aesthetic conventions were widely ignored.

Steel and copper plate engravings and lithographs were the common techniques for illustrations in science and medical textbooks. They made it possible to apply the principle of reduction through a selective representation of the object. A specific quality of traditional science images was their reduction of complex surface structures through techniques of typifying and abstracting. This reduction endowed the science image not only with a desirable degree of substance but associated it also with the ideal of *simplicity*, a central concept of modern aesthetics.[6] A focus on specific details and the exclusion of others and an emphasis on what was considered important produced pictures of terse precision which, at the same time, unintentionally corresponded to an aesthetic ideal.

The invention of the camera initiated a new period of producing science images but did not change the ideal of purity as the basis of the illusion of objectivity. This new imaging technique was functional within a system whose objectivity standards governed scientific inquiry as well as the production of its images. The camera was considered the ideal apparatus for an objective representation of nature interpreted as a collection of objects of scientific investigation. Cameras were soon attached to microscopes and telescopes. In 1839 Daguerre took the first microphotograph of the glands of a spider and in 1854 sunspots were photographed. Photographs served as "vital support," for example, in Robert Koch's, Louis Pasteur's, and other medical researchers' discovery of micro organisms. Their work was heavily based on the belief in the objectivity of photographs as images embedded in the process of scientific investigation. Koch was convinced of the superiority of photography compared with all other techniques of visualizing and considered the photographic process incorruptible. He wrote of the truthfulness of the black and white image that was demonstrated by the genuine chemo-technical process through which the microscopic object represented itself. In 1877 he published the first photographs of bacteria shot with a camera adapted for his specific purposes (Bredekamp and Brons 2004).

In their much quoted essay, Daston and Galison elaborate on a change of the definition of objectivity in the sciences and argue that the nineteenth century turned the earlier ideal of truthfulness to nature to an ideal of objectivity defined by methods of scientific research (Daston and Galison 1992). They call this new ideal, generated within the system of methodical research, *mechanical objectivity*. The term needs qualification. The discontinuity in the definition of objectivity notwithstanding, science photos are indicative of continuities in the concept of science images. Photography seemed to correspond in a perfect manner with this ideal producing a visual environment that created the iconic connotations for making photographic images the ideal of representation for many scientists. It could be interpreted as a technique between the two poles of arts and science and so far as the sciences were concerned, it was believed to be in their *service*. The pioneering visual archive by Hermann Krone listed photographs in the service of archaeology, geology, astronomy, art history, and medicine (Hesse 1998: 152–161, 291). The "pencil of nature" as Talbot's famous title called it, seemed to be the ideal instrument for *serving* the requirements of the sciences. The faculty of the photographic image of assisting the production of scientific knowledge and spreading it fascinated the century of experiments and belief in objective evidence.

> The fact that photography represents every object without interference of subjective views (*Anschauungen*) such as the idealizing pen of an illustrator had to be finally acknowledged and at present all medical disciplines enjoy the vital support of photography...which has become the indisputable and impartial witness of every scientific inquiry made in solitude....(Jankau 1883: 1).[7]

After technical problems of their reproduction were resolved around 1880, photographs in the service of the sciences became essential, providing the basis for the popular image of the sciences in the emerging mass society.

This view of the function of images in the sciences rests on the assumptions, first, that images are documents producing evidence of facts or data produced, in turn, by scientific theory (i.e., that they have a dependant status), and, second, that aesthetic qualities play no role in scientific arguments (i.e., that they have no epistemic function). The objectivity of the science image was identified with the alleged objectivity of the sciences themselves. They produced popular evidence for the image of the sciences through a true, purely technical presentation of their objects. The image of objective science on the macro-level was an extension of the objectivity with which each individual science image was credited. Yet the process of image making was kept separate from the process of producing knowledge. This separation represented remaining skepticism in relation to images based on distrust in the visual and sense perception.

It was a corollary of this ideal that it hides the fact that it was a construction dependent on a range of subjective factors. Science images in the age of photography, it has been demonstrated, were pictures characterized by a carefully planned absence of any trace of their producers and production techniques. Knowledge of the indispensable a priori of the body in the process of observation and visual representation had to be suppressed in favor of an ideal of objective images supposedly free from any trace of subjectivity. The techniques applied in the making of the images were made invisible in the picture. The scientist, his illustrator, and their respective techniques had to disappear completely from the image's construction. The hand of the artist, the artist's aesthetic preferences, and any hint of production techniques remained outside the frame. Through these imaging techniques the individual picture provided visual support for the dominant theory of the objectivity of the sciences. Furthermore, it can be argued that the concept of scientific objectivity depended on this pictorial abstinence.

Pictures of machines or laboratory experiments published in nineteenth-century encyclopedias, textbooks, and journals were realizations of this theory and accordingly they taught ways of seeing these objects of science and technology. Involuntarily they also produced an aesthetic that shaped the image of science and technology in general. Highly stylized images of objects of science and technology designed for public consumption replaced the imagination of earlier science images by conveying cultural ideals such as progress and superiority. To achieve this objective, a fusion of images and text was necessary. The new definition of objectivity notwithstanding, aesthetic values such as symmetry, simplicity, pictorial arrangements, and a general sense of beauty continued to shape science images produced in these modern systems of analogue representation.

The appearance of photography in the sciences was interpreted as support for the objectivity claims. The identification of photography with objectivity made it possible to substitute the photographic image for the reality of nature perceived through the scientist's eyes. As the photographic process itself was considered free of any impact of subjectivity, even traces of the technical process on the surface of the image were not perceived as a distortion of the representation of nature. In contrast to the need for purification in other types of images, they needed no correction. Rather they were read as the visual guarantee of objectivity resulting from the absence of any human interference with the process of representation they could be retained on a picture and interpreted as visual evidence for its truthfulness.

Science photography of the late nineteenth century was the first step on the way of science images to be *seen and believed* by many non-experts and to contribute to a popular image of modern science. Photographing the newly discovered bacteria and later x-ray photography played a significant role in establishing an authoritative position of the photographic image both in medicine and the public (Schlich 1997; Schickore 2002). The impressive progress in hygiene contributed to making these images known outside medical research and associating them with an indisputable authority of the sciences as a system. Each photograph, it can be argued, could be read as an illustration of the triumph of the modern sciences.

The advance of photography in the sciences brought about changes that were more complex than a mechanical interpretation of the photographic picture suggests. Inasmuch as photographic precision followed the ideal of objectivity—a constant characterization of photography in the nineteenth century—the photographic image also served a different ideal of objectivity for which the correspondence with nature was of lesser importance. Resulting from the specific conditions of its technology that represented every detail, important or not, with equal attention and indiscriminating clarity or fuzziness, a new type of science image came into being.[8] More or less important objects as well as object and background were difficult to distinguish. These pictures lacked the precision of the older illustrations. Yet, they introduced a new quality to science images through their ability to make visible what was hidden to the eye. For a while the camera was called the *Daguerre-machine* and its pictures, science images in particular, were understood as products of a production process based upon scientific theory. They could even be considered *more true* than the represented object as long as their details could be justified by theory. This theory based objectivity was important to Helmholtz, Robert Koch, and many contemporary experimental scientists. It differed fundamentally from the ideal of a scientific image's truthfulness to nature. Photography was primarily not an illustration, Koch wrote, but a piece of evidence. It required a different gaze and a specific sort of knowledge about the nature of the technical image. In other words, it had no given referent but required a theory in order to become comprehensible. When in 1881 Koch published a book on micro organisms with numerous photos, he explicitly argued that the photographic image could well be more important than its living object.

For different reasons the trust in science images as evidence of objectivity is vanishing in the current debate on images. We can no longer be certain what the object of scientific imaging is, nor do we know what an image is. Visualization through scientific images is no longer considered free of implications in the material conditions of the production of knowledge. The observation that the new science images make a constitutive contribution to the very generation of knowledge can be generalized. They are changing the way we perceive images in general, also in retrospect. With hindsight we know that this has never been different and that it was a misconception to assume that the process of producing knowledge could be separated from the visual. For us, science images of the nineteenth century are different from these images at the time when they were produced. We approach them with a different set of expectations. What they now make visible may not have possibly been visible previously under different regimes of seeing.

The discovery of new aspects previously unnoticed in familiar images is not a dubious misreading but an implication of the epistemology of seeing images displaced in time or space. The presence of hidden observers, unacknowledged interests, changing conceptions of nature, and various contingencies have been discovered as inescapable qualities of science images precisely of those initially considered the ultimate in objectivity. This understanding of science images is reflected back into its origins in nineteenth-century photography. Among numerous examples, the project of Salpêtrière photos has been critically assessed and Georges Didi-Huberman's criticism is particularly revealing. Based on his critical reading of photographs from the archives, he has demonstrated the illusory character of objectivity claims (Didi-Huberman 1997).[9] Similarly, Hentschel shows in his painstaking analysis of the use of photographs in the documentation of nineteenth-century spectral research the problematic assumption of "mechanical objectivity" in the light of the efforts of practitioners in creating pictures for printing (Hentschel 2005). Closely linked to these changes in the perception of the relation between images and the sciences is a fundamental change of the public image of science.

NEW SCIENCE IMAGES: THE VISUAL AS A GENERAL CATEGORY AND A CHANGING RELATIONSHIP BETWEEN IMAGE AND KNOWLEDGE

The sciences have always employed the visual and images have always been more than pictures illustrating a scientific referent outside and independent of them. There is, however, decisive change. Traditional theories of images are being challenged. Claims that science images are mirrors of facts or produce indisputable evidence are no longer theoretically supported. Images are now perceived as constructions with a considerable degree of contingency. In the late twentieth century, a growing awareness emerged that science images cannot be understood in terms of a relationship to a given reality but must also be read as implicated in science models and theories, and furthermore, that they are not independent of, but fundamentally involved in, the production of knowledge. New techniques of image production demonstrate that science images contribute to a reality of the visual that would not exist without them. The image is, in other words, itself considered an integral element of a scientific process. The disputed status of the image is in a current debate intertwined with a redefinition of knowledge as we have known it from the beginning of the systematic experimental sciences in nineteenth-century laboratories.

This knowledge was reflected in imaging techniques such as radio photography in astronomy, x-ray images and ultrasound pictures that make visible what is imperceptible to the human senses. There is a basic difference between these images and the new computer generated images. The latter are not the product of a supportive technology enhancing the limited human senses. Instead, they create a visual reality without representation that refers to a theoretically supported potentiality. The new images follow a paradoxical logic: the more artificial and complex the apparatus and techniques of imaging are, the more natural their products appear. As a consequence, neither an original of which the image is a picture nor a definite final image exist (for a discussion of the complex relation between original and copy see Fehrmann et al. 2004). These images can be manipulated indefinitely and the point at which a *tentative definite* version is reached is an arbitrary decision. This absence of an original makes it difficult if not impossible to distinguish between corrections and manipulations of an image and its correct or incorrect interpretation. The resulting relativity of criteria for correctness and truth claims has effects in areas where models and simulations are common research tools, as, for example, in theoretical physics or climate research and increasingly also in medical diagnostics.

Visual data produced by imaging machines do not speak for themselves. They are "media without a message but with the potential to become a message," requiring additional input in terms of suppositions regarding the nature of the object; for example, the structure of a tissue and its probable reaction to the magnetic field of the recording process (Schinzel 2006). These images give new importance to the old question of the threshold of the emergence of an image.[10] These products of the most sophisticated technology are *archaic* in the sense that they exist beyond definite shapes, a contourless fog prior to the emergence of separating lines. In order for a picture to emerge, the addition of arithmetic models and algorithms based on medical, statistical, and procedural theories is necessary. The resulting pictures appear surprisingly naturalistic and this impression leads to an underestimation of the function of technical procedures, physiological and mathematical models, and the iconic conventions, which may produce a picture of realist resemblance but, under adverse circumstances, an imprecize or plain false image.

What do images do in the complex process of producing knowledge? Can there be a general theory of the science image? Is modern science conceivable without images? Is it justified to refer to the new techniques of visualization with the word image? While the conventional answer to these questions may have been positive, we now tend to think that the opposite is correct. Terminology is indicative and the semantic field is being re-arranged. Equally, changes referred to as "pictorial turn" are indicative of a new insecurity in relation to definitions of the image-reality relationship.[11] Theories of the sign in the wake of Saussure, Wittgenstein, or Pierce have eroded the theoretical foundation of correspondence theories of truth in images. There is little certainty left regarding the relationship between signs, images, and objects or phenomena (Boehm 1994).[12]

The term *illustration* is no longer frequent and is replaced with *visualization*, which does not refer to a technique of understanding abstract sentences or mathematical equations through illustration. Visualization is not a means for producing images *in the service* of the sciences. Rather, the term refers to a complex operation of creating fields of visual perception engaged in "exploring data and information graphically..." (compare, for example, Earnshaw and Wiseman 1992: 5). This exploration is done like an explorer's voyage. Little is known about its destination and a large portion of the way is concealed like a secret passage.

Whereas illustrating pictures and graphics are concerned with "the communication of information and results that are already understood," the concept of scientific visualization refers to images that make visible what would remain unknown without them. They interfere with processes of numerical simulation and imply "seeking to understand the data" incomprehensible without them (Earnshaw and Wiseman 1992: 5). The new computer generated images are arbitrary in a way comparable to the arbitrariness of linguistic signs. They visualize theories and models of the invisible.[13]

The new technologies for the simulation of visual structures that only superficially resemble conventional images are contributing to a deterioration of common definitions of evidence and objectivity in the sciences and, by implication, to a changing image of the sciences in general. While computer generated images sever the correspondence with the visible reality in the observed world, the supposed autonomy of the sciences and their objectivity claim have been challenged by heterodox theories. Theories uncovering the involvement of images in the scientific process have played a not insignificant role in this growing skepticism regarding the objectivity of modern science. A strong connection between the proliferation of science images, which no longer retain the objectivity claims of Robert Koch and his period, and the public image of science can be observed. Intertwined with the loss of certainty in relation to science images is a growing instability of the sciences as a system and an ambivalence regarding the image of science. Thus, images are a major factor in the crisis of knowledge production and contribute to changing the image of the sciences.

An uneasy co-existence of two popular and mutually exclusive images of the sciences can be observed, one of admiration for their stunning progress and overwhelming achievements and a simultaneous image of fundamental uncertainty and distrust. Both are supported by an abundance of science images published in the print and electronic media. These popular images are the focus of public debates regarding ethical issues and may lead to jeopardizing credibility and public support of the sciences (see Lenhard et al. 2007).

READING SCIENCE IMAGES OUT OF CONTEXT: THE FUNCTION OF THEIR HIDDEN AESTHETIC DIMENSION

Science images are made with a specific intention and for a specific audience. It seems that it is the intention of both producing and reading them that, more than anything else, sets them apart from other images (Rheinberger 2001). They are published in specific places designed for a specific and small market of experts in scientific disciplines and perceived within the framework of these media. When science images cross the dividing lines that separate distinct fields of seeing and are moved (e.g., from narrowly defined textbooks to the public media), they become different images. What happens in the zone of transference between these spheres?[14] What are the dynamics when science images are not read as images made for intrinsic purposes but as material for the creation of popular images?

Science images share with other types of images qualities that can be identified as aesthetic. It is an indication of their changed status in the scientific process that historians of science and of art have begun to search for a common language capable of accounting for both discursive content and the aesthetic value.[15] Computer generated images of fractals have initiated a discussion on the relationship between science and art. Titles such as *Fractals and Art for the Sake of Science* (Mandelbrot 1989) or *Fractal Expressionism—Where Art Meets Science* (Taylor 2002) reflect this. Likewise the images produced by astronomers with the help of the Hubble telescope and computers to add arbitrary choices of color have achieved the status of popular pictures that are viewed and downloaded from the Internet for aesthetic rather than scientific purposes.[16]

Cultural studies have placed emphasis on the "polysemy of the text" (Fiske 1987: 16) and the composition of images, particularly new digital science images, is similarly polyvalent. This polyvalence has been made strikingly obvious in recent years, largely as a result of computer generated images with an open range of meanings. White noise produced by the interaction of object and image machine offers an unlimited potential for processing the underdetermined data it is made up of. Producing an image through medical imaging systems is the active process of creating clusters of such machine generated data and composing them to a preliminary image with the potential to reveal information. This information must not

be understood as representing a reality outside the range of machine generated data. It has been observed that the new computer generated images
create an effect complementary to their origin in technology and in order
to become comprehensible, they require a considerable involvement of the
imagination (Weigel 2004).[17] Computer generated digital images displaying
abstract shapes and colors randomly attributed to them, are examples of
potential arrangements of data towards an image that is not final but retains
the potential for multiple combinations. They speak to cognitive faculties
as much as to the imagination by offering the possibility of aesthetic readings. These images are accessible not only to skilled medical experts but
also to non-experts by opening a space for the non-scientific imagination.
Computer generated images more than traditional images of representation
blur the dividing line between imitation and simulation, between arts and
science, between the real and the imaginary, and it is this *blur* that can be
considered symptomatic of science images.

It can be argued that these highly technical images correspond to the
aesthetic imagination to an extent that no previous science image ever did.
In spite of their dependence on theoretical models and algorithms, the new
science images are characterized by a high degree of indeterminacy and
potentiality. Their indeterminacy, which appeals to the scientific imagination, also suggests a hidden secret that opens a space for playful combinations not dissimilar to pictures in the history of the artistic avant-garde.
This ambivalence may well be the origin of their popularity.

Images used in a scientific treatise visualizing the structure of an atom
or the neuronal connections in the brain can straddle scientific and public
discourses and be reproduced in an art book, a newspaper, or any other
medium for public consumption. The same image published in different
contexts and perceived by readers with an expectation different from the
one it was initially made for will gain different and often unpredictable
qualities. Products of complex and highly abstract processes of visualizing
computer technology have found their way into popular media and adver-
tizements, undoubtedly for their particular aesthetic appeal and the unintended enigma of their iconic message (Borck 2006). Furthermore, highly
sophisticated image machines and their technologies have been transferred
into the realm of art and fused with aesthetic practices. For over twenty
years now, installations and artistic performances have incorporated
adapted apparatus and procedures from advanced scientific experiments.
Not much is known about this collaboration of the arts and the sciences.[18]
By appealing to the aesthetic imagination, these post-modern science images
contribute to the perception that a large number of non-experts have of the
respective scientific disciplines as well as of science in general.

Once moved to contexts of meaning outside the field of its intended
significance, the science image gains new meaning. A new image emerges
when it enters into a different sphere of reception. Once connections are
made between a scientific event and the looking upon the science image as

an artistic composition, the image is not only meaningful as the construction of factual *evidence* but derives sense (Sinn) from the interplay between imagery and different subsystems of society.[19] Transferred to an artistic or a teaching environment, it makes sense in terms of aesthetic associations with other works of art or aesthetic practices or of an instrument for didactics.[20] In aesthetic contexts the science image is released from the false impression of being un-constructed. On the contrary, it then performs in contexts that do not create the illusion of images as a duplication of nature. We know of productive misreadings of the coded information contained in images in terms of scientific innovation and, more prominently, of artistic creativity stimulated by science images.[21] Looked at out of their intended context, the new science images are capable of addressing and stimulating the aesthetic mind, setting in motion the imagination and creating aesthetic pleasure (for a general discussion of this issue see the still relevant essay by Stuart Hall 1980: 128–138; Krohn 2006). Heterogeneity, indicative of producing and reading images, opens science images to a variety of readings and calculated as well as involuntary misreadings, resulting in aesthetic innovations.

It is safe to suggest that the line separating the arts and the sciences is currently shifting. The arts are being reconstituted with an eye on the sciences; at the same time, the sciences are undergoing fundamental change by reassessing the visual for their episteme and are moving closer to artistic practices. Contemporary art, it can be summarized, is an important medium for popularizing the sciences and, by implication, contributes to their changing popular image. Changes of the science images are connected to a changing art world and also to a change in the popular image of the sciences.

POPULARIZATION OF SCIENCE IMAGES

"Science is the most public and the most private of activities" (Galison et al. 2005: 332). In contradistinction to secrecy, hermetic traditions, and divination of a very long past, modern science laid claim to producing true knowledge accessible to every one. This picture was never entirely true; the spatial seclusion of laboratories, competitive or proprietary secrecy, and a language barrier kept the large majority of the population at a considerable distance from science and research. While the sciences benefited from the image of an inaccessible holy grail of the modern period, this was also an impediment. A major strategy to overcome its isolation can be called popularization. Apart from public lectures, a central element has always been the transfer of science images to the sphere of public spectatorship.

Among several nineteenth-century projects, the controversy surrounding Charles Darwin's theory of evolution was the most spectacular example of popularization. Supporters as well as opponents made extensive use of images as visual evidence for their respective positions that led to bitter public controversies. Darwin's student Ernst Haeckel, a professor of anatomy at Jena University published a book in support of evolutionary theory (Haeckel 1868, 1874).[22] It went through many print runs and became the most popular catechism of evolutionary theory. The ensuing controversy was based on a series of plates depicting stages of the embryonic development from simple forms of life to the human fetus. More than the book's representation of Darwinian theory, the images guaranteed a wide distribution and gave rise to one of the first public battles on scientific issues involving mass media. This war of words included accusations of fraud based on the book's illustrations. The plates applied the technique of eliminating details and simplifying complex forms customarily used in science images but now interpreted by opponents as forgery and pseudo-science. This early example of popularization through images was symptomatic in many respects and demonstrated that popularization is not a one-way avenue but has repercussions for the definition of the sciences and sciences' image of themselves. (A fascinating example of science fiction's impact on the sciences is Gugerli 2002.) What is accepted as true knowledge ultimately depends not exclusively on truth claims negotiated among experts but requires public mediation. Science images in conjunction with a popular, generalized image of science have made significant contributions to this complex and opaque process of mediation.

The term *popularization* has been central to defining the program of cultural studies and in recent debates the complex process of popularization has attracted renewed attention in science studies. It was observed that common theories of popular culture were often based on the assumption of a dependency of the popular on high culture and consequently led to pejorative value judgments (compare Ruchatz 2005). Until the late twentieth century, the popular was customarily defined through voids and deficiencies. In relation to the authentic and original it was believed to be a derivative that could be explained by reference to its lack of scientific precision. Popularization was conceived as a category of negativity and identified with creating codes on low cognitive levels. Its social place was held in low esteem as that of the uneducated and underprivileged, a deplorable result of mass culture. In more recent approaches this negative value judgment has been reversed. and a changed perception of the popular has placed emphasis on popularization as a means for "democratizing information" (Deilmann 2004: 90). The popular is then not reduced to a position of deficiency and subjectivity but conceived as a category of *difference* (Stäheli 2005). A theory of functional differentiation can account for the popularization of science images as an element of cultural practice that privileges neither the rationality of the sciences nor a prerogative of the popular.

The complexity of modern science creates a problem for democratic societies. If democracy is to be practiced in the age of science and technology, complex matters of science and technology need to be simplified and popularization of science images can be considered a means to this end. It is also a means for selection. From the late nineteenth century on, the production of knowledge has grown exponentially. Popularization is a way of making decisions about what is considered worthy of attention outside the circles of experts. It is a selection process that serves the maintenance of a collective memory and contributes to making decisions for future developments. In societies increasingly characterized by asymmetries in the distribution of knowledge, popularization can be considered a precondition for social integration through creating an order of knowledge by assessing and selecting information.

In analogy to linguistic transfers between scientific and non-scientific contexts popular science images have been explained as an operation of *translation* between codes of distinct spheres of reception (Pörksen 1986).[23] However, translation presupposes an original text to which a transfer from one idiom in another idiom must be truthful. The popular science image does not fit into this hierarchical dependency. Popularization, it has been argued, should be understood as a replacement rather than a translation (Link 2005). Semantic constructions used in chemical laboratories of the nineteenth century, Jürgen Link argues, became part of public language. As a result, inherited semantics were substituted for the language of the sciences. In the late nineteenth century, private as well as public language couched in the code of chemistry became a common system of symbols.

Similar replacements can be observed on the market of images. Popularizing science images can be understood as a cultural technique devised for *formatting* knowledge in such a way that it can move between different subsystems. It creates a mechanism for distinct spheres of modern societies to enter into communication and create non-hierarchical inter-discursivity. It is a universalizing mode of social communication aimed at the inclusion of non-experts by overcoming exclusions resulting from knowledge hierarchies. The transformation of specialized science images to a general image of the sciences is part of this social communication. The direction of the information flow from the closed circles of scientists to the open space of public discourse, including engineers and amateur inventors, can be reversed. It is not surprising that scientists are not immune to the impact of popular images of science. In recent years, the traditional media for popularization have been extended by a new strategy that is particularly efficient in fields such as computerization, artificial intelligence, or applied mathematics, called *demo*. They make results of scientific and technological research publicly accessible to experts and a broader group of interested non-experts. These demonstrations of new technology and programs are explicitly targeting experts who find it difficult to keep pace with the rapid innovation in their fields. They make an attempt to find a solution for the problem of communication between experts and the public as well as among experts. They are the latest variety of popularization and are binding the "making and the marketing of science and technology" (Rosenthal 2005: 348).[24]

The popularization of science images can be a powerful factor in decision making processes both outside and within the sciences and technology. As the construction of science images is intermeshed with the creation of the popular image of the sciences, the definition of criteria for financial and mental support for scientific projects is also a result of their reception. On the level of conscious planning, scientific communities are known to involve the public in their deliberations when it suits their purposes, be it to settle internal disputes, or to secure support for their research (Hilgartner 1990; Bucchi 1996). These exchanges are not limited to linguistic discourse but extend to the visual media. Furthermore, it can be assumed that they have a much more subtle impact on pre-conscious attitudes towards the sciences. Transfer of information among scientists and between scientists and non-scientists is of utmost importance and images are deeply involved in this process of communication that has no discernable agents. Science fiction films such as *Star Wars*, or *The Fly*, which is analyzed by Bruce Clarke in this volume, have not only shaped public images of the sciences but have also stimulated the imagination of scientists and engineers, leading them to design machines and even entire research programs. It is for these reasons that the transformation of images *in* the sciences to images *of* the sciences needs to be explored from a variety of perspectives including epistemology, science policy and ethics.

So far as image transfer or replacement of science images—including such elements as charts, graphs, diagrams, or digital simulations—are concerned, we want to understand and describe the rules that govern the process. Yet, this replacement also requires a critical perspective and needs to be interpreted as a consequence of power relations. The creation and dissemination of popular science images are inevitably involved in political decision making, marketing, and a partisan manipulation of public opinion. It must be seen as a cultural practice of industrialized societies in which the general public has influence that used to be the privilege of small elites. In modern mass democracies popular images assume an importance they have never had before. This burdens the popularization of images with problems of appropriate information and criteria for defining competence.

POPULAR IMAGES OF SCIENCE

Shifting the perspective to images that are not produced in science, but by film-makers, PR agents, and authors of fiction, one gets an impression of the popular images of science. Representations of science in the media, it can be assumed, are reflections of popular views and prejudices of science. In what ways does this contribute to the "reality" of images, to the construction of a mental image of science and the scientist in public media?[25] There is a covert coherence in what past periods were and what our present is attracted by. Science images never appear in isolation and each cluster activates memories of previous, similar images, which together form a generalized image of science, often compressed to a stereotype not dissimilar to an archetypical image. It connects most of us by shaping an imaginary space that differs from those of politics or the arts. There is a need to explore the logic by which this collective imaginary space of science is formed and its virtual topography.

Most of the media products that convey images of science are produced by commercial enterprises, which implies that they compete for public attention. Few specialize and try to find niches. Common to all of them is that knowledge of their respective "publics" is superficial and, as empirical studies have demonstrated, the result of stable, preconceived views. Producers of films, TV series, or print media have only a faint knowledge of their audiences. They cater to a public of their own imagination. Movie scripts may appear original but they are original on a superficial level only, because they are made to appeal to the audience by reflecting the audience's beliefs, expectations, and anxieties. Surprising settings, characters, and plots are a layer underneath which stable and predictable patterns are hidden. Scientists who aim to "improve" the image of science in popular media often ask the question of how to make it more "realistic." The question itself attests to the fact that self-image and popular image do not coincide. It also reflects a lack of understanding on the part of scientists of both the media's independence and the nature and origin of popular images of science. There is no political or scientific authority for surveying and controlling popular images of science. Scientists have little influence on the selection and design of images of science. They are observers of ill-understood changes of these imaginary spaces with no power to interfere. Popularization of science may create a mythology for the digital age that resists any control of image production and places the abstract, technical image at its core.

While the popular image of science is to a large degree a continuation of pre-modern and mythical images, it is not immune to the impact of modern science images. The question must be raised how a combination of popular culture, rooted in irrational images, and advanced science images is constructed? Has the "reality" of science *contained* or *misconstrued* in science images any resemblance with its popular images produced by the media? If the image of science is the product of a complex combination of pre-modern mythology and recent science images, is it possible to uncover a logic of this fusion?

CONTRIBUTIONS TO THIS BOOK

The essay by Nikolow and Bluma provides the historical background for our volume as far as research traditions are concerned. They present the different, still separate strands of discussion, about visualization and the role of images *in* science, on the one hand, and about popularization and the role of publics and their images *of* science, on the other. Their plea is that both strands should be merged under the impression of technologies of image production and the erosion of boundaries between expert communities and the general public, between the scientifically "objective" and the "popular."

Mitchell's essay raises the fundamental question as to whether a "science of images" is conceivable. Images are a matter of interpretation, social construction, and performance and, as images in the sciences, they are customarily understood as tools or media in the service of their logical operations. Referring to iconology as an attempt to liberate the image from the categories of art history, Mitchell's essay makes a new and encompassing attempt in theory design, pursuing the project on the two levels of its position within the system of academic disciplines and its constitution. It would have to be a cognitive science, based on testing and experimentation, including the physiological investigation of human perception in terms of making and processing images by the brain and the mind. This would need to be complemented by a psychology that is capable of addressing the role of the subconscious in creating and perceiving images. Such a science needs to be "released from the tyranny of the physical eye" since images can be visual and mental, are subject to transformations, and move between a range of spheres not always accessible to the eye. The essay addresses the logical and diagrammatical relation under the category of a mathematics of images; images as immaterial and imaginary constructions require a concrete, material medium to qualify for communication and scientific investigation. This he calls the physics of the image. Third, a biology of the image would constitute its object as a "metapicture" of images perceived as organisms. This adaptation of the methodology of the life sciences to a scientific theory of images might lead to a "natural history" of the temporal and spatial conditions of the making and circulation of images.

The essay by Schummer and Spector, in particular, points to an *Urbild* of the modern scientist by presenting the iconographic variations of the laboratory chemist. From the early story of Dr. Faustus on, it has been ingrained as a central element into the public image of science and it still provides a powerful image for the conception of science.

Popularization works as a *spatial* displacement of images. They are moved among simultaneous regimes of visions. Our reading of science images can greatly benefit from focusing on a displacement in *time* by subjecting images separated by chronology to a gaze of synchronicity. When Colin Milburn and Bernd Hüppauf read late nineteenth-century images from the perspective of recent image theory, they emancipate them from the stigma of an irrationalism of a bygone past and make them contemporaneous. This reading against the flow of time uncovers dimensions of these pictures hitherto unnoticed but present as soon as we are prepared to perceive them. Framed by this knowledge, reading a science image of the past is not concerned with obsolete practices but can reveal a great deal about the ways in which these images involuntarily contributed to the construction of the image of the sciences in general. Science images and the popular images of science are still entangled in these twisted ways.

Charlotte Bigg's careful analysis of a sequence of photographs shot by the photographer Nadar on the occasion of French organic chemist Michel Eugene Chevreul's hundredth birthday in 1886 and published in a periodical with a wide distribution, points the way to better understanding this fusion. Her essay investigates how images of an individual scientist were publicly launched in such a way that the particular could be turned into the general. In contrast to conventional images produced at the official celebrations, Nadar's photos turned readers into participant observers, and pretended thereby to give intimate access to the great scientist's personality and mind. Nadar's combination of the still limited resources available to photography in the late nineteenth century can be read as paradigmatic and a model for future transformations of specific images into a general image of modern science suited for identification.

Mersch's essay argues that visualizations should be understood as "visual arguments." With Wittgenstein he holds that all methods of illustrating, the iconic as well as the linguistic, are equal and a "genuine method" of creating knowledge which implies that the iconic correlations of identity and difference, equivalence and contrast are central to scientific reasoning. An algebraic equation is no less an image than its diagrammatic transfer in a two-dimensional figure. Evidence demonstrated by visualizations is inextricably tied to the truth claims of discursive practices. The new technical image is a diagrammatic hybrid and constitutes a new genre, which belongs neither to the pictorial nor to the written. Diagrammatics signify the *visibility of a thought*.

No technical image can be deciphered through its visibility alone. It emerges from theory and requires theory as a frame for reading. Thus, we are confronted with the dilemma of an impossibility of distinguishing between the image as an instrument and as an argument. This indeterminacy constitutes the precarious status of images in the age of the technical image.

The science images, Mersch argues, have not freed themselves from an aesthetic dimension. Instead they have subjected themselves to new dependencies by leaving the creativity of the gaze to apparatuses and replacing the subjective observer with technology. However, their "work" is no less opaque as production and circulation follow the logic of an anonymous structure, which answers to other than scientific interests.

The essay by Cartwright and Alač is a contribution to the long tradition of a medical history of the imagination. It focuses on images created in magnetic resonance laboratories, not, as one could assume, on the images produced by this technology, but on the mental images produced by the patient and the researcher/medical technician during the diagnostic process. How is the situation experienced by the individuals involved? The situation, the paper argues, *is* an image and its experience requires an active involvement of the imagination. The imagination as the originator of images is defined as an intersubjective field made up of the visual and acoustic percepts of the laboratory which are, partly consciously and partly subconsciously, perceived and evaluated by the persons involved in the interactive situation. The paper's approach is to reconstruct these images not through empathy and speculation nor through interviews but through observation of bodies. Thus, the invisible mental images become accessible through reading the non-verbal body language.

As Peter Weingart's essay about myths and stereotypes of science in fiction films demonstrates, the similarities are obvious. Movies continue a literary tradition that leads back to the beginning of modern science, and they convey a deep ambivalence about the production and application of new knowledge, varying the mythological figure of Prometheus (for a summary of the literary lineage of the ambivalence towards science see Haynes 1994). The essays by Petra Pansegrau and Erika Flicker also show that the stereotypes of scientists are remarkably stable and simple. They have a long lineage often reaching back to the Middle Ages and are saturated with images from the period of alchemy.

Bruce Clarke and Lutz Koepnick probe into two themes that occupy center stage in films about science. Clarke looks at Cronenberg's remake of *The Fly*, the original being an icon of movies about scientists because it was the first to turn away from the previous "mad scientist" image so familiar in the Frankenstein tradition. But the theme of *The Fly* is the same "perennial mythos of bodily metamorphosis." Clarke's treatise, however, focuses on *The Fly* as a "mutating narrative structure whose literal transformations, like the fantastic bodily changes under narration, are driven by the extensions of media devices."

Koepnick addresses the man–machine interaction and the hopes and worries associated with new technologies. His focus is on science fiction films which are considered a seismograph of public anxieties about the present and projections into the future. In contrast to previous science fiction films more recent ones are no longer concerned so much with future technologies reshaping "the material architectures of human life" but how they "expand our present drastically and collapse the boundaries between past and future." The case in point are the electronic media and their role in expanding memory. The movies which are his points of departure include John Dahl's *Unforgettable* (1996), Tom Tykwer's *Winter Sleepers* (1997), Wim Wenders's *Until the End of the World* (1991), and Christopher Nolan's *Memento* (2000).

The close relationship between contemporary art and the neural sciences is reflected in a dialogue between Berlin based artist Gabriele Leidloff and neural scientist Wolf Singer. Contemporary art is engaged in searching for ways of rendering visible what is invisible to the human eye and fascinated with advanced research into the biology of perception. From the invention of x-ray photography immediately following Roentgen's discovery in 1895, to recent advanced computer assisted magnetic technology, the history of modern medical diagnosis is saturated with techniques for making images of the invisible. Like every representation of reality, these images require construction and have meaning only through interpretation. Leidloff's work is an example of contemporary art's attempt to make this process transparent and accessible to a wide spectatorship.

NOTES

1. In current debates two schools can be distinguished, one focused on material conditions of research and the other one on changing ideas or "paradigms" (Kuhn) as the engine of scientific innovation. The essays of this volume are committed to the former position.
2. A closer look reveals that the situation is even more complex. In their contribution to this volume Joachim Schummer and Tami Spector distinguish between the public self-image of science, which refers to the image that scientists wish to project to the public, and the self-image of science that is private and idiosyncratic, dependent on the individual scientist or a specific scientific discipline.
3. Pioneers of popularizing the sciences were chemists, physiologists, and physicists of the late nineteenth century such as Helmholtz, Ernst Haeckel, Ludwig Boltzmann, and Ernst Mach. Popularization of the sciences, Helmholz argued, requires an artistic talent that is contrary to scientific thinking. He did not specifically refer to images as the vehicle for transferring abstract scientific theories to popular knowledge. But his view was representative of the perceived divide between scientific theory and the non-scientific thought which, he thought, was required for making the sciences widely understood (Helmholtz 1884: 355).
4. The four volumes of the pharmacist Albertus Seba's *Locupletissimi rerum naturalium Thesauri* (1734–1765) combined *accurata descriptiones* with *iconibus artificiosissimis expressio* and included realist representations of mostly exotic flora and fauna with animals of the imagination such as a double-headed monster from Africa, a seven-headed serpent, or bats with characteristics of the human body. The question of whether they were works of artistic imagination or true to nature was hotly debated and settled in favour of the latter.
5. Arnheim elaborated on a theory of vision centered on the term *visual thinking* and arguing that perceiving an image is predicated upon a visual language (Arnheim 1969). A provocative position is developed in an essay by Ronald Giere who, based on Alfred Wegener's theory of continental drift, attempts to demonstrate that a scientific theory is constructed more like an image than vice versa (Giere 1966).
6. The more substance (*Prägnanz*) an image has the closer it is related to a diagram. The difference between a diagram and a simplified naturalistic image is "more a difference of degree than a difference of kind" (Hall 1996). The history of simplicity resulting from reduction in modern art has not been written. It would reveal a strong connection between the aesthetics of the visual arts and science illustrations. The beginning can be seen in literary theory, specifically Charles Batteux (1748). In the twentieth century, the aes-

thetics of simplicity through reduction led to the ideal of functional transparency (Le Corbusier) and the economy of abstraction. For a survey of the origins see Henn (1974).

7. "Die Thatsache aber, dass die Photographie uns jeden Gegenstand naturgetreu wiedergibt, d.h. ohne dass subjective Anschauungen wie z.b. durch den idealisirend wirkenden Zeichenstift des Zeichners hineingetragen werden können, musste doch endlich durchdringen, und es erfreuen sich heute alle medicinischen Disciplinen der regen Unterstützung der Photographie, welche die allgemeine unbestechliche Protokollführerin der Naturwissenschaften, der unanfechtbare, unparteiische und sichere Zeuge für jede, besonders in der Einsamkeit gemachte wissenschaftliche Untersuchung genannt werden muss" (Jankau 1883).

8. In recent years the topic has attracted considerable attention in publications and exhibitions; see, for example, the pioneering exhibition Beauty of Another Order. Photography and Science organized by The National Gallery of Canada in 1997, catalogue ed. by Ann Thomas; and the exhibition Wahr-Zeichen. Fotografie und Wissenschaft, organized by the Technische Sammlungen, Dresden; catalogue in two volumes edited by Krase and Matthias (2006).

9. The quote from Albert Londe, "La photographie médicale," is taken from Didi-Huberman's book.

10. They can be seen as examples of Plato's *apeiron* (Philebos 23Cff). The question as to when indeterminate and obscure shadows cross the line and become an image is addressed in contemporary photography's strong predilection for fuzziness; see for example Stefan Heyne's photographs: catalogue with essays by Tannert and Gamm (2005).

11. Among the first to respond to this new condition were Mitchell (1986); Belting (1990, 2001), Boehm (1994), Jones and Galison (1998), Elkins (1999). The ensuing debate has led to numerous essays and books. Compare the *Journal of Visual Culture* (2000); Bredekamp et al. (2003); Holländer (2000). The debate was represented in journals and books (see Maar and Burda 2004). Comprehensive collections reflecting recent positions are *Visual Culture: The Reader* (Evans and Hall 1999); Dikovitskaya (2005) (with seventeen interviews); Heßler (2006).

12. An earlier and very influential attempt to frame the issues was Mitchell (1986). In the meantime, this debate has produced a number of readers, among them Maar and Burda's (2004) *Iconic Turn. Die neue Macht der Bilder.*

13. They are produced for specific purposes and therefore need to be decoded in terms of an abstract message. The expert's eye is trained to perceive and decode these images as signs in congruence with the ideals of photographic realism and the objectivity claim of modern science.

14. Not much is known about the transfer between the spheres of producing and reading images. Problems of reading a video produced as forensic evidence and later presented to a public audience are discussed in Hüppauf (2006).

15. See the contributions to the conference Observing Nature—Representing Experience at the Max Planck Institute of History of Science, January 2005.

16. http://hubblesite.org/gallery/album/solar_system_collection/

17. She refers to the traditional *facultas fingendi* as a faculty now required by computer simulation (Weigel 2004). See also: Introduction to Heintz and Huber (2001). They argue that for the new images to become legible, a return of the despised imagination is needed.

18. A noteworthy exception was the exhibition Wahr-Zeichen (see note 8). A complete section at the Altana Gallery is devoted to the sciences in contem-

porary art. A catalogue with high quality reproductions and short essays on fourteen artists is edited by Andreas Krase and Agnes Matthias, Dresden (2006).

19. Gottlob Frege observed in his distinction between "Sinn" and "Bedeutung" that the former denotes a sphere of surplus that transcends the immediate context required for the generation of meaning attached to a sign (word or image). The science image, removed from its intended context and re-placed in a different context gains the potential of creating different meaning(s), unintended and unforeseen by its originator.

20. It needs to be added that the transfer of science images to the sphere of the popular lacks the liberating effect which cultural studies associated with the discovery of the polysemy of texts. The suggestion of a value-free description of such transfers is made by Kaspar Maase (2002). While Maase is not concerned with science images, his suggestion can be read as paradigmatic for the general issue of transfer between the *serious* and the *popular*.

21. Examples are numerous from early photos of the moon and x-ray images of the late nineteenth century to micro photography and computer art.

22. The images of these publications were designed to demonstrate that the human ontogenesis was a recapitulation of the phylogenesis.

23. Uwe Pörksen refers to two stages of translation, first from Latin to national vernacular and then from scientific to popular languages.

24. In spite of a universal tendency, it is important not to overlook, however, that many organizations and experts have no interest in popularising science images. See Peter Galison: "...The classified universe is...on the order of five to ten times larger than the open literature..." (Galison 2005: 591). Interests in keeping scientific research secret are manifold, including that of governments and the military. Uncompromising believers in the mission of modern science despise the popular as the unprofessional version of the genuine image. They have scruples at opening the closed circles of the sciences and are content, as Mesopotamian diviners or medieval alchemists were, to communicate among themselves and stay close to their heroes and deities. This is a common experience for every one who made an attempt to bring together image theorists and engineers or scientists experienced in working with imaging technology.

25. A recovery of "reality" after post post-structuralism and the demise of deconstruction has been on the agenda of an emerging international discourse for a few years now. Important for the issues raised in this book is: Latour and Weibel (2005). The book is based upon an exhibition at the ZKM, Karlsruhe, 20 March to 3 October 2005.

BIBLIOGRAPHY

Arnheim, Rudolf (1969) *Visual Thinking*, Berkeley and Los Angeles: University of California Press.

Baigrie, Brian S. (1996) "Introduction", Brian S. Baigrie (ed.), *Picturing Knowledge. Historical and Philosophical Problems Concerning the Use of Art in Science*, Toronto: University of Toronto Press.

Batteux, Charles (1748) *Cours de belles lettres*, vol. 3.

Belting, Hans (1990) *Bild und Kult. Eine Geschichte des Bildes vor der Kunst*, München: Beck.

Belting, Hans (2001) *Bildanthropologie*, München: Beck.

Boehm, Gottfried (ed.) (1994) *Was ist ein Bild?* München: Wilhelm Fink Verlag.

Borck, Cornelius (2006) *Maschinenbilder des Geistes. Zur Visualisierung in den Neurowissenschaften*, paper presented at the IWT, Bielefeld, April 27, 2006, unpublished manuscript.

Bredekamp, Horst and Franziska Brons (2004) "Fotografie als Medium der Wissenschaft—Kunstgeschichte. Biologie und das Elend der Illustration", in Christa Maar und Herbert Burda (eds), *Iconic Turn. Die neue Macht der Bilder*, Köln: Brons: 365–381.

Bredekamp, Horst, Angela Fischel, Birgit Schneider, and Gabriele Werner (2003) "Bildwelten des Wissens", *Bildwelten des Wissens. Kunsthistorisches Jahrbuch für Bildkritik*, 1, 1: 9–20.

Bucchi, Massimiano (1996) "When scientists turn to the public: Alternative routes in science communication", *Public Understanding of Science*, 5: 375–394

Daston, Lorraine and Peter Galison (1992) "The image of objectivity", *Representations*, 37: 67–106.

Daum, Andreas (1998) *Wissenschaftspopularisierung im 19. Jahrhundert: Bürgerliche Kultur, naturwissenschaftliche Bildung und die deutsche Öffentlichkeit 1848–1914*, München: R. Oldenbourg Verlag.

Deilmann, Astrid (2004) *Bild und Bildung. Fotografische Wissenschafts- und Technikberichterstattung in populären Illustrierten der Weimarer Republik (1919–1932)*, Tönning: Der Andere Verlag.

Dikovitskaya, Margaret (2005) *Visual Culture. The Study of the Visual after the Cultural Turn*, Cambridge, MA: The MIT Press.

Earnshaw, Rae A. and Norbert Wiseman (1992) *An Introduction to Science Visualization*, Berlin: Springer.

Elkins, James (1999) *The Domain of Images*, Ithaca, NY, and London: Cornell University Press.

Evans, Jessica and Stuart Hall (eds) (1999) *Visual Culture. The Reader*, London: Sage Publications.

Fehrmann, Gisela, Erika Linz, and Eckhard Schumacher (eds) (2004) *Originalkopie. Praktiken des Sekundären*, Köln: DuMont.

Fiske, John (1987) *Television Culture*, London, New York: Routledge.

Galison, Peter (2005) "Removing knowledge", in Bruno Latour and Peter Weibel (eds.), *Making Things Public. Atmospheres of Democracy*, Cambridge, MA: The MIT Press: 590–599.

Galison, Peter, Robb Moss and Students (2005) "Wall of science", in Bruno Latour and Peter Weibel (eds), *Making Things Public. Atmospheres of Democracy*, Cambridge, MA: The MIT Press: 332–333.

Giere, Ronald (1996) "Visual models and scientific judgment", in Brian S. Baigrie (ed.), *Picturing Knowledge: Historical and Philosophical Problems Concerning the Use of Art in Science*, Toronto: University of Toronto Press: 269–302.

Gugerli, David (2002) "Der fliegende Chirurg. Kontexte, Problemlagen und Vorbilder der virtuellen Endoskopie", in David Gugerli und Barbara Orland (eds.), *Ganz normale Bilder. Historische Beiträge zur visuellen Herstellung von Selbstverständlichkeit*, Zürich: Chronos: 251–270.

Haeckel, Ernst (1868) *Natürliche Schöpfungsgeschichte*, Berlin: Verlag Georg Reimer.

Haeckel, Ernst (1874) *Anthropogenie oder Entwicklungsgeschichte des Menschen*, Leipzig: Verlag Wilhelm Engelmann.

Hall, Bert S. (1996) "The didactic and the elegant", in Brian S. Baigrie (ed.), *Picturing Knowledge. Historical and Philosophical Problems Concerning the Use of Art in Science*, Toronto: University of Toronto Press: 9.

Hall, Stuart (1980) "Encoding/Decoding", in Stuart Hall (ed.), *Culture, Media, Language. Working Papers in Cultural Studies*, London: Hutchinson: 128–138.

Haynes, Roslynn D. (1994) *From Faust to Strangelove: Representations of the Scientist in Western Literature*, Baltimore and London: Johns Hopkins University Press.

Heintz, Bettina and Jörg Huber (2001) "Der verführerische Blick. Formen und Folgen wissenschaftlicher Visualisierungsstrategien", in Bettina Heintz and Jörg Huber (eds), *Mit dem Auge denken*, Wien, New York, Zürich: Springer, Edition Voldemeer: 9–40.

Helmholtz von, Hermann (1884) *Vorträge und Reden, zugleich dritte Auflage der* Populären wissenschaftlichen Vorträge *des Verfassers*, Braunschweig: Vieweg.

Henn, Claudia (1974) *Simplizität, Naivität, Einfalt. Studien zur ästhetischen Terminologie in Frankreich und Deutschland 1674–1771*, Zürich.

Hentschel, Klaus (2005) "Wissenschaftliche Photographie als visuelle Kultur. Die Erforschung und Dokumentation von Spektren", *Berichte zur Wissenschaftsgeschichte*, 28: 193–214.

Hesse, Wolfgang (ed.) (1998) *Hermann Krone. Historisches Lehrmuseum für Photographie. Experiment. Kunst. Massenmedium*, Dresden: Verlag der Kunst.

Heßler, Martina (ed.) (2006) *Konstruierte Sichtbarkeiten. Wissenschafts- und Technikbilder seit der frühen Neuzeit*, München: Wilhelm Fink Verlag.

Hilgartner, Stephen (1990) "The dominant view of popularization: Conceptual problems, political uses", *Social Studies of Science*, 20: 519–539.

Holländer, Hans (2000) *Erkenntnis, Erfindung, Konstruktion. Studien zur Bildgeschichte von Naturwissenschaften und Technik vom 16. bis zum 19. Jahrhundert*, Berlin: Gebr. Mann Verlag.

Hüppauf, Bernd (2006) "Ein Grab wird geöffnet. Was zeigt ein Video?", in Jürgen Matthäus und Klaus-Michael Mallmann (eds), *Deutsche, Juden, Völkermord. Der Holocaust als Geschichte und Gegenwart*, Darmstadt: Wissenschaftliche Buchgesellschaft: 311–325.

Jankau, Ludwig (1883) *Die Photographie in der praktischen Medicin*, München.

Jones, Caroline and Peter Galison (eds) (1998) *Picturing Science, Producing Art*, New York, London: Routledge.

Journal of Visual Culture (2000) ed. Marquard Smith, London: Kingston University.

Krase, Andreas and Agnes Matthias (eds) (2006) Wahr-Zeichen. Fotografie und Wissenschaft. Museen der Stadt Dresden, Technische Sammlungen Dresen, Dresden.

Kretschmann, Carsten (ed.) (2003) *Wissenschaftspopularisierung. Konzepte der Wissensverbreitung im Wandel*, Berlin: Akademie Verlag.

Krohn, Wolfgang (ed.) (2006) *Ästhetik in der Wissenschaft. Interdisziplinärer Diskurs über das Gestalten und Darstellen von Wissen*, Hamburg: Felix Meiner Verlag.

Latour, Bruno and Peter Weibel (eds) (2002) *Iconoclash. Beyond the Image Wars in Science, Religion and Art*, Cambridge, MA: The MIT Press.

Latour, Bruno and Peter Weibel (eds) (2005) *Making Things Public. Atmospheres of Democracy*, Cambridge, MA: MIT Press.

Lenhard, Johannes, Günter Küppers, and Terry Shinn (eds.) (2007) *Simulation. Pragmatic Constructions of Reality*, Sociology of the Sciences Yearbook, Vol. 25, Dordrecht: Springer.

Link, Jürgen (2005) "Aspekte 'molekularer' Popularisierung von Wissenschaft durch Kollektivsymbolik und Interdiskurs. Am Beispiel der sozialen Chemie im 19. Jahrhundert", in Gereon Blaseio, Hedwig Pompe, und Jens Ruchatz (eds), *Popularisierung und Popularität*, Köln: DuMont: 199–216.

Maar, Christa and Hubert Burda (eds) (2004) *Iconic Turn. Die neue Macht der Bilder*, Köln: DuMont.

Maase, Kaspar (2002) "Jenseits der Massenkultur. Ein Vorschlag, populäre Kultur als Massenkultur zu lesen", in Udo Göttlich, Winfried Gebhard, und Clemens Albrecht (eds), *Populäre Kultur als repräsentative Kultur. Die Herausforderung der Cultural Studies*, Köln: Halem Verlag: 79–104.

Mandelbrot, Benoît B. (1989) "Fractals and an art for the sake of science", *Leonardo*, 2: 21–24.

Mitchell, W.J.T. (1986) *Iconology. Image, Text, Ideology*, Chicago: University of Chicago Press.

Mondzain, Marie José (2006) *Was ist: Ein Bild sehen?* in Bernd Hüppauf and Christoph Wulff (eds), *Bild und Einbildungskraft*, München: Wilhelm Fink Verlag.

Observing Nature—Representing Experience (2005) Conference at the Max Planck Institute of History of Science, Berlin, January 2005.

Oels, David (2005) "Wissen und Unterhaltung im Sachbuch, oder: Warum es keine germanistische Sachbuchforschung gibt und wie eine solche aussehen könnte", *Zeitschrift für Germanistik*, 15: 8–27.

Papp, Charles S. (1968) *Scientific Illustration*, Dubuque, IA: W.M.C. Brown.

Pörksen, Uwe (1986) *Deutsche Naturwissenschaftssprachen. Historische und kritische Studien*, Tübingen: Narr.

Rheinberger, Hans-Jörg (2001) "Objekt und Repräsentationen", in Bettina Heintz und Jörg Huber (eds), *Mit dem Auge denken. Strategien der Sichtbarmachung in wissenschaftlichen und virtuellen Welten*, Wien, New York: Springer Verlag: 55–61.

Rosenthal, Claude (2005) "Making science and technology results public", in Bruno Latour and Peter Weibel (eds), *Making Things Public. Atmospheres of Democracy*, Cambridge, MA: MIT Press: 346–348.

Ruchatz, Jens (2005) "Der Ort des Populären", in Gereon Blaseo, Hedwig Pompe und Jens Ruchatz (eds), *Popularisierung und Popularität*, Köln: DuMont.

Schickore, Jutta (2002) "Fixierung mikroskopischer Beobachtungen. Zeichnungen, Dauerpräparat, Mikrophotografie", in Peter Geimer (ed.), *Ordnungen der Sichtbarkeit. Fotografie in Wissenschaft, Kunst und Technologie*, Frankfurt am Main: Suhrkamp: 285–310.

Schinzel, Britta (2006) "Wie Erkennbarkeit und visuelle Evidenz für medizintechnische Bildgebung naturwissenschaftliche Objektivität unterminieren", in Bernd Hüppauf and Christoph Wulff (eds), *Bild und Einbildungskraft*, München: Wilhelm Fink Verlag: 354–370.

Schlich, Thomas (1997) "Repräsentationen von Krankheitserregern. Wie Robert Koch Bakterien als Krankheitsursache dargestellt hat", in Hans-Jörg Rheinberger, Michael Hagner, und Bettina Wahrig-Schmidt (eds), *Räume des Wissens. Repräsentation, Codierung, Spur*, Berlin: Akademie Verlag: 165–190.

Stäheli, Urs (2005) "Das Populäre als Unterscheidung", in Gereon Blaseo, Hedwig Pompe, und Jens Ruchatz (eds), *Popularisierung und Popularität*, Köln: DuMont: 146–167.

Tannert, Christoph and Gerhard Gamm (2005) *Katalog — Stefan Heyne*, Köln: Salon Verlag.

Taylor, Richard (2002) *Fractal Expressionism—Where Art Meets Science*, Santa Fé Institute.

Topper, David (1996) "Towards an epistemology of scientific illustration", in Brian S. Baigrie (ed.), *Picturing Knowledge. Historical and Philosophical Problems Concerning the Use of Art in Science*, Toronto: University of Toronto Press: 215–249.
Weigel, Sigrid (2004) "Das Gedankenexperiment: Nagelprobe auf die *facultas fingendi* in Wissenschaft und Literatur", in Thomas Macho and Annette Wunschel (eds), *Science & Fiction. Über Gedankenexperimente in Wissenschaft, Philosophie und Literatur*, Frankfurt am Main: Fischer Taschenbuch-Verlag: 183–205.

2 Science Images between Scientific Fields and the Public Sphere
A Historiographical Survey[1]

Sybilla Nikolow and Lars Bluma

In the famous photograph of James Watson, Francis Crick, and the double helix, the two future Nobel laureates pose around the object of their desire (on the history of this photograph see de Chadarevian 2003). The photograph was taken in 1953, the year of their discovery of the DNA double helix structure; it went on public display only 15 years later in Watson's best-selling book, *The Double Helix* (Watson 1968). Today this picture still stands for the successful representation of the DNA molecule in the construction of its model as the key event of their discovery. Together with Watson's book, this visualization contributed in its own way to the process by which Watson and Crick's version of events entered public consciousness and by which other research paths were cast into the shade, notably the crucial preliminary work of their fellow researchers. In the scientific practices of the young molecular biology of the 1950s and 1960s, model making became the ultimate research tool. Molecular models and their images circulated between laboratories, were deployed in lectures and exhibitions, became commercially produced and distributed. Simultaneously, they determined the public image of molecular biology/genetics in the media then and now.[2]

The Crick-Watson photograph continues nineteenth-century traditions of modern scientific portraiture, which had scientists display themselves confidently to the public in their workspace and surrounded by their instruments, models, and objects (Jacobi and Schiele 1989; Jordanova 2000; Werner 2001; Sichau 2004).

At the same time this image of the DNA model also belongs to another context of visual representations of scientific objects, namely to the sketches and drawings of molecular structures that usually accompany scientific notes and publications.[3] What for molecular biologists chiefly posed a new form of representing a knowledge acquired in other scientific contexts, giving them insight into the structure of a key molecule, came to be paraded to the readers of Watson's book as authentic evidence of scientific practice, to whom it was presented as an image *of* science, a picture that was compatible with popular myths of discovery, calculated to strengthen public trust in such scientific undertakings.

Meanwhile the gene and with it the structural image of DNA has become one of the cultural icons of the twentieth century (Nelkin and Lindee 1995; de Chadarevian and Kamminga 2002). Predictably it leapt from the covers of the magazines *Science* and *Nature* in February 2001, when the renowned biochemist J. Craig Venter and his colleagues and the International Human Genome Project published their drafts of the human genome in *Science* and *Nature*, respectively (Venter et al. 2001; International Human Genome Sequencing Consortium 2001). While *Nature* used a conventional schematic illustration of the model, in *Science* several portraits—meant to evoke different human races and ages—coiled up along an imaginary helical structure. Watson and Crick's model construction kit may bear little comparison with the computer-supported deciphering industry of a Craig Venter, yet the publicly circulated picture of the discipline remains the same. To the genome researchers of today, who seem to spend more time in the media than in their laboratories, it offers a conciliatory tradition; and it allows their audiences to participate through visual stereotypes.

Whether a science image is perceived as a scientific image or as a public image of science not only depends on the original context of its production and intended use. It also depends on who has seen it. In view of the openness of the process of reception, we want to make a plea here for science images to be considered as objects that occupy a space between scientific practices and their respective publics, a space that is not defined in advance and that can be traced only through an analysis of the very circumstances of their reception. Historical examples in particular—such as the rise of the visual representations of the DNA-double helix in scientific fields and in the public sphere—provide insight both into which scientific and technological products were made public and the precise fashion in which this occurred. Science images interest us here with respect to their function as a means of communication not just within science but also vis-à-vis a public that participates in scientific results.

Images generated within a process of scientific production differ from those generated for public consumption both in respect to their context of origin and the context of their use. Nevertheless, it is worth noting that their strategic deployment at the boundary between science and the public does not occur without mutual references. Of interest from the perspectives of science studies and the history of science are precisely those kinds of transformation where the life of images extends from the process of their creation within a particular discipline right through to their subsequent public use as proof, evidence, or argument for a specific culture of science. The success of science images in both science-internal and public communication seems to depend on the degree to which they, as particularly mobile objects, offer semantic flexibility and identities for different groups of recipients. At the same time they need to be sufficiently robust to prevail over rival images with their respective arsenal of interpretative trajectories.[4]

Science images thus should be regarded as specific forms of representation and production which fluctuate between the scientific and the public sphere and acquire their meaning only through this process of exchange.

Especially images like the model of the double helix, which reach the public arena as scientific representations and therefore have bearings on the issue of the relationship between science and the public, do not fit a taxonomy that neatly separates "purely" scientific images from "purely" public or popular ones (for discussion, see Hüppauf and Weingart's classification in this volume). Their peculiarity lies in the fact that, while originally produced in a context of scientific inquiry, they gain an additional dimension of meaning as public images of science. The alternative, namely a classification that insists on a strict separation between scientific and public images, is itself potentially vulnerable to the charge of serving political goals in the negotiations between science and the public. For this reason we want to argue for a history of visual representations that includes the reception of science images in the public sphere. In the following we suggest how questions about science images can be fruitfully explored by linking up the results of visualization studies with those of popularization research.

IMAGES AND VISUAL CULTURE STUDIES

The use of images, especially science images, as source material for historical research may have become the norm by now. However, considering the wealth of visual sources it has taken academic historiography surprisingly long to pay analytical attention to these cultural productions. To be sure, in museums the portraits of physicians, technologists, and scientists, along with images of their models, instruments, and working environments (and visual and graphical representations of the results of their researches and measurements) have traditionally been treated as evidence of past scientific cultures. In historical narratives on the other hand, visual materials have more often than not been deployed in a peculiarly unreflected manner (similarly Pang 1997a). In many ways this served to enforce the protocols of glorification and hero worship by which an "older" historiography reproduced (not least visually) the vested interests of the "scientific community." Yet what historians of science and technology ended up underestimating along the way was the value of images as historical sources. The reason for this lies in their disciplines' old alliance with the history of ideas and its entrenched habit of giving sole recognition to texts as serious sources. In this tradition, visual evidence has seldom been credited with meaning-making or cognitive qualities. Only in recent years, with the rise of cultural history in the humanities in general, has this process of rethinking reached science studies (e.g., Jäger 2000; Burke 2001; Roeck 2003; Heßler 2005; Tucker 2006).

Apart from art history it was in "cultural studies" that scholarly debate about images began to flourish from the mid-1970s (in particular Berger 1972; Mulvey 1975). With the 80s and 90s finally came a sustained boom of "visual culture readers" (see, among others, Evans and Hall 1999; Cartwright and Sturken 2001; Howells 2003; Jones 2003). The research trajectory that formed under the label of "visual culture" picked up numerous influences from structuralism and post-structuralism. For instance, it became one of the goals of "visual studies" to overcome the distinction between "high" and "low cultural form" that was advocated by traditional history of art, and to pay increased attention to images of everyday life. Likewise familiar from cultural studies was the call to transcend the boundaries of academic discourse. Visual studies made it their premise that the analysis of visual representations must occupy a central role in the interrogation of Western culture. Categories such as gender and race came to count as images of produced cultural taxonomies that are open to reconstruction through semiotic, linguistic, and psychoanalytic methods. Accordingly, the meaning of an image was no longer assumed to reside solely in the image; rather, it must be derived from its "sociocultural environment." This deconstructivist attitude of "visual culture studies" expresses itself in that central basic oppositions that determine Western culture (mind/body, subject/object, male/female, self/other, etc.) are seen as visually constructed and therefore accessible to political and ideological criticism—notably through visual studies.

Mitchell, whose seminal notion of a "pictorial turn" governed the debate in the 1990s (Mitchell 1992), has formulated a comprehensive criticism of "visual culture studies" in his most recent work (Mitchell 2005: ch. 16). He takes issue above all with the assumption that an image analysis can be reduced entirely to a semiotic-linguistic analysis that, in quasi-scientific fashion, ultimately disposes of the magical-mystical dimension of the image. Against the premise that the mythico-magical content of an image is open to complete rationalization, Mitchell offers the thesis that the magical effects of images continue to be constitutive for modernity. Even if images communicate social relationships, he argues, seeing is not just a form of cultural construction; it is also a non-cultural, in other words, natural activity.

SCIENCE IMAGES AND SCIENCE STUDIES

For a long time, images produced by science and technology played only a marginal role in "visual culture studies." Historical image studies did not really pick up on the visual orders of the sciences unless they patently represented race or gender constellations. Likewise, in science studies the demand for a "pictorial turn" (Gugerli 1999) became an option only with the so-called "practical turn," following which attention came to be directed to practices as locally situated scientific and technological systems of action (see Latour 1987; Lynch and Woolgar 1990; Rheinberger et al. 1997). Now, scientific representations and inscriptions are seen as resulting from a complex interplay of instruments, experiments, measurements, representational technologies, and rhetorical strategies, an interplay to be decoded in detail. Visualizations thus appear as embedded in the process of the production and stabilization of scientific truths and objectivity (Knorr-Cetina 1999). As a result, visualizations of knowledge moved from forming a subspecies of representations in scientific practice to the very center of attention for science studies. In the past, it must be said, visual studies had all too often followed a mimetic theory of representation, conceding images no independent role in the production of knowledge. Only the practical turn in science studies made it clear that scientific images do not simply reproduce/represent knowledge: they themselves produce knowledge. Images thus occupy a double function as knowledge representation and as knowledge production. In view of their status in the process of producing knowledge, Knorr-Cetina proposed a terminological innovation: owing to the predominant association of scientific "discourse" with verbal communication, discourse should be replaced by "viscourse." Her coinage means to emphasize the striking significance of visualization with respect to that which is rendered an object of science in the "laboratory" and which can then be carried out into the public sphere.

Interest in science images is no doubt fed also by the contemporary insight that imaging technologies have gained ground in all disciplines across the map. There is talk of a "pictorialisation of knowledge" (Heintz and Huber 2001) and discussion has begun of what this would imply for knowledge production. Yet from a historical perspective it can hardly come as a surprise that the development of new technologies of representation and their introduction should have brought radical departures for scientific practices.

When asking what can be seen in an image, we must also ask what is not shown and what therefore remains hidden by the image. Processes of visualization shift the boundaries of what henceforth counts as visible and invisible. They constitute (re-)"orderings of the visible" (Geimer 2002), and can be interpreted as "constructed visibilities" (Heßler 2006). The meaning of an image is the outcome of a dialectical relationship between the visible and the invisible, of visual representation and visual absence. What becomes the visible of a science image emerges in a process of selection that is dependent on the manner of visual design, the aesthetic experience of the scientist, the reception on the part of the public, and the representational options and norms of a given scientific discipline. It can, for instance, in no way be taken for granted that the atoms of nanotechnology be displayed as movable and arbitrarily manipulable balls. Entirely different modes of representation would have been entirely possible, representations that do not correspond to our culturally and historically burdened ideas of manipulable atoms (Hennig 2004).

The first results of empirical work on the visualization of science can be summed up as follows: images pose a form in its own right for knowledge representation and knowledge production. As such, their epistemic status deserves attention similar to that accorded to texts. For a professional study of images we must draw not only on the methods of science studies, but also on those of art history and communication and media studies. The analysis of the production and use of images in scientific practice lends itself in a special way for the exploration of the concrete intersections between culturally and scientifically informed ways of perceiving nature and society on the one hand, and of the technologies of image production on the other. Yet such a project cannot prosper unless image analysis is subjected to the strict standards of source criticism that are already applied to textual sources.

The thesis defended here, namely that images pose a form of knowledge representation and knowledge production in its own right, supports no ad hoc primacy vis-à-vis other forms of representation and production. Rather, it transpires that different knowledge cultures have generated their own specific forms of representation in which the significance and role of imaging technologies received a variety of evaluations. In this context Hentschel speaks of "visual cultures of science," a language he applies to the histories of single disciplines specifically where visualizing practices in research converge with those of teaching and of the scientists' *Lebenswelt* (Hentschel 2005: 193–194). From a historical perspective this gives rise not just to such questions as which visual forms acquired or failed to acquire epistemological significance in which discipline and at what time, but also how it should be evaluated when visual forms of expression appear; for example, to devolve again to a lesser status compared to other forms during specific phases of disciplinary development.

Above all there is an opportunity here to overcome unnecessary disciplinary boundaries between an iconographical art history that focuses on the aesthetic function of picture genres and their genesis; a historiography of science intent on tracing the positive knowledge embedded in images; and a history of technology that has engaged with the history of image-creating devices all along, but which still too often uses images of technological artifacts in a merely descriptive way. Recent examples from the history of photography show how fertile a joint approach can prove to be for an analysis of technologies of visual representation and their products (see Tucker 1997; Geimer 2002). When inquiring further about the circumstances under which particular objects of a culture become visible and in what ways their visibility can be considered as an expression of either science or art, such disciplinary boundaries emerge as historically contingent (Jones and Galison 1998).

The potential aim of a productive collaboration and mutual complementation could be the development of integrative methods for a conceptualization and historicization of visual orders. So far, the heterogeneity of both method and content of the extant case studies leaves it open whether historians of science and technology ought to follow the battle cry for a comprehensive "science of images" (Belting 2001), for an "iconic turn" (Boehm 1994) or "pictorial turn" (Mitchell 1992; Gugerli 1999) in cultural studies and history. Already beyond dispute though is that such a project cannot proceed without addressing the "visual worlds of knowing" (Bredekamp et al. 2003) in the processes of research, artistic creation, or popularization. Science-historical approaches can illuminate the specificity, peculiarity, and transformation of science images. In turn they stand to benefit from the current discipline-transgressing re-evaluation of images as specific cognitive products. Interest in visualizations has even reached literary studies and linguistics, where infographics, computer images, and pictograms in commercials and the mass media are already being addressed (Pörksen 1997; Gerhard et al. 2001).

HISTORICAL STUDIES ON SCIENCE IMAGES

Recent interdisciplinary work in these directions reveals three distinct approaches in the history of science. First, a number of authors have focused on the development of and the practices associated with particular technologies of visual representation such as photography (Tucker 1997; Geimer 2002), radiography (Dommann 2003), microscopy (Schickore 2002), endoscopy (van Dijk 2001; Gugerli 2002), the graphical recording of data (Hankins 1999), and the popular genre of the scientific portrait (Jacobi and Schiele 1989; Fara 1998; Jordanova 2000; Werner 2001; Sichau 2004), to name a few striking examples. Their analysis of the respective scientific practices addresses the relationships between the scientific production of knowledge and the artistic, technological, and craft components within scientific image production, as they came to be applied in, for example, scientific experimentation.

Secondly, for about twenty-five years, the study of visualization in the history of particular disciplines has moved beyond the classical concerns of the history of mathematics with geometrical objects and images, of the history of geography with the production of maps, and of the history of technology and architecture with drawings. Studies on medicine (Reiser 1978; Eckart 1980; Jordanova 1990; Maehle 1993; Cartwright 1995; van Dijk 2005); geology (Rudwick 1976, 1992); anatomy and natural history (Daston and Galison 1992); botany (de Chadarevian 1993a; Secord 2002); bacteriology (Schlich 1995, 1997); immunology (Cambrosio et al. 1993); theory of evolution (Prodger 1998; Voss 2007; Hopwood 2004, 2005, 2006); physiology (de Chadarevian 1993b); thermometry (Hess 2002); brain research (Hagner 1996, 2002, 2006; Borck 2002, 2005); astronomy (Lynch and Edgerton 1988; Pang 1997b; Schaffer 1998; Hentschel 2002); nano-science (Hennig 2004); physics (Galison 1997; Wiesenfeldt 2001);

engineering (Baynes and Pugh 1981; Ferguson 1992; Henderson 1995; Lefèvre 2004); and economics (Klein 1997; de Marchi and Goodwin 1999; Tanner 2002) offer noteworthy examples. This disciplinary historiography foregrounded the significance of image production in the epistemic process and in discipline-internal communication. The questions put to disciplinary pasts here targeted discipline-specific visual cultures and visual languages. The status occupied by different procedures of representation and images in the disciplines in question formed a special concern, along with the issue of how these procedures came to prevail over previous and rival forms of knowledge representation, and of how disciplinary standards, practices, and discourses changed in the course of time. Furthermore the decisive role of visual argument in scientific controversy was stressed.

Third, there has been a move to explore the visual orders of particular epochs and their strategies. For instance, a "visual culture" has been diagnosed for the period around 1800 and explored by interrogating this period's typical modes of perceiving nature and society (Stafford 1994; Dürbeck et al. 2001). These manifested equally in artistic and scientific productions, and therefore must be regarded as trans-disciplinary. Moreover, renewed attention has been drawn to the aesthetic dimensions of the sciences (Krohn 2006). Surely the issue of the aesthetics of science images can no longer be dismissed as an external or secondary aspect, nor can it simply be reduced to audience taste.

Orthogonal yet no less revolutionary to these historical traditions, the specific visual resources of the image of objective knowledge have become a focus of historical analysis. In their study of scientific atlases from the late seventeenth to the early twentieth century, Daston and Galison identified the practices of truth and objectivity that became characteristic of their age and the shifts representations (and thus also conceptions) of objectivity underwent along the way (Daston and Galison 1992; see also Galison 1998; Daston 1999). Since approximately the middle of the nineteenth century, new imaging technologies, for instance photography, promised to deliver objective images that promoted an ethos of non-interference. These technologies appeared to guarantee that scientists' personal idiosyncrasies did not contaminate scientific representation. Yet the rhetoric of keeping one's prejudices and theories out of data and images that are difficult to produce and even more difficult to interpret can be read also as a response to growing public skepticism regarding the trustworthiness and objectivity of a scientific enterprise that was forced to revise itself at ever shorter intervals.

THE RELATION BETWEEN SCIENCE AND PUBLIC

Science images have been singled out for the visual obviousness that allows scientific-technological visualizations to be read by the public as "perfectly normal images" (Gugerli and Orland 2002). This has led to a call not to overlook the manifold conditions that preceded such perception. We welcome this broadening of perspective for its inclusiveness regarding the processes of circulation of science and technology in the public sphere. Inquiry into the functions of science images and their actions must remain incomplete as long as the analysis remains confined to the laboratories of image production. It would seem advisable also to pay closer attention to the places where science images are unquestioningly taken as the hallmark of scientificity and objectivity or truth. This makes the history of the public reception of science images a key concern. Studies on the practices of science popularization offer promising opportunities to engage with concrete questions, such as which images came to be produced by the sciences, and how and why they were produced and successfully received at a particular time.

Popular forms and media of knowledge entered the gaze of science studies and the history of science about twenty years ago (Whitley 1985; for a summary see Felt et al. 1995: ch. 9). The resulting case studies lead to a profound revision of formerly dominant views on popularization work (e.g., Hilgartner 1990; Weingart 2001: ch. 6). It is a matter of consensus now that popularization by no means forms a derivative and secondary activity that takes place quite separately from the process of discovery and to which scientists generally attribute little significance. On the contrary, these aspects of one social phenomenon, science and the public, intimately intersect in both knowledge production and in its reception (for the history of technology see also Bluma et al. 2004).

From the perspective of science studies, popularization forms an intrinsic component of the process by which facts are scientifically produced because popularizations play a crucial role in what, at the end of a complex process of communication and negotiation, is termed scientific knowledge and accepted as such by the public. Considering this, the sweeping dismissal of public or popular representations as a lower form of supposedly higher scientific knowledge appears as a rhetorical strategy through which scientific experts or their representatives strive to exercise greater interpretative power over science in the public sphere. As a number of studies have shown, these attempts serve to maintain or even to reinforce a social hierarchy that stratifies producers and consumers of knowledge (Hilgartner 1990; Lubar 1995; Bensaude-Vincent 1997; Weingart 2005).

In two respects, the research path advocated here offers new impulses for the analysis of science images between scientific practice and the public sphere. First, in asking about the conditions under which knowledge is received in public (Brecht and Orland 1999) and about how these conditions are tied to the media of communication (Weingart 2005), it makes it possible to avoid a reproduction of the scientist's gaze that still tends to inform studies of science visualization. Instead, the aim should be a symmetrical analysis of the relations between science and the public, an analysis in which both sides pose potential resources for one another (Shapin 1990; Nikolow and Schirrmacher 2007). Secondly, attention to the images themselves can help overcome the textualist bias of traditional popularization research. Instead, the goal should be to take the significance of images for the communication between science and the public more seriously (see also Cooter and Pumpfrey 1994: 255; Evans and Priest 1995: 332, for a topical case study see Deilmann 2004).

HISTORICAL CASES ON SCIENCE IMAGES AND THEIR PUBLIC

To illustrate the potential of the approach presented here we have chosen five case studies from the history of science and technology.[5] In the following, images are discussed as means and as media of communication between scientific, technological, or medical practice and their respective publics. Detailed attention is paid to the ideas about disciplinary practices that are transported in the respective visual sources.

Anke te Heesen's case study (te Heesen 2002) of an eighteenth-century copperplate engraving uses the image of a physician and electrician in his workspace for an exercise in the history of iconography. Placing the portrait in question in the context of the miraculous healings of Christ, her argument forges a promising link to the history of art. Beyond this she embeds her findings in the history of Enlightenment communication and educationism, accentuating the tactile and material qualities of an image that appeared in a picture encyclopedia (te Heesen 1997). From her analysis this etching at once emerges as evidence of the public image of representatives of a scientific discipline that saw themselves as enlightened and rational, and as an example for contemporary traditions as to how knowledge was collected, prepared, and ordered for educational purposes.

In the images analyzed by Gerhard Wiesenfeldt, we see instruments that were used in late eighteenth-century experiments on the phenomenon of electricity, notably the famous electrifying machine in the Teyler Museum in Haarlem. In this manner we acquire an impression of the contemporary settings of experimentation (Wiesenfeldt 2002). However, the actors themselves appear to have become invisible in the production of this image. Only detailed comparisons of the genesis of different representations that accompanied the publication of experimental results produced by the use of the machine make it possible for us to look behind the trappings of the work with the apparatus in the museum. These comparisons reveal the variability with which the idea of what made a realistic representation of a machine expressed itself in the perspectives of those involved (experimenters, sponsors, architects, copperplate engravers, draughtsmen, and instrument makers), and of how the director of the cabinet of instruments exercised control over which aspects of his experimental practice would reach the public in his museum. The images of the Teylerian machine illustrate not only a particular historical practice of experimentation, they are also an important component of a specific experimental politics.

In turn, Sabine Höhler has shown with the example of ocean maps from the *Deutsche Atlantische Expedition* of 1925 to 1927 how a political public came to interpret the graphical representations of the results of deep-sea explorations as authentic evidence for a political conquest through scientific means, and what share marine scientists had in the creation of this reading (Höhler 2002a, 2002b). From the measuring activities with novel instruments, the use of the resulting data in the production of a first deep-sea profile, to the construction of a deep sea map whose relief illustration entered popular imagination about the deep sea alongside a model of the research ship *Meteor*, Höhler traces the gradual transformations through which knowledge about unknown geographical spaces had to pass in order to achieve a visual obviousness which had the power to serve the creation of national identity outside the laboratory.

Lars Bluma (2002) discusses the use of block diagrams in the processes of technical design that became an established scientific practice in the context of the systematic reconfiguration of the engineering sciences in the aftermath of the Second World War. With the aid of these diagrams, systems engineers did not just render something visible, they simultaneously rendered something else hidden. Block diagrams allowed for the processual functionality of complex technological systems to be visualized on a very high level of abstraction, but they let the materiality of technological components disappear out of sight. Their deployment in job advertisements shows that the block diagram acquired iconic status in the broader context of engineers' public self-representation. These block diagrams exemplify how science images can shift the boundaries between scientific fields and the public sphere. In this case, block diagrams became symbolic of the taming—at the hands of systems engineers—of complex technological and political problems in the Cold War. In order to emerge as such they had to undergo a transformation from science-internal and epistemic objects to generally comprehensible and symbolic objects.

To what extent scientific images can perform identity-making functions through their public circulation has been shown by Sybilla Nikolow on the example of statistical images of populations in the first half of the twentieth century (Nikolow 2006; see also Brecht and Nikolow 2000). In contrast to Höhler's case the focus here is on images that were intended for public consumption all along. Through them the latest results of demography were communicated, and population was meant to be transported into public consciousness as a "scientific object" by means of specific statistical practices of visualization. The statistical images analyzed here pose complex objects of knowledge insofar as they simultaneously communicated knowledge about population as a scientific object *and* statisticians' scientific practice. Their comparison shows how statistical images of communities became specific instruments to secure authority and credibility in the public sphere for particular political ideals about population.

These examples illustrate the importance of visual analyses which address the "double-bind" of scientific image and public image of science, the double-bind that determines what circulates as a science image and therefore equally as a scientific image and as an image of science in the public sphere.

DISCUSSION

As the recent literature discussed here shows, today the historical analysis of images is in many respects practiced much more than it used to be and in almost all historical disciplines across the board. Even so, many problems endure. What is missing particularly with respect to the question of the relationship between science and the public are, for example, studies that provide a comparative analysis of the formation of different cultures of representation in the sciences in different political systems and different social orders.[6] One may surmise that visual representations in democratic societies differ from those of repressive systems, owing to the differences in their respective conceptions of the public, of politics and of scientific expertise. Studies along these lines promise to lead to interesting collaborations between the history of science or technology and political history. Such collaborations easily seem more realistic than the demand for a science of images, which often pretends that the historically grown boundaries between natural and cultural science can be ignored at will.

Equally, within the history of science, too—and despite many isolated results—the implications of a renewed examination of visual forms of knowledge representation and knowledge production would seem a long way from being exhausted. Only recently Norton Wise has pointed out that attention to images could facilitate a departure from such ingrained historiographical dichotomies as, for example, art versus science, museum versus laboratory, or geometrical versus algebraic methods. It is necessary to take images seriously as arguments, he contends, in order to develop a "materialized epistemology" (Wise 2006).

It would furthermore be desirable to question the dominance of visual representations within the body of non-textual sources and to explore the relationships of scientific images with other forms of scientific representation. This could take the form of addressing the interplay within the fabric of that which can be seen and that which can be said in the analysis of knowledge production and public representation (see Jenkins 1987; Schäfer 2004). The question about nonverbal forms of communication should open our minds also to the other senses of scientists (sound, taste, smell) whose histories still remain largely unwritten.

NOTES

1. This paper was translated from the German by Anna-K. Mayer. This argument was already developed in Nikolow and Bluma 2002. The version presented here substantially expands and updates our earlier survey.
2. On the scientifico-technological practices of model making and their public representations, see de Chadarevian (2004).
3. The inaugural publication of Watson and Crick in April 1953 in *Nature* contained a sketch drawn by Crick's wife, see de Chadarevian (2003: 94).
4. For the history of engineering see Henderson (1995: 214). It is not mere coincidence that Star and Griesemer developed their influential concept of a boundary object on the example of the history of negotiations between science and the public in a museum of natural history (Star and Griesemer 1989).
5. This selection does not claim to be representative. A list of works that offer a fruitful combination of the histories of popularization and visualization as we conceive of it must include at least Borck (2002); de Chadarevian (2003, 2004); Hennig (2004); and Müller (2005).
6. The history of political propaganda and of the making of national identities by means of symbols, rituals, and myths has produced several studies that could prove relevant for science studies. A survey of them in general history would go beyond the scope of this essay. For a sound introduction to this theme (with a bibliography) see, for example, Paul (2004).

BIBLIOGRAPHY

Baynes, Ken and Francis Pugh (1981) *The Art of the Engineer*, Guildford, Surrey: Overlook Press.

Belting, Hans (2001) *Bild-Anthropologie. Entwürfe für eine Bildwissenschaft*, München: Wilhelm Fink Verlag.

Bensaude-Vincent, Bernadette (1997) "In the name of science", in John Krige and Dominique Pestre (eds), *Science in the Twentieth Century*, Amsterdam: Harwood: 319–338.

Berger, John (1972) *Ways of Seeing: Based on the BBC television series with John Berger*, London: British Broadcasting Corp.

Bluma, Lars (2002) "Das Blockdiagramm und die 'Systemingenieure'. Eine Visualisierungspraxis zwischen Wissenschaft und Öffentlichkeit in der US-amerikanischen Nachkriegszeit", *Internationale Zeitschrift für Geschichte und Ethik der Naturwissenschaften, Technik und Medizin (NTM)*, 10: 247–260.

—— Karl Pichol and Wolfhard Weber (eds) (2004) *Technikvermittlung und Technikpopularisierung. Historische und didaktische Perspektiven*, Münster: Waxmann.

Boehm, Gottfried (ed.) (1994) *Was ist ein Bild?* München: Wilhelm Fink Verlag.

Borck, Cornelius (2002) "Das Gehirn im Zeitbild. Populäre Neurophysiologie in der Weimarer Republik", in David Gugerli and Barbara Orland (eds), *Ganz normale Bilder. Historische Beiträge zur visuellen Selbstverständlichkeit*, Zürich: Chronos: 195–225.

—— (2005) *Hirnströme. Eine Kulturgeschichte der Elektroenzephalographie*, Göttingen: Wallstein.

Brecht, Christine and Sybilla Nikolow (2000) "Displaying the invisible. 'Volks-krankheiten' on exhibition in imperial Germany", *Studies in the History and Philosophy of Biomedical and Biological Sciences*, 31: 511–530.

—— and Barbara Orland (1999) "Populäres Wissen. Editorial", *Werkstatt Geschichte*, 23: 4–12.

Bredekamp, Horst, Angela Fischel, Birgit Schneider, and Gabriele Werner (2003) "Bildwelten des Wissens", *Bildwelten des Wissens*, 1, 1: 9–20.

Burke, Peter (2001) *Eyewitnessing. The Uses of Images as Historical Evidence*, London: Reaktion.

Cambrosio, Alberto, Daniel Jacobi, and Peter Keating (1993) "Ehrlich's 'Beautiful Pictures' and the controversial beginnings of immunological imagery", *Isis*, 84: 664–699.

Cartwright, Lisa (1995) *Screening the Body. Tracing Medicine's Visual Culture*, Minneapolis, MN: University of Minnesota Press.

—— and Marita Sturken (eds) (2001) *Practices of Looking. An Introduction to Visual Culture*, Oxford: Oxford University Press.

de Chadarevian, Soraya (1993a) "Instruments, illustrations, skills and labora-tories in nineteenth-century botany", in Renato G. Mazzolini (ed.), *Non-Verbal Communication in Science Prior to 1900*, Florenz: Leo S. Olschki: 529–562.

—— (1993b) "Graphical method and discipline. Self-recording instruments in nineteenth century physiology", *Studies in History and Philosophy of Science* 24: 267–291.

—— (2003) "Portrait of a discovery. Watson, Crick, and the doppelhelix", *Isis*, 94: 90–105.

—— (2004) "Models and the making of molecular biology", in Soraya de Cha-darevian and Nick Hopwood (eds), *Models. The Third Dimension of Science*, Stanford, CA: Stanford University Press: 339–368.

—— and Harmke Kamminga (eds) (2002) *Representations of the Double Helix*, rev. ed., Cambridge, UK: Whipple Museum of the History of Science.

Cooter, Roger and Stephen Pumpfrey (1994) "Separate spheres and public places. Reflections on the history of science popularization and science in public culture", *History of Science*, 32: 237–267.

Daston, Lorraine (1999) "Objectivity versus truth", in Hans Erich Bödecker, Hans Reill und Jürgen Schlumbohm (eds), *Wissenschaft als kulturelle Praxis 1775–1900*, Göttingen: Vandenhoeck & Ruprecht: 17–32.

—— and Peter Galison (1992) "The image of objectivity", *Representations*, 40: 81–128.

Deilmann, Astrid (2004) *Bild und Bildung. Fotographische Wissenschafts- und Technikberichterstattung in populären Illustrierten der Weimarer Republik (1919–1932)*, Osnabrück: Der andere Verlag.

van Dijk, José (2001) "Bodies without borders. The endoscopic gaze", *International Journal of Cultural Studies*, 4: 219–237.

—— (2005) *The Transparent Body. A Cultural Analysis of Medical Imaging*, Seattle, WA: University of Washington Press.

Dommann, Monika (2003) *Durchsicht, Einsicht, Vorsicht. Eine Geschichte der Röntgenstrahlen 1896–1963*, Zürich: Chronos.

Dürbeck, Gabriele, Bettina Gockel, Susanne B. Keller, Monika Renneberg, Jutta Schickore, Gerhard Wiesenfeldt, and Anja Wolkenhauer (eds) (2001) *Wahr-nehmung der Natur. Natur der Wahrnehmung. Studien zur visuellen Kultur um 1800*, Dresden: Verlag der Kunst.

Eckart, Wolfgang Uwe (1980) "Zur Funktion der Abbildung als Medium der Wis-senschaftsvermittlung in der medizinischen Literatur des 17. Jahrhunderts", *Berichte zur Wissenschaftsgeschichte*, 3: 35–53.

Evans, Jessica and Stuart Hall (eds) (1999) *Visual Culture: The Reader*, London: Sage.

Evans, William and Susanna Hornig Priest (1995) "Science content and social context", *Public Understanding of Science*, 4: 327–340.

Fara, Patricia (1998) "Images of a man of science", *History Today*, 48, 10: 42–49.

Felt, Ulrike, Helga Novotny, and Klaus Taschwer (eds) (1995) *Wissenschaftsforschung. Eine Einführung*, Frankfurt am Main: Campus.

Ferguson, Eugen S. (1992) *Engineering and the Mind's Eye*, Cambridge, MA: MIT Press.

Galison, Peter (1997) *Image and Logic. A Material Culture of Microphysics*, Chicago: Chicago University Press.

—— (1998) "Judgement against objectivity", in Caroline A. Jones and Peter Galison (eds), *Picturing Science. Producing Art*, London: Routledge: 327–359.

Geimer, Peter (ed.) (2002) *Ordnungen der Sichtbarkeit. Fotografie in Wissenschaft, Kunst und Technologie*, Frankfurt am Main: Suhrkamp.

Gerhard, Ute, Jürgen Link, and Ernst Schulte-Holtey (eds) (2001) *Infographiken, Medien, Normalisierung. Zur Kartografie politisch-sozialer Landschaften*, Heidelberg: Synchron Wissenschaftsverlag der Autoren.

Gugerli, David (1999) "Soziotechnische Evidenzen. Der 'pictorial turn' als Chance für die Geschichtswissenschaft", *Traverse*, 6, 3: 131–159.

—— (2002) "Der fliegende Chirurg. Kontexte, Problemlagen und Vorbilder der virtuellen Endoskopie", in David Gugerli and Barbara Orland (eds), *Ganz normale Bilder. Historische Beiträge zur visuellen Selbstverständlichkeit*, Zürich: Chronos: 251–270.

—— and Barbara Orland (eds) (2002) *Ganz normale Bilder. Historische Beiträge zur visuellen Selbstverständlichkeit*, Zürich: Chronos.

Hagner, Michael (1996) "Der Geist bei der Arbeit. Überlegungen zur visuellen Repräsentation cerebraler Prozesse", in Cornelius Borck (ed.), *Anatomien medizinischen Wissens. Medizin, Macht, Moleküle*, Frankfurt am Main: Fischer: 259–286.

—— (2002) "Mikro-Anthropologie und Fotografie. Gustav Fritschs Haarspaltereien und die Klassifizierung der Rassen", in Peter Geimer (ed.), *Ordnungen der Sichtbarkeit. Fotografie in Wissenschaft, Kunst und Technologie*, Frankfurt am Main: Suhrkamp: 252–284.

—— (2006) "Bilder der Kybernetik: Diagramm und Anthropologie, Schaltung und Nervensystem", in Martina Hessler (ed.) *Konstruierte Sichtbarkeiten. Wissenschafts- und Technikbilder seit der Frühen Neuzeit*, München: Wilhelm Fink Verlag: 383–404.

Hankins, Thomas L. (1999) "Blood, dirt, and nomograms. A particular history of graphs", *Isis*, 90: 50–80.

te Heesen, Anke (2002) "Elektrisieren und Heilen. Vier verschiedene Betrachtungen zu einem Kupferstich der Aufklärungszeit", *Internationale Zeitschrift für Geschichte und Ethik der Naturwissenschaften, Technik und Medizin (NTM)*, 10: 209–221.

—— (1997) *Der Weltkasten. Die Geschichte einer Bildenzyklopädie aus dem 18. Jahrhundert*, Göttingen: Wallstein; trans. as *The World in a Box. The Story of an Eighteenth-Century Picture Encyclopedia*. Chicago: Chicago University Press.

Heintz, Bettina and Jörg Huber (eds) (2001) *Mit dem Auge denken. Strategien der Sichtbarmachung in wissenschaftlichen und virtuellen Welten*, Wien: Springer Verlag.

Henderson, Kathryn (1995) "The visual culture of engineers", in Susan Leigh Star (ed.), *The Cultures of Computers*, Oxford: Blackwell: 196–218.

Hennig, Jochen (2004) "Vom Experiment zur Utopie. Bilder in der Nanotechnologie", *Bildwelten des Wissens*, 2, 2: 9–18.

Hentschel, Klaus (2002) *Mapping the Spectrum. Techniques of Visual Representation in Research and Teaching*, Oxford: Oxford University Press.

—— (2005) "Wissenschaftliche Photographie als visuelle Kultur. Die Erforschung und Dokumentation von Spektren", *Berichte zur Wissenschaftsgeschichte*, 28: 193–214.

Hess, Volker (2002) "Die Bildtechnik der Fieberkurve. Klinische Thermometrie im 19. Jahrhundert", in David Gugerli and Barbara Orland (eds), *Ganz normale Bilder. Historische Beiträge zur visuellen Selbstverständlichkeit*, Zürich: Chronos: 159–80.

Heßler, Martina (2005) "Bilder zwischen Kunst und Wissenschaft. Neue Herausforderungen für die Forschung", *Geschichte und Gesellschaft*, 31: 266–292.

—— (ed.) (2006) *Konstruierte Sichtbarkeiten. Wissenschaft- und Technikbilder seit der Frühen Neuzeit*, München: Wilhelm Fink Verlag.

Hilgartner, Stephen (1990) "The dominant view of popularisation. Conceptual problems, popular uses", *Social Studies of Science*, 20: 519–539.

Höhler, Sabine (2002a) "Profilgewinn. Karten der Atlantischen Expedition (1925–1927) der Notgemeinschaft der Deutschen Wissenschaft", *Internationale Zeitschrift für Geschichte und Ethik der Naturwissenschaften, Technik und Medizin (NTM)*, 10: 234–246.

—— (2002b) "Depth records and ocean volumes: Ocean profiling by sounding technology, 1850–1930", *Technology and Culture*, 18: 119–154.

Hopwood, Nick (2004) "Plastic publishing in embryology", in Soraya de Chadarevian and Nick Hopwood (eds), *Models. The Third Dimension of Science*, Stanford, CA: Stanford University Press: 170–206.

—— (2005) "Visual standards and disciplinary change. Normal plates, tables and stages in embryology", *History of Science*, 43: 239–303.

—— (2006) "Pictures of evolution and charges of fraud. Ernst Haeckel's embryological illustrations", *Isis*, 97: 260–301.

Howells, Richard (2003) *Visual Culture*, Cambridge, UK: Polity Press.

International Human Genome Sequencing Consortium (2001) "Initial sequencing and analysis of the human genome", *Nature*, 409: 860–921.

Jacobi, Daniel and Bernd Schiele (1989) "Scientific imagery and popularized imagery: Differences and similarities in the photographic portraits of scientists", *Social Studies of Science*, 19: 731–753.

Jäger, Jens (2000) *Photographie: Bilder der Neuzeit. Einführung in die Historische Bildforschung*, Tübingen: Edition Diskord.

Jenkins, Reese V. (1987) "Words, images, artifacts and sound: Documents for the history of technology", *British Journal for the History of Science*, 20: 39–56.

Jones, Amelia (ed.) (2003) *The Feminism and Visual Culture Reader*, London and New York: Routledge.

Jones, Caroline A. and Peter Galison (eds) (1998) *Picturing Science. Producing Art*, London: Routledge.

Jordanova, Ludmilla (1990) "Medicine and the visual culture", *Social Studies of Medicine*, 3: 89–99.

—— (2000) *Defining Features. Scientific and Medical Portraits 1660–2000*, London: Reaktion Book.

Klein, Judy L. (1997) *Statistical Visions in Time. A History of Time Series Analysis, 1662–1938*, Cambridge, UK: Cambridge University Press.

Knorr-Cetina, Karin (1999) "'Viskurse' der Physik. Wie visuelle Darstellungen ein Wissenschaftsgebiet ordnen", in Jörg Huber und Martin Heller (eds), *Konstruktionen. Sichtbarkeiten*, Wien: Springer Verlag: 245–263.

Krohn, Wolfgang (ed.) (2006) *Ästhetik in der Wissenschaft. Interdisziplinärer Diskurs über das Gestalten und Darstellen von Wissen*, Hamburg: Felix Meiner Verlag.

Latour, Bruno (1987) *Science in Action. How to Follow Scientists and Engineers Through Society*, Cambridge, MA: Harvard University Press.

—— (1990) "Drawing things together", in Michael Lynch and Steve Woolgar (eds), *Representation in Scientific Practice*, Cambridge, MA: MIT Press: 19–68.

Lefèvre, Wolfgang (ed.) (2004) *Picturing Machines. 1400–1700*, Cambridge, MA: MIT Press.

Lubar, Steven (1995) "Representation and power", *Technology and Culture*, Supplement to volume 36, 2: 54–81.

Lynch, Michael (1998) "The production of scientific images. Vision and re-vision in the history, philosophy, and sociology of science", *Communication & Cognition*, 31, 2/3: 213–228.

—— and Samuel Y. Edgerton (1988) "Aesthetics and digital image processing. Representational craft in contemporary astronomy", in Gordon Fyfe and John Law (eds), *Picturing Power. Visual Depiction and Social Relations*, London: Routledge and Kegan Paul: 184–220.

—— and Steve Woolgar (1990) *Representation in Scientific Practice*, Cambridge, MA: MIT Press.

Maehle, Andreas-Holger (1993) "The search for objective communication. Medical photography in the nineteenth century", in Renato G. Mazzolini (ed.), *Non-Verbal Communication in Science Prior to 1900*, Florenz: Leo S. Olschki: 563–586.

de Marchi, Neil and Craufurd D.W. Goodwin (eds) (1999) *Economic Engagement with Art*, Durham, NC, and London: Duke University Press.

Mitchell, W.J.T. (1992) "The pictorial turn", *Artforum*, March: 89–94.

—— (2005) *What Do Pictures Want? The Lives and Loves of Images*, Chicago: University of Chicago Press.

Müller, Falk (2005) "Zwischen Bilderbuch und Meßgerät. Der elektronenoptische Blick auf die Realstruktur von Festkörpern", in Martina Heßler (ed.), *Konstruierte Sichtbarkeiten. Wissenschafts- und Technikbilder seit der Frühen Neuzeit*, München: Wilhelm Fink Verlag: 75–98.

Mulvey, Laura (1975) "Visual pleasure and narrative cinema", *Screen*, 16: 3, 6–18.

Nelkin, Dorothy and M. Susan Lindee (1995) *The DNA-Mystique. The Gene as Cultural Icon*, London: Freeman.

Nikolow, Sybilla (2006) "Imaginäre Gemeinschaften. Statistische Bilder der Bevölkerung", in Martina Heßler (ed.), *Konstruierte Sichtbarkeiten. Wissenschafts- und Technikbilder seit der Frühen Neuzeit*, München: Wilhelm Fink Verlag: 263–278.

—— and Lars Bluma (2002) "Bilder zwischen Öffentlichkeit und wissenschaftlicher Praxis. Neue Perspektiven für die Geschichte der Medizin, Naturwissenschaften und Technik", *Internationale Zeitschrift für Geschichte und Ethik der Naturwissenschaften, Technik und Medizin (NTM)*, 4: 201–208.

—— and Arne Schirrmacher (eds) (2007) *Wissenschaft und Öffentlichkeit als Ressource füreinander. Studien zur Wissenschaftsgeschichte im 20. Jahrhundert*, Frankfurt am Main: Campus Verlag.

Pang, Alex-Sooyung Kim (1997a) "Visual representation and post-constructivist history of science", *Historical Studies in the Physical and Biological Sciences*, 28: 139–171.

———— (1997b) "Stars should henceforth register themselves. Astrophotography at the early link observatory", *British Journal for the History of Science*, 30: 177–202.

Paul, Gerhard (2004) *Bilder des Krieges—Krieg der Bilder. Die Visualisierung des modernen Krieges*, Paderborn: Verlag Ferdinand Schöningh.

Pörksen, Uwe (1997) *Weltmarkt der Bilder. Eine Philosophie der Visiotype*, Stuttgart: Klett-Cotta.

Prodger, Phillip (1998) "Illustration as strategy in Charles Darwin's *The Expression of the Emotions in Man and Animals*", in Timothy Lenoir (ed.), *Inscribing Science. Scientific Texts and the Materiality of Communication*, Stanford, CA: Stanford University Press:140–181.

Reiser, Stanley Joel (1978) *Medicine and the Reign of Technology*, Cambridge, UK: Cambridge University Press.

Rheinberger, Hans-Jörg, Michael Hagner, and Bettina Wahrig-Schmidt (eds) (1997) *Räume des Wissens. Repräsentation, Codierung, Spur*, Berlin: Akademieverlag.

Roeck, Bernd (2003) "Visual turn? Kulturgeschichte und die Bilder", *Geschichte und Gesellschaft*, 29: 294–315.

Rudwick, Martin J.S. (1976) "The emergence of a visual language for geological science 1760–1840", *History of Science*, 14: 149–195.

———— (1992) *Scenes from Deep Time. Early Pictorial Representations of the Prehistoric World*, Chicago, IL: University of Chicago Press.

Schäfer, Armin (2004) "Das Gewebe aus Sichtbarem und Sagbarem", *Zeitschrift für Ästhetik und allgemeine Kunstwissenschaft*, 49, 2: 281–294.

Schaffer, Simon (1998) "On astronomical drawing", in Caroline Jones and Peter Galison (eds), *Picturing Science. Producing Art;* London: Routledge: 441–474.

Schickore, Jutta (2002) "Fixierung mikroskopischer Beobachtungen. Zeichnung, Dauerpräparat, Mikrofotographie", in Peter Geimer (ed.), *Ordnungen der Sichtbarkeit. Fotografie in Wissenschaft, Kunst und Technologie,* Frankfurt am Main: Suhrkamp: 285–310.

Schlich, Thomas (1995) "Wichtiger als der Gegenstand selbst. Die Bedeutung des fotographischen Bildes in der Begründung der bakteriologischen Krankheitsauffassung durch Robert Koch", in Martin Dinges and Thomas Schlich (eds), *Neue Wege in der Seuchengeschichte*, Stuttgart: Franz Steiner Verlag: 143–174.

———— (1997) "Die Representation von Krankheitserregern. Wie Robert Koch Bakterien als Krankheitsursache dargestellt hat", in Hans-Jörg Reinberger, Michael Hagner, and Bettina Wahrig-Schmidt (eds), *Räume des Wissens. Repräsentation, Codierung, Spur*, Berlin: Akademieverlag: 165–190.

Secord, Anne (2002) "Botany on plate. Pleasure and the power of pictures in promoting early-nineteenth-century scientific knowledge", *Isis*, 93: 28–57.

Shapin, Stephen (1990) "Science and public", in R.C. Olby, G.N. Cantor, J.R.R. Christie, and M.J.S. Hodge (eds), *Companion to the History of Modern Science*, London: Routledge: 990–1006.

Sichau, Christian (2004) "Wissenschaftliche Instrumente und das Bild der Wissenschaft", *Bildwelten des Wissens*, 2, 2: 37–45.

Stafford, Barbara (1994) *Artful Science. Enlightenment, Entertainment and the Eclipse of Visual Education*, Cambridge, MA: MIT Press.

Star, Susan Leigh and James R. Griesemer (1989) "Institutional ecology. Translations and boundary objects. Amateurs and professionals in Berkeley's Museum of Vertebrate Zoology. 1907–39", *Social Studies of Science*, 19: 387–420.

Tanner, Jakob (2002) "Wirtschaftskurven. Zur Visualisierung des anonymen Marktes", in David Gugerli and Barbara Orland (eds), *Ganz normale Bilder. Historische Beiträge zur visuellen Selbstverständlichkeit*, Zürich: Chronos: 129–158.

Tucker, Jennifer (1997) Photography as witness, detective, and impostor. Visual representation in Victorian science", in B. Lightman (ed.), *Victorian Science in Context*, Chicago: Chicago University Press: 378–408.

────── (2006) "The historian, the picture, and the archive", *Isis*, 97: 111–120.

Venter, J. Craig et al. (2001) "The sequence of the human genome", *Science*, 291: 1304–1351.

Voss, Julia (2007) *Darwins Bilder. Ansichten der Evolutionstheorie 1837–1874*, Frankfurt am Main: Fischer.

Watson, James (1968) *The Double Helix. A Personal Account of the Discovery of the Structure of DNA*, London: Weidenfeld & Nicolson.

Weingart, Peter (2001) *Die Stunde der Wahrheit. Zum Verhältnis der Wissenschaft zu Politik, Wirtschaft und Medien in der Wissensgesellschaft*, Weilerswist: Velbrück Wissenschaft.

────── (2005) *Die Wissenschaft der Öffentlichkeit. Essays zum Verhältnis von Wissenschaft, Medien und Öffentlichkeit*, Weilerswist: Velbrück Wissenschaft.

Werner, Gabriele (2001) "Das Bild vom Wissenschaftler. Wissenschaftler im Bild", *Kunsttexte.de, Sektion Bild/Wissen/Technik*, No. 1, http://www.kunsttexte. de

Whitley, Richard (1985) "Knowledge producers and knowledge acquirers. Popularisation as a relation between fields and their publics", in Terry Shinn and Richard Withley (eds), *Expository Science. Forms and Functions of Popularisation*, Boston, MA: Reidel: 3–28.

Wiesenfeldt, Gerhard (2001) "Säkularisierung der Naturerkenntnis. Zur bildlichen Darstellung von Experimenten in Lehrbüchern des 18. Jahrhunderts", in Gabriele Dürbeck, Bettina Gockel, Susanne B. Keller, Monika Renneberg, Jutta Schickore, Gerhard Wiesenfeldt, and Anja Wolkenhauer (eds.), *Wahrnehmung der Natur. Natur der Wahrnehmung. Studien zur visuellen Kultur um 1800*, Dresden: Verlag der Kunst: 103–116.

────── (2002) "Politische Ikonographie von Wissenschaft. Die Abbildung von Teylers 'ungemein großer' Elektrisiermaschine, 1785/87", *Internationale Zeitschrift für Geschichte und Ethik der Naturwissenschaften, Technik und Medizin (NTM)*, 10: 222–233.

Wise, Norton (2006) "Making visible", *Isis*, 97: 75–82.

Part II
Towards a Science of Images

3 Image Science

W.J.T. Mitchell

Everyone knows that science uses imagery, both verbal and visual, as an essential part of its quest for ever more accurate accounts of material reality. Models, diagrams, photographs, graphs, sketches, metaphors, analogies, and equations (the whole Peircean family of icons or signs by resemblance) are crucial to the life of science. They introduce whole ways of seeing and reading, particularly in dazzling figures of thought such as string theory, which seeks an elegant universe, a multiverse of parallel worlds and supple spaces, folding in upon themselves into wormholes, sparticles, and gravitrons. And these images do not remain in the sphere of technical science, but quickly circulate into popular culture, especially cinema and special effects video (as in the PBS Nova series). As an institution, science is quite gifted at presenting itself in mass media, in popular writing as well as visual media. From the paleontologists' reconstructions of extinct life-forms like the dinosaur, to the model of the atom, to the speculative images that circulate across the borders between science and philosophy, science and science fiction, science and poetry, reality and mathematics, science is riddled with images that make it what it is—a multi-media, verbal-visual discourse that weaves its way between invention and discovery.

But amid all the proliferation of scientific images there is one conspicuously missing component, and that is a scientific focus on images themselves. I don't mean to suggest that scientists fail to examine images critically, trying to sort out the true from the false, the misleading or fanciful from the verifiable and accurate. I'm thinking of a more general problem, a science of images or *Bildwissenschaft* as such, which would treat images as *objects* of scientific investigation and not merely as useful tools in the service of scientific knowledge. So I would like to take the topic of this conference and turn it "inside out," as it were, and reframe the discussion of "images of science and the scientist" inside some reflections on the science of images. In particular, I would like to raise the following questions:

1. Is a "science of images" even conceivable, or are images, as social and cultural constructions, simply outside the domain of the sciences in the usual sense of the word, so that their proper domain is that of the

arts and humanities, the realm of interpretation, appreciation, and performance, rather than that of empirical investigation and abstract, rational, or even mathematical modeling? If your answer to this question is no, then you should probably stop reading now, because I want to proceed on the assumption that the answer is yes, a science of images is conceivable (and in fact there are a number of researchers already committed to the idea, which has been around in one form or another, sometimes under the rubric of "iconology," or the theory of images, as distinct from "iconography," that lexical sorting of different kinds of images).

2. If there is a science of images, what kind of science would it be? Would it be an experimental science like physics or chemistry, or a historical science like paleontology and geology? What would be its relation to mathematics, and specifically to the role of geometry, diagrams, and other graphic expressions as instruments of mathematical thinking? How would a science of images align itself with divisions between the "physical" and the "biological" sciences? Would it be a theoretical, speculative science or a practical and technical discipline such as medical science? How would a science of images distribute itself between the "hard" and "soft" sciences?

3. If there were a science of images, what use would it be to the other sciences? Would it provide a method of separating true and verifiable images from false and misleading ones? Would it help to settle what Peter Galison describes as the ongoing battle between pictures and logic, images and algorithms in scientific procedures, a battle that has its counterpart in the arts in the phenomenon Leonardo da Vinci called a *paragone*, the contest of words and images, poetry and painting?

The ever-elastic concept of science itself as a synonym for knowledge of any sort, embracing the human as well as natural sciences, seems on the one hand to empty the idea of science of any specificity, so that a science of images would just be any knowledge of images whatsoever. On the other hand, and in quite the opposite way, we seem to be trapped at the outset by the very terms of our topic—"images of science and the scientist"—that is, by stereotypes and caricatures of white-coated technicians, gleaming research laboratories, supercomputers, supercolliders, and super-geniuses who are depicted either as madmen who threaten to destroy the world, or as benign sages who will show us how to save it. We are also led astray by a mainly experimental-quantitative picture of science which portrays it as a rather mechanical activity of proof and demonstration, gathering of data, and establishing of certainty and positive knowledge.

As for what might be called the "unscientific" or "soft" sciences, there the name of science is generally taken to be a meaningless courtesy. Only the Germans seem comfortable with attaching *wissenschaft* to words like *Kultur* and *Bilder*. The English-speaking scientist, hard-headed and empirical, tends to contrast "real science" with the world of culture and images, the arts and letters. In the domain of images and culture, presumably, opinion and ungrounded speculation rule, empiricism is a dirty word, data are haphazardly gathered or ignored, and impressionistic, unreliable results are acceptable if they are intuitively pleasing. Our stereotype of the social scientist is thus of someone who is gathering dubious data based in subjective opinions in order to confirm what is already common sense knowledge. Our stereotype of the political scientist is of someone who has almost nothing to do with science at all. And the economist, as everyone knows, is a practitioner of "the dismal science," filled with tedious charts and statistics that produce "predictions" that rival the readings of tea leaves and animal entrails. Our stereotype of the human scientist is of an absent-minded drudge who labors endlessly in the dusty stacks of the library, only to emerge with a set of "findings" that are of interest to absolutely no one else, or with a wild theory that is attractive precisely in proportion to its implausibility and perversity. Our picture of C. P. Snow's "two cultures" of science and the humanities, in other words, seems to dictate from the outset that the study of images would belong in one domain, science in the other. Images might be instrumental to the pursuit of science, then, but they would be at best incidental, ornamental, or functional, not an essential part of, much less the target of, science. Their proper home would be in the domain of culture and psychology, in the realm of subjective perceptions, and poetic associations.

Rather than settle for this impossible dilemma in which, on the one hand, any cognitive activity takes on the name of science, or a rigid stereotype of a specific form of science dominates our thinking, I want to think about the range of scientific models that might be brought to bear on the problem of images and ask which ones seem especially promising—or not. What would such a science be? How would one move from the image considered as the instrument or medium of science, to the image as the object of science, something to be tested, experimented with, described, explained, in accordance with the most rigorous scientific methods? What would it mean to run an experiment on an image? Would it involve investigation of the material particles in an oil painting? (NYU's Institute of Fine Arts actually requires considerable training in chemistry if one wants training in restoration and connoisseurship). Insofar as images are material things, objects in the world, chemistry and physics have something to contribute to their understanding, but mainly at the level of forensics, the detection of forgeries, the examination of the physical body or support in or on which the image appears. The science appropriate to the study of images has to include this level, but it also has to be an optical science, one that pays attention to visual perception and imagination, to optical illusions, reflections, transparency, and translucence. The science of images, then, would have to consider not just material objects, but the spaces between objects, and the light that is transmitted from one thing to another. Insofar as images appear in non-visual media like language, it would have to engage with linguistics, with psycho-linguistics, and with the study of logical as well as spatial relations. Since images are generally defined as "icons," or signs by similarity, it would have to be a science of similitude, analogy, and likeness, as well as of dissimilarity, adjacency, and difference.

The science of images would also have to be a historical science, one that deals with the spatial and temporal circulation of images, their migrations from one place or epoch to another. It would have to be a science that looked at images as groupings, families, classes, linked by resemblance to other things in the world and to each other. It would have to be a science that registered the capacity of images to represent reality, but which also recognized that images can be highly misleading and deceptive, precisely because they have such a capacity to engage our trust with their seemingly immediate testimony to the visible, the palpable, the concrete world of experience. (A science of images would have to address the very divided reputation of images within science, the tendency of scientists, as Peter Galison has observed, to divide themselves into iconoclastic and iconophobic camps, with some researchers putting great stock in the usefulness of images, others regarding them as mere distractions that prevent the hard questions from being addressed.)

Image science would have to be, and has been, a cognitive science, an empirical study of the conditions of human perception, of the centers of pattern recognition, image formation, and transformation in the brain and the mind. Given the affective potency of images in human consciousness, however, it seems clear that a cognitive science would not be enough by itself. It would have to be complemented by a psychology that reckoned with the unconscious effects of imagery, their uncanny ability to lure, seduce, and even traumatize the beholder. The science of imagery could not just be about pattern recognition, but about mis-recognition, fantasy, dreaming, and hallucination. It would have to be about memory images, but also about *false* memory, screen memory, and the dubious status of "recovered memory."

If science in general uses images as part of its cognitive apparatus, then it seems clear that a science of images would also have to use images to get at its subject, but it would then be in the curious position of doubling that subject, of working with *images of images*, or what I have called "metapictures" to capture its object. Metapictures are pictures that show us what pictures are, how they function, where they are located, and they are seen most literally in the familiar sight of pictures-within-pictures, a "nesting" of a picture inside another picture. A science of images, then, could try to be rigorously iconoclastic, and limit its representations of images to non-pictorial, non-graphic forms (sentences, descriptions, equations, etc.) or it could accept the inevitability of the metapicture, and grapple with the vertiginous abyss of infinite regress that it seems to open up when we use pictures to understand pictures.

My own impulse, of course, is to hurl myself into the abyss. And not into the abyss of culture, society, or politics, where the "soft" science of images seems most at home, but into the hard sciences of mathematics, physics, and biology, places where images abound, but where a science of images has yet to emerge.

DIAGRAMMATOLOGY

Peter Galison's essay on the *paragone* between the iconophiles and iconoclasts of mathematics provides a metapicture of the role of images in the abstract science of numbers, functions, and logic. Galison reconstructs this struggle at many levels: as a debate between an "intuitive" approach to mathematics, stimulated by visual, pictorial, and sculptural model-building, and an approach governed by logic, calculation, and demonstration. It is also a debate between analog and digital media, between "concrete" and "abstract" renderings of problems, between an "eyes open" and "eyes shut" approach to problem solving. As should be clear from this partial inventory, there is something quite slippery and even incoherent about this array of oppositions. Galison notes that, despite the vehement emotions often expressed, the battle lines are never very well-defined, and that figures such as Poincaré and David Hilbert, who seem to occupy opposite sides, often wind up defecting to the enemy camp at critical moments. Galison concludes that there is something illusory about the whole struggle:

> At the heart of the scientific image is the search for rules: at the heart of the logical-algorithmic has been a hunt for the recognition that is the eternal promise of representation. Said another way: the impulse to draw the world in its particularity never seems to be able to shed itself of the impulse to abstract, and that search for abstraction is forever pulling back into the material-particular. (Galison: 2002: 302)

The question arises, then, is the difference between images and logic an illusory frontier, a poorly analyzed distinction that, on reflection, dissolves into a general science of signs, a semiotics that would transcend all the superficial barriers between words and images, numbers and diagrams, the abstract and the concrete? Galison's own rhetoric—"the eternal promise of representation" that is "forever pulling back into the material-particular"—is quite equivocal, as if he were describing something between a religious-metaphysical utopia (an "eternal promise") and a physical force, a gravitational or electromagnetic field "forever pulling." His historical account of the *paragone* of image and logic in twentieth-century science, on the other hand, suggests that the invention of a new entity, the so-called "digital image," has overcome the ancient division and rendered crossovers and conversions routine:

> But now, ever more intensively, the routinization of analog-to-digital and digital-to-analog conversions has made the flickering exchanges

routine: image to non-image to image. No longer only set in motion at moments of crisis, we find that ordinary, every-day science propels this incessant oscillation: "Images scatter into data, data gather into images." (Galison: 2002: 322)

Is the ancient quarrel, with its eternal promises and endless tug of war about to be settled once and for all? Has the image finally been tamed by the rule of the computer, the digitization of the analog sign? Galison does not answer this question, and perhaps it has no answer, though intuitively, I think, we want to respect both the philosophical and historical impulses in his discussion. That is, there seems to be something *both* eternal and historical about this problem, as if every solution simply reintroduces the problem at a new level.

This impasse is, in my view, precisely what necessitates a science of the image, rather than a merely instrumental use of images as unexamined instruments in getting at other kinds of reality. The first step in this science, however, has to be some definition of its object. What is an image that it can, as Galison argues, "scatter into data" or (conversely) appear as a "gathering" of data? I have written on this question elsewhere, and simply want to rehearse my conclusions briefly here. First, I take as axiomatic the intuition of C.S. Peirce that an image is an icon; that is, a sign by resemblance. This means, of course, that the whole notion of the specifically "visual image," and the accompanying language of intuition, concreteness, perceptual immediacy, graphic arrays, and so forth, has to be put into question. One of Peirce's deepest insights, often forgotten, was that the algebraic equation is no less an icon than its diagrammatic rendering in two dimensional space. The equals sign (=) is itself an indexical sign, like the brackets of set theory, or other relational pointers (< as "less than"; > as "greater than"; congruence, similarity, and the signs for operations such as addition, subtraction, division, etc). At the heart of logic and mathematics, then, the iconic relationships of identity and equivalence, similitude and difference, are lurking. These relationships also obtain in some pictures and diagrams, as well as in language, so that we speak of "verbal icons" and mean by that expression both the names and descriptions of objects, and the figurative comparison of one object to another. Both names and metaphors are "verbal images," but in very different senses of the word *image*. An image is a double sign, then, naming something we see, like a portrait or landscape or graph, and something we comprehend as a signifying relation to something beyond itself: *this* portrait represents *that* person. Our ordinary language captures this double relation when we say that a portrait "looks like" the person it represents. Looking and likening, seeing and similitude are fused, and *con*fused in ordinary representational images, which is why we run into difficulty when we encounter an image that doesn't look like anything, or doesn't look like what it represents. The scientific image of the atom as a kind of miniature solar system, for instance, is widely understood to be a completely false picture of the atom, *if* we take it as a picture that is supposed to "look like" actual atoms. It is rather a model that attempts to capture some features of the atom that are graspable in other, mainly quantitative terms. It is surely an image, and a visual image, but not one that looks like what it represents.

If we are going to have a science of images, then, the first step is to release it from the tyranny of the physical eye, literally understood, and understand that images (as icons) circulate through many domains: there are mental, and mathematical, and verbal images as well as pictorial and visual images. The images we should be concerned with in science are *not* just the pictures, graphs, and physical models, but also the metaphors that provide pictures of a domain of research—the universe "as" a heat-engine, or a clock, or a ball of string-pictures that need not be made visible or drawn graphically. Images are not medium-specific, though they are never encountered outside of some medium or other. An image can move from one medium to another, appearing now as an equation, now as a diagram, now as a figure in a narrative, and now as a figure in a narrative painting. Panofsky called the image a "motif" to emphasize its repeatability in many different pictures, but he failed to draw the obvious conclusion that resides in our ordinary language of talking about images: the image and the picture are very distinct, yet intimately linked entities. The English language (but not German) registers this distinction when we say that we can hang a picture, but it would seem odd to speak of hanging an image. The picture is the material support, the physical medium (whether stone or paint, metal or electromagnetic impulses) in which the image appears. But the image "as such," if we can speak of such a thing, is not itself a material thing though it must always appear in or on some material support—a statue, an embodied perceiver. An image is a relationship and an appearance: it might be better, in fact, to think of images as events or happenings than as objects, in order to register their often fleeting temporality (appearing and disappearing, going in and out of focus, or, in Galison's lovely metaphor, scattering and gathering). We might then want to speak, with phenomenologists like Bachelard and Merleau-Ponty, of the "onset" of the image, or with Wittgenstein about the "dawning of an aspect." But only an immaterial, fantasmatic conception of the image that treats its being in what Derrida described as a "hauntology" can capture its spectral nature.

THE PHYSICS OF THE IMAGE

I realize that the foregoing remarks will convict me of being thought an idealist or worse, and that this is an unfashionable position in a time when invocations of materiality and embodiment are de rigueur. But a monistic materialism can never grasp the specific materiality of the image. For that, we need a dialectical materialism of the sort that led Marx to solve the riddle of the commodity: "So far," Marx points out, "no chemist has ever discovered exchange-value either in a pearl or a diamond." (I. 83) Exchange value is not a physical property of objects, but of the circulation and exchange of objects, their alienation from use, their abstraction from their concrete, material properties. Images are another form of the exchange value of things, operating primarily at the perceptual and cognitive level, though of course the commodification of images is a familiar enough phenomenon, and the fetishism of commodities marks precisely that moment when the spectral, fantasmatic character of images seems to settle like an aureole around the physical body of an object.

But an even simpler demonstration of the peculiar physics of the image can be glimpsed if we raise the question of their destruction. Iconoclasm is the effort to destroy images, usually for political or religious reasons, though Galison's account of twentieth-century science makes it clear that there can be professional and epistemological motives for the effort to banish images. But his story also makes it clear that the image invariably comes back in a kind of "return of the repressed," even in the thinking of the most relentless iconophobes. This situation could have been predicted, of course, by a historical reflection on the age-old crusade to stamp out idolatry, to purge the world of graven images, and even of verbal images. This is not simply a matter of grinding or melting down a Golden Calf, scattering it on water, and forcing the Israelites to drink it, as in the Exodus story. (As we know, this sort of materialist effort to destroy an image always fails: the Calf survives as a verbal image in the very narrative that tells of its destruction, and is reborn as a visual, graphic image in scores of Renaissance paintings). So the effort to destroy images cannot rest with their mere sculptural or graphic rendering. The most persistent effort to achieve utter annihilation of images in words and even in thought is the commentaries of Maimonides' *Guide for the Perplexed*, which finds that even the language of the Bible itself is riddled with misleading metaphors and concrete words that attribute things like a body, a face, hands, feet, and spatial location to the invisible, unrepresentable deity. The mandate of iconoclasm is, finally, not just the destruction of graven images, but the purging of words and ideas as well, to arrive at a purified language and consciousness that is capable of thinking about god without thinking about anyone or anything. This is, of course, an impossible, unattainable state, but it has the virtue of revealing just how impossible it would be to follow or enforce the second commandment, prohibiting the making of images of any kind and of any thing. The destruction of images, as Michael Taussig has argued, is a sure way of guaranteeing them an even more potent presence in memory, or as reincarnated in new forms.

So a fundamental law of the physics of images is: images cannot be destroyed. The picture or physical support in which they appear can be destroyed, but the image survives destruction, if only as a memory in the mind—that is, the body—of the destroyer. The question arises, then: if we are going to pursue the metaphor of a physics of the image, is it subject to a "law of conservation" similar to the one that governs matter and energy in the physical world? That is, should we say that images can neither be created nor destroyed? It is easy to see why it is so difficult to destroy images, but creation seems to be another matter. Surely new images are always being created by artists and scientists, as well as by ordinary people, from the child's first drawing to the everyday snapshot. Here I think we have reached a boundary of our understanding, but my intuition (if that will be allowed) is that images cannot be created, at least not *ex nihilo*. Insofar as images are always images *of* something, then what they are images of must always logically and temporally precede them. We say that a child is the image of its parent, by which we mean that there is a discernible family resemblance, even though we also know that in some other respects the child looks nothing like its parent. The image, then, is not the bearer of the new, the different in the child, but of what was already present in the parent. The rule of likeness is a conservative rule, defying innovation and insisting on the return of the similar. This is true, I suspect, even when we are attempting to create a totally new, original image, and explains why it is so difficult, if not impossible, to imagine what it would mean to create a radically new image. The efforts of the surrealists are especially instructive in this regard, for their wildest innovations are invariably discovered to be conjunctions of already existing images in novel combinations. One could call these "new images," but it would have the same kind of force that is entailed in uttering a "new sentence" in a language, where the sentence entirely consists of words that are themselves not new. Perhaps we should say, then, that new combinations of images can be created, and even new images in the sense that we speak of "coinages" in language (which are invariably old recognizable words morphed into something new). But if an image (or a word) were *completely* new, how would we recognize it? It is this moment of recognition that makes the image readable as such, and that provides the thread of continuity with variation, deviation, and difference that makes it possible to see images "morphing" (as in a Michael Jackson video) from one identity to another. This morphing would be purely abstract and non-referential if it did not pass through moments of stillness—"freeze frames" as it were—in which the multiple identities were registered as *this* image or *that* image. And even if we imagine the morphing of an abstract image—a very concrete and technical possibility—the various stages of morphogenesis would each have a specific gestalt as *this* form or *that* form. Perhaps the only sense in which a new image could appear would be in some composite or synthetic or transitional appearance, as in the Galtonian photographs that blur together several portraits to produce a portrait that looks "strangely familiar," but is of no individual who ever existed.

THE BIOLOGY OF IMAGES

I warned you at the outset that the search for a science of images might lead us into an abyss of speculation, and I hope that so far you have not been disappointed. We have traced the mathematics of images as diagrammatic and logical relations, and the physics of images as immaterial, fantasmatic entities that require a physical medium to make their appearance. But what about the life sciences: Could there be a "natural history" of images built around a metapicture of images as organisms or life-forms? This question can be broached by returning to the question of the Galtonian photograph as a "new" or "created" image. The reason this Galtonian image is "strangely familiar"—both old and new—is that it broaches the question of the image as a *type* or "typical" representation, rather than as a representation of an individual. We are familiar enough with this phenomenon in the realm of stereotypes and what might be called "reductive" or "schematic" images. The smiley face on the bumper sticker is recognizable *as a face*, but not as any particular face. In fact, when we speak of "recognizing" what an image represents, the form of re-cognition involved can be quite general and abstract: it can amount to seeing something as a face or a body without seeing it as any particular face or body, just as, in geometry we can recognize something as a square or circle without thinking of it as a unique, particular entity. The specific drawing or diagram may function as the *token* of a *type*, a concrete embodiment of a quite general and generic image, one that can be translated into an algebraic icon in Peirce's sense (x=y, in the case of the square; C=πr² in the case of the circle).

But this generalizing property of images is exactly what links them to the life sciences, and quite specifically to the concept of the species and the specimen. One thing we could not account for in the physical model of the image was the question of morphing, transformation, and the genesis of "family resemblance" across a series of images. But the metaphor of the image as life-form brings this process into focus, at the same time that it raises a whole new set of difficulties. If images are like living things (rather than the spectral, ghostly entities we encountered in the realm of physics) then surely they can be created and destroyed. But here we must remind ourselves that we are constructing a multi-tiered stack of analogies in which material objects are to apparitions as pictures are to images, as specimens are to species. I have never argued that a picture or a material object cannot be created or destroyed. This painting, that statue, this manuscript with those equations and diagrams can surely be created and destroyed. But think of what it would mean to destroy a species, rather than a specimen. Not impossible, perhaps (and quite a realistic prospect in an era that is defined by its consciousness of "endangered species" and extinctions). So perhaps the move from the physics to the biology of the image does reveal a level of our science that it was not possible to address within the sphere of physical, inanimate matter. It is in the sphere of the life sciences that our science of images would confront the problem of the *reproduction* of images, their *mutations* and *evolutionary transformations*. If the image is to iconology what the species is to biology, then pictures (in an extended sense that includes sculpture and other material constructions or "situations") are the specimens in a natural history of images. This natural history is, of course, also a cultural and social and political history, but it is one that is focused on the "second nature" that we have created around ourselves—the entire image-repertoire of human consciousness and civilization. We have always understood that the arts were, as Aristotle insisted, "imitations of nature," and that this meant not just that they represented or resembled the natural world, but that they themselves were a kind nature "in process," as expression of the species identity of human beings. Now we live at a moment of crisis when the human is regarded by some as an extinct or endangered species, when the "posthuman" looms as the horizon of speculative thought. At the same time, we are told that the ancient, indestructible domain of images has been mastered finally by the "digital," and that numbers, calculations, and mechanical operations will finally replace us in an infinite circuit of information. (Kittler 1999). Neither posthumanism nor the digital image seem to me especially coherent or promising as concepts, but they belong together, I think, as symptoms of a failure to think historically or philosophically, and to take refuge in a curious kind of post-historical presentism. And they also help us to see why the two most conspicuous and highly publicized "natural images" of our time dominate our picture of the fate of our own species. I'm thinking of the fossil and the clone, and I'll conclude with a brief reflection on their meaning.

FOSSIL AND CLONE

The destruction of a species is not necessarily the destruction of its image. On the contrary, extinction of a species is a pre-condition of its resurrection as image in the form of fossil traces. The fossil record is a material and pictorial record, a vast iconic and indexical archive of extinct species that have inhabited this globe. And of course fossils are not the only image-traces that we have to reconstruct the evolution of life-forms. Contemporary paleontology regards birds as the descendants of dinosaurs, placing the reptilians in quite another class. (In that sense, the analogy of image and species needs to be qualified, because higher level taxa such as phyla also have their clusters of attributes, their family resemblances, their Galtonian composite images. There is also, though I don't have time to address it, the crisis within biological theories of taxonomy as such, in which the rise of cladistics has made the whole notion of species itself an object of controversy. How this will play out for iconologists is beyond the scope of this paper.)

The fossil image is what survives the death of a species, just as the corpse is what survives the death of the individual specimen. The sciences of natural history are the species equivalent of the rituals of mummification and preservation of effigies of the dead that find their place in the ethnographic wing of the natural history museum. Both are sciences of resurrection and re-animation, an effort of life-forms, bodies (our own) to manage mortality by means of images. But that is exactly what makes them so uncannily similar to the other great break-through in the life sciences in our time, the DNA revolution epitomized by the clone. The clone is the obverse of the fossil in every way. It epitomizes the hope for species immortality, in the promise of therapeutic cloning to "scrub" all birth defects from our DNA, to produce replaceable organs and ever-improved specimens. It also signifies the hope for the immortality of a singular, individual specimen in the utopia of reproductive cloning, where exact duplicates of parent organisms can be produced.

The fossil and the clone, then, play the role of end-point species for both the image and the human. Both are, quite precisely, image "families" or classes: the fossil the product of a slow process of petrification (taphonomy: science of petrifaction), reversed by resurrection and re-animation in the paleontological imagination (it is no accident that most paleontologists have highly developed visual acuity and that many of them are artists and image-processors). The fossil is also an allegorical image, a premonition of our own species mortality. It is thus what Walter Benjamin called a "dialectical image," capturing history at a standstill— in this case the "deep time" of the geological record projected backward through the entire history of life on earth, and forward toward the specter of our own extinction.

The clone, by contrast, is the technical, biocybernetic chimera of our time, and is thus generally pictured as a monstrosity, an unnatural and sterile freak. It personifies and incarnates in living flesh the anxiety about images that pervades hypericonoclastic critiques such as Baudrillard's precession of simulacra: Copies without an original; indistinguishable copies, the horror of repetition and indefinite sameness; the fear of the double; homophobia and heteronormativity; reproduction without difference; confusion of identity and similitude. The clone then is also a dialectical image. It points forward to a utopian or dystopian future in which the rule of the "spitting image," the exact simulacrum, is extended to an unprecedented degree. It points backward toward our most archaic fantasies about images: that they are imitations of life in a more than figurative sense, that some of them possess "aura" (literally, the breath of life), that they look back, and have desires and agency. All the taboos about image-making are revived around the clone, and strange political alliances emerge between eco-activists, Greens, and fundamentalist Christians. Notions such as the "circulation" or "mobility" of images in an age of globalization and genetic engineering are clearly insufficient. We need to think instead of a *migration* of images, in which their movements are incessantly regulated, prohibited, or accelerated by fantasies of contamination, plague, and purification (see my essay, Mitchell 2004). With the clone the metaphor of the life-form as image and vice versa seems to be literalized, and rendered reversible. Is it that the image is like a life-form, or the reverse?

The figures of the clone and the fossil merge, as it happens, in *Jurassic Park*, one of the great cinematic spectacles of the early nineties, a period that now seems like a time of innocence. For just an instant (captured in this still) a velociraptor is caught in the beam of a film projector that is projecting the DNA sequence that made it possible to clone a living dinosaur from its fossil traces. Let this "digital dinosaur" stand as a nexus point for these speculations on the science of images. It is, first, a science fictional image, a speculative projection of what the convergence of paleontology and genetic engineering might produce. It is also a technical, cinematic image, an early example of the revolution in digital animation that has ushered in a whole new era in the relation of animated and live action images. Within the life sciences this image has to be dismissed as a fantasy, a biological impossibility. But within the science of images, it is a crucial specimen, a kind of rare "missing link" in the evolutionary record of these strange, fantasmatic likenesses and apparitions. In the narrative of the film, this animal is breaking into the computer control room of the park and threatening to devour the controllers. Perhaps it is an allegory for our hope that the digitizing of the image is a way of controlling the wild kingdom of images, making peace between scientific logic and fleshly, concrete pictures, annihilating the Golden Calf once and for all: A likely story.

BIBLIOGRAPHY

Galison, Peter (2002) "Images scatter into data, data gather into images", in Bruno Latour and Peter Weibel (eds), *Iconoclash: Beyond the Image Wars in Science, Religion and Art*, Cambridge, MA: MIT Press: 300–323.

Kittler, Friedrich A. (1999) *Gramaphone, Film, Typewriter*, Stanford, CA: Stanford University Press.

Mitchell, W.J.T. (2004) "Migrating images: Totemism, fetishism, idolatry," in Petra Stegmann and Peter C. Seel (eds), *Migrating Images*, Berlin: House of World Cultures: 14–24.

4 Popular Images versus Self-Images of Science

Visual Representations of Science in Clipart Cartoons and Internet Photographs[1]

Joachim Schummer and Tami I. Spector

INTRODUCTION

The public image of almost anything is substantially a visual image. Public discourses are visually mediated, and for most people images sustain when words have long been forgotten. Visual images communicate more readily to the public than other media, and even the written or spoken word translates into visual images in the human imagination.

Although it has become the subject of scholarly studies about science only recently, visualization has accompanied science at least since medieval times. Many alchemical texts and Renaissance textbooks of practical knowledge were packed with images, setting the stage for the later tradition of textbook illustrations. Late medieval attempts to classify knowledge were illustrated by woodcuts of the arts, thus establishing emblems of the disciplines. The front page of Renaissance science books typically presented a portrait of its author in his characteristic setting, on which later traditions of portraiture could draw. The satirical literature of the fifteenth and sixteenth centuries, which was richly illustrated with woodcuts, did not spare fields like alchemy, pharmacy, and astronomy, nor did the later genre paintings. All these images contributed essentially to the public image of science in their time, and continue to do so today.

Any study of the contemporary public visual image of science is faced with two problems: the sheer mass of existing pictures, and the gap between pictures as the objects of a study and language as the medium of communicating that study. Margaret Mead and Rhoda Métraux in 1957 and David Chambers in 1983 avoided both problems by letting people, in the first case, describe in words and, in the second case, draw on paper what a typical scientist looks like. Such studies revealed many stereotypes, for example, "a man who wears a white coat and works in a laboratory [...] is elderly or middle aged and wears glasses [...] wear[s] a beard, may be unshaven and unkempt [...] is surrounded by equipment [...] and spends his days doing experiments," is a loner with "no social life, no other intellectual interests, no hobbies or relaxations," or associated the scientist with a magician, alchemist, and mad scientist. Another option is to focus on a defined set of images as Marcel LaFollette (1988) did for U.S. magazine illustrations and Peter Weingart (2002) for film, and then analyze their visual content according to chosen categories.

In this paper we present a different approach. We take digital images from databases that can be searched by keywords and analyze them both quantitatively and qualitatively to identify the stereotypes, emblematic objects, and typical gestures and elements used to image science. In addition, we examine the differences between the scientific disciplines, their relative visibility, and their characteristic visual representations. Most importantly, however, we distinguish between the *popular image* of science and the *public self-image* of science. The popular image of science depicts how non-scientists see science; in contrast, the public self-image of science communicates how scientists visually represent science to the public.[2] While both images are public images, the public self-image of science is at the interface between science and the public. As with the linguistic interface between science and the public, the way scientists openly depict their field is based on an intricate compromise. They want to describe how it "really" is, but also want to look better than what they perceive as their public image. They want to illustrate the complexity of their work and possibly correct misleading clichés of the public image, but need to draw on simple visual elements and metaphors, because they want to be understood. The public self-image of science thus both responds and adapts to the popular image in subtly differentiated ways. And because there are many scientific disciplines and different institutions that represent science, the responses are necessarily varied.

While discussions of the public image of science have been largely motivated by scientists' complaints about their allegedly bad public repute, our approach is primarily motivated by the need for *understanding* the image of science, whether the popular image or the public self-image. What exactly is the popular visual image composed of and where does it come from? How do scientists intuitively respond to this popular image? Do they, by their public self-images, correct or reinforce the popular image of science? And, if their intent is corrective, does the public self-image have any impact on the popular image of science?

Our approach requires that we first explore the popular image of science, which we elaborate in the next section through the analysis of clipart images. In the following section we focus more deliberately on chemistry and physics and investigate how and in what ways chemists and physicists respond and adapt to their popular image through their visual self-representations in Internet photographs.

THE POPULAR IMAGE OF SCIENCE IN CLIPART CARTOONS

Cartoons are humorous or satirical drawings that present their subject matter in a very reduced, stereotypical manner. They depict and compose only the most essential characteristics, so that the subject matter is easy to recognize and the image memorably humorous. Unlike our conception of what fine artists do, cartoonists visually analyze and reproduce clichés and stereotypes that are part of the cultural heritage of both visual and literary images. Cartoons are an extremely popular visual medium that use an artistically simplistic style to communicate deep seated cultural assumptions; and, therefore, cartoons of science are an ideal source for analyzing the popular image of science, its cultural clichés, and visual stereotypes.

Clipart cartoons and the methodology of quantitative image analysis

Cartoons are now commercially available in digital form in huge searchable databases of clipart for illustration of virtually any topic in print and electronic media. The Internet has made clipart cartoons the most popular image source for private and professional use. Because clipart databases are searchable by keywords, they are an ideal source for analyzing visual stereotypes both qualitatively and quantitatively. Qualitative analysis selects a set of cartoons by keyword search and analyzes the visual content according to standard procedures from art history or visual studies. Since the results are supposed to be expressed in linguistic form rather than in images, the analysis includes the crucial step of image interpretation that translates visual content into linguistic form. That procedure is faced with the two main problems of any visual study: the subjectivity of image interpretation and the practical limitation regarding the number of images that can be analyzed in reasonable time.

Quantitative analysis avoids both problems by focusing on the keywords alone. If the image content has been professionally analyzed by database managers and encoded by keywords, such that database users can easily find their desired motifs, the analysis can be performed at the linguistic level of keywords on any number of images within seconds. The procedure of image analysis is then similar to co-keyword analysis, as familiar from bibliometrics. A set of images selected by one keyword can be analyzed according to the co-occurrence of other keywords or of combinations of keywords. Once the set of images representing science is identified, the frequency of co-occurring keywords provides a quantitative measure for visual associations with science. Measuring the strength of visual associations is the key to quantitative visual studies. In our study it allows us not only to measure the dominant associations with science but also the relative visibilities of disciplines and their respective emblematic objects.

For our study we have used the searchable online database http://www.clipart.com by Jupitermedia. As is true of any database, clipart.com does not meet all the ideal requirements for research. In particular, keywords have not been assigned in the same systematic way for all the images, because the database combines images from more than ten primarily U.S. based clipart publishers, each combining images drawn by a variety of cartoonists. Yet, the difference in keyword assignment and possible distortions by selective image input are presumably leveled out in most cases due to the number and diversity of original sources and the large number of images. Indeed, clipart.com had more than 2.1 million clipart images at the time of our analysis (June 2004).

The relative visibility of science and its disciplines

Database analysis, when analyzing a clipart database, requires extensive pre-studies and qualitative checks of the images retrieved by means of the keywords. Indeed, the unmodified keyword "science" provides mostly (78 per cent) cartoons of animals in anthropomorphic gestures, reminiscent of the pre-scientific medieval tradition in which the "animal kingdom" served to illustrate moral fables, which survives in modern comic strips. Since these images are not associated with specific scientific keywords, like "biology," we have excluded them in the following analysis, as we have similar cartoons of flowers and trees. In addition, we have excluded all science images associated with the keyword "technology," although the overlap between science and technology is surprisingly low (10 per cent). Thus refined, the search provides 1,360 cartoons about science,[3] which represents 0.6 per cent of the overall clipart database. If this number measures the relative visibility of science in the popular visual culture, it suggests that science plays only a very marginal role, compared for instance to the much more visible field of technology (3 per cent).

Our first analysis of the set of science clipart images compares the relative visibility of the disciplines. About three quarters of the images are keyword-related to at least one discipline, but the distribution reveals a clear disciplinary focus (Figure 4.1). Indeed, more than 40 per cent are related to the discipline of chemistry, demonstrating that chemistry dominates the popular visual stereotype of science overall. Next comes physics with only 16 per cent. Apart from that, only five other disciplines play a visible role. The combined field of the biomedical sciences is represented in 12 per cent of the images.[4] The relatively strong presence of "rocket science" (11 per cent) suggests a clear U.S. origin of the cartoons, where space engineering has strongly influenced the popular view of science since the Apollo program—and has generated "rocket science" as an ironic idiom for any "endeavor requiring great intelligence or technical ability" (*The American Heritage Dictionary of the English Language* 2000), usually, however, in the negative (i.e., "It isn't rocket science"). In visual images, anatomy (9 per cent) is distinguished from the biomedical sciences, as is astronomy (7 per cent) from physics. Although strictly speaking not a science, mathematics follows with 4 per cent, whereas all the other real sciences are virtually absent in the visual stereotypes of science.

The co-occurrence of disciplinary keywords provides further insight into how the visual image of science is structured. Overall, there is little

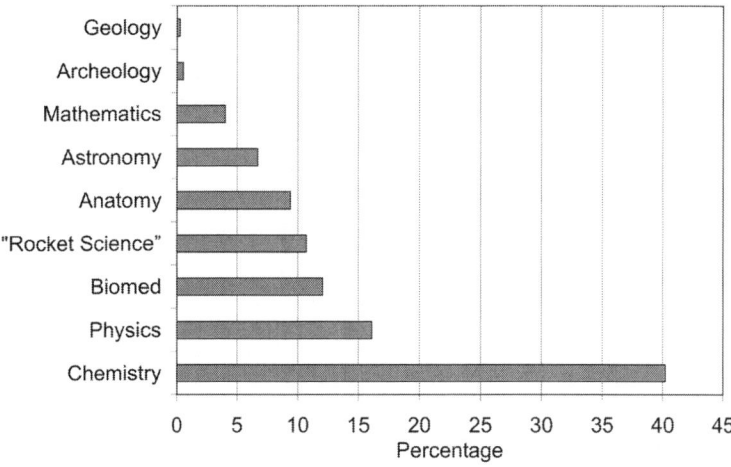

Figure 4.1 The relative visibility of disciplines in clipart figures on science.

overlap (about 5 per cent), which suggests that each of the disciplines has a relatively clear visual identity. Apart from some minor overlap between astronomy and rocket science, anatomy and biomedical sciences, physics and mathematics, and physics and chemistry, only the biomedical sciences stand out because 28 per cent of their cartoons are also related to chemistry. This overlap is related to the relative weakness of the emblematic object of the biomedical sciences (see below).

Several approaches explain the relative visibility of the various disciplines and why the weight of each discipline's visibility does not necessarily accord with its impact on the actual modern-day research landscape. There are of course historical reasons, which we explore in the sub-section concluding this chapter, but there are also specific visual reasons that we investigate in the next section, thereby illustrating the potential of our approach for quantitative emblematic studies.

The emblematic objects of scientific disciplines

The visual stereotypes of science contain emblems that stand for science or for a scientific discipline and which must be simple enough to be easily recognized. A discipline without an emblem hardly exists in the popular visual image of science. Of the seven visible disciplines in Figure 4.1, six have strong emblematic objects (Figure 4.2). Glassware such as beakers, flasks, or test tubes epitomizes chemistry; the microscope, the biomedical sciences; a rocket, the popular idea of "rocket science"; bones stand in for anatomy; the telescope, astronomy; and mathematics finds representations either by means of formulas (algebra) or a pair of compasses (geometry). Physics is the exception to the rule because it has no such popular emblem, although it shares with chemistry to some degree the atom. Indeed, the cartoons that are keyword-related to physics consist largely of images of rather unspecified experimental settings, which suggest that the keyword "physics" is rather understood in terms of its pre-modern meaning where it was used to denote the natural sciences overall.[5] One of the reasons why physics has no clear visual emblem might be that its more abstract subject matter has resisted the visual imaginability of the popular culture.

Figure 4.2 illustrates that the emblems can be research instruments, objects, or graphical languages, but each must be easily recognized by anybody. Compared to the other emblems, the main emblems of chemistry (beakers, flasks, and test tubes) stand out because they have the simplest graphical structure and can even be drawn with a single line. This points to a possible visual reason for chemistry's dominance in the visual image of science; the elegance of the emblems enables them to serve as emblems of science overall.

The emblematics of the popular visual image of science obviously does not reflect the actual instrumentally based scientific practices of today. For instance, much of the emblematic glassware portrayed in the database was previously used by many scientific disciplines but is currently outdated. And although chemists historically used microscopes and occasionally still do today, that instrument is now much more important in other fields. Yet, the popular visual culture has its own rules for selecting emblems, which seem to draw more on the history of science rather than on modern methodology.

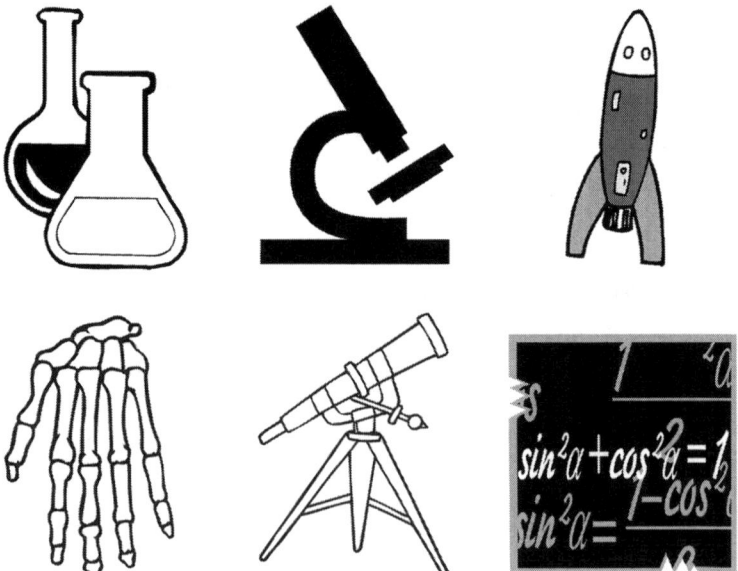

Figure 4.2 Emblematic objects of the disciplines chemistry, biomedical science, "rocket science," anatomy, astronomy, and mathematics.

The clipart database is an excellent research tool for quantitative emblematic studies. Based on the assumption that a visual element is an emblem of a field if the element occurs frequently in visual representations of the fields, we can make two key distinctions. First, we can distinguish between *weak* and *strong* emblems, depending on how frequently the element appears in representations of the field compared to those of other fields (i.e., on the percentage of co-occurrences of element and field keywords based on the total number of element keyword occurrences). Second, we can distinguish between *important* and *less important* emblems, depending on how many representations of the field contain the emblematic elements (i.e., on the percentage of co-occurrences of element and field keywords based on the total number of field keyword occurrences).

Let us clarify these distinctions: the Bunsen burner, beaker, flask, and test tube are all strong emblems of chemistry because they almost exclusively occur in representations of chemistry (Figure 4.3). However, the Bunsen burner, which is the strongest emblem, is also the least important one because only 4 per cent of the chemistry images contain it, compared, for instance, to 44 per cent of the chemistry images that are keyword-related to glassware.[6] Surprisingly, the microscope is a slightly stronger emblem of chemistry than of the biomedical sciences (Figure 4.3). However, half of the chemistry images, including the microscope, are also related to the biomedical images, which explains the disciplinary overlap mentioned above; and since 35 per cent of all the biomedical images show a microscope, it is the most important emblem of this field. Overall, the most important visual emblems of science are glassware (beaker, flask, and test tube; present in 18 per cent of all science images) and the microscope (10 per cent). Although these percentages based on science overall are not very high due to the visual diversity of the disciplines, they are the strongest emblems of science.

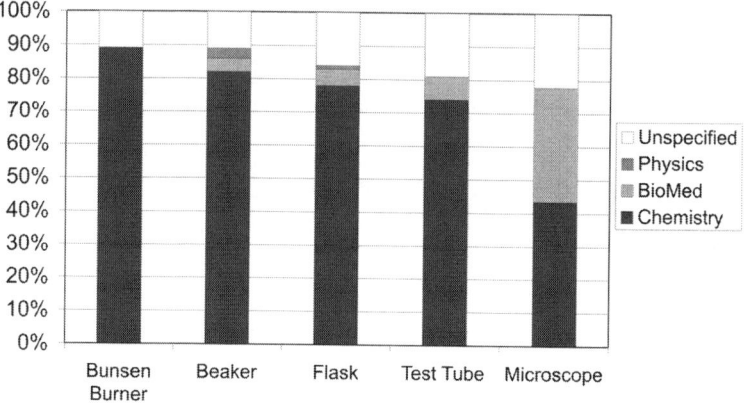

Figure 4.3 Identification of emblematic laboratory instruments by association with different disciplines.

The laboratory as the archetypical location of science

The previous two sections supply foundations for a co-keyword analysis of the inner structure of the popular visual image of science; that is, of which disciplinary elements and of which emblematic objects it is composed. Co-keyword analysis also allows for the investigation of the outer structure of the popular visual image of science, giving us a perspective on the broader symbolic associations of science. Before doing this, however, we focus on the semantic field of "research" and its relation to the various disciplines.

In the public view, research is not confined to science (Figure 4.4, first column). Clipart demonstrates that, visually, people associate research equally with scientific and other disciplines: with test tubes and with reading books and writing texts (39 per cent).[7] They assume, correctly, that scholars in the humanities, lawyers, journalists, bankers, and so forth, all do research. As one might expect, however, if we focus on the keyword "scientific research," reading and writing shrinks dramatically and the dominant research field is again chemistry (Figure 4.4, second column). In addition to the emblematic objects of chemistry, the microscope is a strong visual emblem of research, both with and without biomedical associations, whereas all the other scientific disciplines and their emblematic objects are virtually absent. The reason for this becomes obvious in the third and forth column of Figure 4.4. In the popular image, the characteristic activity of scientists is laboratory research, and the stereotypical laboratory is equipped with glassware—the emblematic objects of chemistry. This stereotype is so strong that 95 per cent of all cartoons depicting laboratory research are keyword-related to chemistry and only to chemistry. A cor-

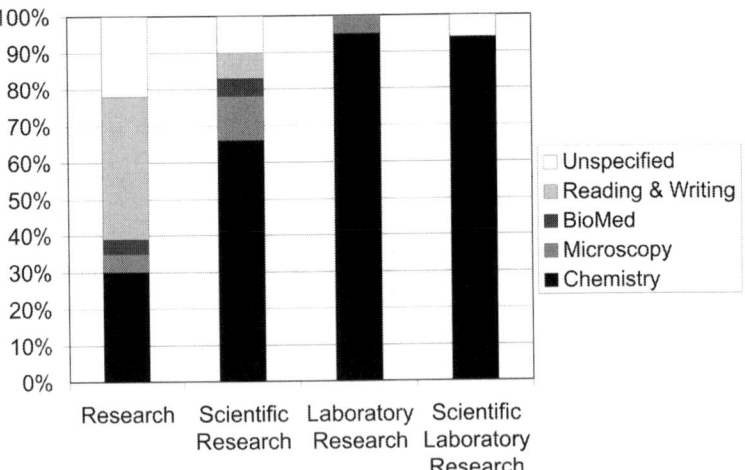

Figure 4.4 Visible associations with research types (note that other disciplines, like physics and mathematics, are hardly associated with any of these research types).

responding analysis shows that chemistry is also the archetypical field of experimentation.

This suggests another reason for the dominance of chemistry in the visual image of science. Apart from their elegant graphic structure, depictions of glassware are the simplest visual elements to indicate science: a room equipped with some glassware turns into a laboratory; a person holding a test tube is a scientist. In sum, glassware epitomizes scientific research, which in the visual popular image means conducting chemical experiments in a laboratory. This popular image, while at odds with actual contemporary scientific activity, is historically linked. Before the early nineteenth century, the laboratory *per se* was a chemical or alchemical laboratory and experimental research approximated chemical research (see Nye 1996: 9ff.). The clipart image featured in Figure 4.5, which presents a typical modern cartoon of laboratory research, is a legacy of this tradition. The object that reveals the cartoon's alchemical legacy most overtly is the flask in the foreground: shaped like the retort used by alchemists, it unconsciously assimilates the historical references included in these images.

Figure 4.5 Typical clipart cartoon of laboratory research.

The mad scientist and other visual associations

Keyword analysis allows us to study the broader visual associations with science simply by counting the frequencies of all the keywords assigned to cartoons of science. However, keyword frequencies are generally lower than the actual occurrences of the corresponding elements in the cartoons, because background elements and details are frequently not considered in the keywords; as a result, the frequencies have only relative significance. Instead of presenting a long list of keyword frequencies, we link the keywords to semantic groups. For instance, the educational context is indicated by the set of keywords "class, school, teacher, pupil, student, learning, education." In addition, since chemistry strongly dominates and appears to embody the popular visual image of science, we have confined the analysis to cartoons that are keyword-related to chemistry in order to capture a clear visual character. In principle, however, the analysis can be performed for each discipline for which enough images are available. In the following portrait of chemistry we highlight not only the strong characteristics but also aspects of chemistry that seem to be underrepresented in the popular visual image compared to the actual practice of modern chemistry.

Chemistry is clearly viewed as a science where people (32.2 per cent) do experiments (36.7 per cent) with various instruments and tools (59.2 per cent) in a laboratory (22.5 per cent). Two thirds of the people dealing with chemistry are male,[8] their experiments include reactions and the visual inspection of liquids rather than measurements, and their instruments are mostly glassware. Chemistry is more associated with the biomedical sciences including pharmacy (7.4 per cent) than with physics (2.7 per cent) or mathematics (0.2 per cent). Despite their potentially symbolic use, models of atoms and molecules rarely appear (2.9 per cent), such that the theoretical side of chemistry is hardly visible. Equally uncharacteristic are books (2.3 per cent), computers (1.2 per cent), and diagrams (1.0 per cent, including the Periodic Table of Elements). Apart from research, people associate chemistry visually primarily with education (19.1 per cent) and rarely with industry (0.8 per cent), technology (0.8 per cent), or business (0.6 per cent), although chemical technicians are not unknown (3.5 per cent). Also, toxicity (2.5 per cent), explosions (1.2 per cent), fire (1.0 per cent), and other hazards are rarely represented in cartoons of chemistry, whereas specific cartoons of chemicals indicate such dangers more frequently (8 per cent).

Cartoons of the "mad scientist" deserve particular attention for two reasons. First, scientists often fear that this aspect would dominate their public image. This fear is unfounded, however, because only 2 per cent of all science cartoons depict a "mad scientist." Second, these cartoons illustrate how popular visual culture has incorporated elements from other media.[9] Originally, the figure of the "mad scientist" was developed by early nineteenth-century writers to portray chemists specifically (see Schummer 2006). And, indeed, half of the "mad scientist" cartoons are clearly recog-

Figure 4.6 Typical mad scientist cartoon.

nizable as chemists through their emblematic glassware (Figure 4.6). Yet, the visual image of the "mad scientist" was shaped by movies, particularly by those that adopted and transformed Mary Shelley's *Frankenstein* (Toumey 1992), from which it moved to cartoons. Mad scientist cartoons thus illustrate the power of the public domain to hijack historical artifacts and, via media as popular and accessible as clipart, to elide the history from which they have been "clipped," and to preserve them, fossilizing a history while simultaneously sustaining an image in the corporate imagination.

Conclusion: The conservativeness of the popular visual culture

Because they capture visual stereotypes, cartoons are an important source for the study of popular visual culture in general and the popular image of science in particular. Collected in huge searchable databases, cartoons allow quantitative investigations that are otherwise not viable in visual studies. If performed with due care and an understanding of the keyword assignments, co-keyword analysis is a powerful tool for the investigation of the inner and outer visual structure of a field. Combined with the qualitative analysis of pictorial elements, new types of visual arguments can be developed that the relatively new field of visual studies deeply needs.

Because the popular visual culture has incorporated historical elements, such analysis must also be historically informed. This is particularly true of the popular visual image of science, which has conserved age-old stereotypes. In fact, the cartoons contain few elements of actual science from the last two centuries, but refer instead to a period before the nineteenth-century professionalization of the sciences.

Perhaps most striking is the relative invisibility of science overall, which corresponds to a historical period when only a few amateurs were doing science in private, unlike the public funded "Big Science" that has emerged since the nineteenth century. Furthermore, only the sciences that developed a disciplinary character before the nineteenth century are specifically visible in the cartoons, while the visually unspecific character of physics reflects the premodern meaning of "physics" as the generic term for the natural sciences. The emblematic objects go either back to antiquity (glassware, bones, mathematical formulas, and a pair of compasses) or are inventions of the early seventeenth century (microscope and telescope). Only rockets are new, but "rocket science" is neither a science nor a developed engineering discipline outside the United States, which suggests that powerful national publicity can under certain conditions impact the visual culture of science in a country.[10]

The characteristics of the popular image of chemistry as the visually dominant discipline provide further support for the extremely conservative nature of the visual culture. Although chemistry is still by far the biggest scientific discipline,[11] its visual role as epitomizing laboratory research and experimental science goes back to the period before the early nineteenth century, when chemistry was *the* prototypical experimental science. The absence of measurement instruments and experiments, which have dominated chemistry since the late eighteenth century, the virtual lack of theoretical aspects of chemistry (except a few images of the atom and the Periodic Table), and the neglect of industrial chemistry, which has otherwise come to the public attention at least through environmental problems, all reveal the pre-nineteenth-century origin of the popular visual image. This is also true of the emblematic glassware which harkens back to early alchemy.[12] The only component of the popular visual image from the nineteenth century seems to be the "mad scientist," but that figure also has its origin in medieval and early modern portraits of the "mad alchemist" both in writing and painting (see Schummer 2006). Finally, the archetypical image of a chemist, which is a person gazing at a flask that contains some liquid, goes back via sixteenth-century iatrochemistry to early fourteenth-century medicine, where it represented uroscopy as the emblematic gesture of medicine for at least two centuries (Figure 4.7).[13] Although this motif has been carefully avoided in any of the portraits of famous chemists since at least the eighteenth century, it has persisted in the popular visual culture up to the present day.

All these findings suggest that the popular visual culture of science is extremely conservative, and has not readily incorporated new visual elements for centuries. Cartoons are expected to convey older stereotypes, but it may be surprising that their visual components go as far back as the fourteenth century. Cartoons like those found in clipart are a source for

Figure 4.7 Left: Fourteenth-century illustration in Avicenna's Canon medicinae. It depicts uruscopy that soon became an emblem of medicine. Right: Modern cartoon of a chemist (Source: Gerard of Cremona, 1283, The Hague, MMW, 10 B 24).

considerable amusement, but they also conserve historical visual traditions that are sometimes no longer part of explicit public knowledge.

In the second half of this paper, we explore a different component of the public image of science, namely, how scientists visually represent themselves to the public. One of the guiding questions of our analysis is whether or not they follow the conservative visual tradition of the popular culture.

THE PUBLIC SELF-IMAGE OF SCIENCE IN INTERNET PHOTOGRAPHS

The preservation of age-old stereotypes in clipart images of science suggests that popular visual culture is not very susceptible to new pictorial input. If such input exists, it most likely comes from the visual self-image of science; that is, from how scientists and science-related institutions present themselves to the public in pictures.

The public self-image of science at the interface of science and the public

In the visual regime, public self-images of science are at the interface between science and the public. On the one hand, they try to communicate to the public visual aspects of science that scientists think are either important or required to correct or enrich the popular image of science, while on the other hand, they need to adopt symbolic elements from popular visual culture for effective visual communication. For instance, if a self-image is meant to communicate the research strength of an institution, it might employ popular visual emblematics for research, even if that reinforces undesirable stereotypes. Their mediating capacity and multifunctionality make the public self-images of science particularly interesting for a comparative visual study with the popular images of science.

Unfortunately, such comparative studies are faced with several methodological problems, particularly if performed on a quantitative level. In the ideal world, there are two databases of images, one for popular images and one for public self-images of science, with systematic keywords that allow for comparative co-keyword analysis. Yet, in the real world, there is neither a database of public self-images of science nor are there any keywords. Moreover, scientists tend not to represent themselves in cartoons, but in photographs, which are different kinds of images. First, unlike clipart, the primary purpose of photographs is not humor. Photographs are also more authoritative, putting more emphasis on detail, nuance, and authenticity rather than on general impressions and stereotypes; photographs have a variety of creators, including both scientists and professional photographers and are legitimated by a variety of sources: the person being photographed, the person commissioning the photograph (which may be one and the same with the person photographed); and the procedure of selecting and publishing photographs differs considerably from commissioning clipart cartoons for a database. All of this makes it difficult to make a comparison between the two types of images.

The source of public self-images of science that comes closest to the ideal requirements for our study are photographs posted on Internet websites from scientists and science-related institutions such as universities and research institutes. With about a billion users per day,[14] the Internet provides a means by which scientists can put out their self-image more energetically and purposively to the broadest possible public. Compared to images in printed material, these images can easily be retrieved by Internet search engines and quantitatively analyzed in large numbers. In addition, while images in printed material are usually processed by professional designers, Internet photographs are frequently self-made by scientists (or at least self-selected) and thus come closer to their unmediated public self-image. These images may be retrieved by selecting relevant science-related websites and then collecting all their photographs; however, this procedure provides a heterogeneous mixture of images, many of which are difficult to interpret as self-images of science. We found it more effective, though still not satisfactory, to search for science-related photographs using an appropriate Internet search engine and then select the relevant photographs from science-related websites. Even then, using the Google image search tool with science-related search terms tends to provide images that can hardly be defined as self-images of science, since the search engine relates images to search terms only because they both appear somewhere on the same web page. The most effective, though very restrictive, method is to search for images whose file names consist of the keyword in question (e.g., "chemistry.jpg" or "chemist.jpeg"). Although science websites use images with different file names to represent themselves, it is almost certain that an image called "chemist.jpeg" on a chemistry website is meant to represent the visual self-image of chemists.[15] The shortcomings of this approach are the limited number of images that meet the formal conditions and the limited number of legitimate keywords.

In the following we explore the feasibility of this approach by combining quantitative with qualitative analysis. We assume that the public visual self-image of science involves a complex response to the popular visual image of science and that different disciplines and different science-related institutions respond differently to, and interact differently with, popular visual culture. Thus, after surveying the relative visibility of disciplines on the Internet compared to those in the clipart database, we analyze the characteristic styles of disciplinary and institutional self-representation and how these intersect with the popular stereotypes represented by clipart images. Because chemistry and physics dominate both the popular image and the self-image of science, and because these disciplines exemplify two opposed styles of representation, we focus on the self-images of these disciplines. And as is the case with the clipart images, our Internet sources are from the English-speaking world and predominantly from the United States.

The relative visibility of disciplines and their different styles of self-representation

To obtain an overall picture of the relative visibility of the various sciences on the Internet, we have used in each case the search terms for the discipline and the corresponding scientist (e.g., "chemistry" and "chemist"; Figure 4.8). Of course, the results do not show the extreme visual dominance of chemistry, as in clipart images, while smaller, more esoteric disciplines are at least visually present, if only in small proportions. Though less than in clipart, chemists (33 per cent) dominate the portraits of scientists, followed by astronomers, geologists, biologists, and physicists with only 7 per cent of the total scientists' images. On the other hand, images that illustrate the scientific disciplines are dominated by physics (25 per cent), closely followed by chemistry (24 per cent) and then by biology, geology, and astronomy. The almost reverse order between the disciplines and the corresponding scientists and, particularly, the low visual presence of physicists compared to physics calls for explanation.

An examination of the actual images and their websites reveals that the disciplines are primarily linked to educational institutions, including uni-

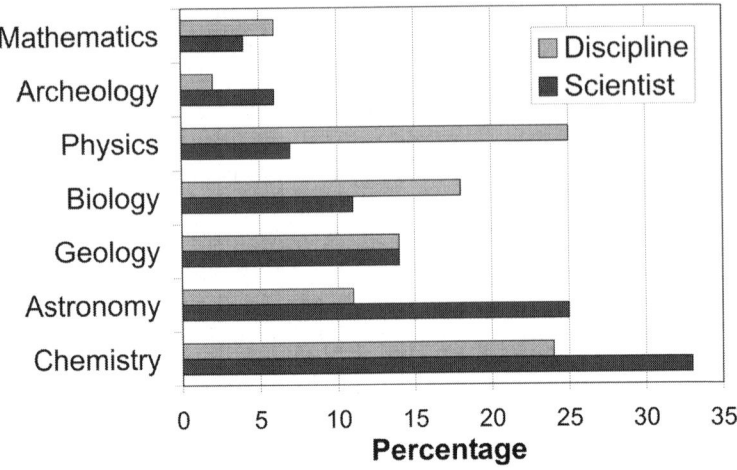

Figure 4.8 Relative number of images of different disciplines from Google image search, each for search terms for the discipline and the scientist (e.g., for "chemistry" and "chemist").

versity buildings, teaching labs, and textbook covers. Thus, the order of the visual presence of the disciplines corresponds in some degree to the number of their corresponding institutions, which for chemistry and physics is about the same. In contrast, the images of scientists differ greatly by discipline. Most "astronomers" and many "chemists" are posted on non-scientific or historical websites, which indicates that their visibility reflects only the popular visual image of a generalizable science found in clipart images. If displayed as self-images by science-related websites, the many images of chemists, astronomers, and biologists portray unknown scientists in their research settings surrounded by emblematic objects. Moreover, the prototypical photograph of a chemist shows the historically linked image of a person gazing at a flask of colored liquid. In stark contrast, about half of the relatively few images of physicists depict theoretical physicists, frequently famous people such as Albert Einstein, Richard Feynman, and Enrico Fermi.[16] As self-images of physics, and in direct contrast with chemistry, such photographs of almost mythic physicists translate *the person portrayed* into an emblem of the discipline, rather than their research tools. Apart from chalkboards with equations on them or bookshelves, almost all are devoid of emblematic objects and other scientific or disciplinary indicators.

In their images of scientists, chemists and physicists exemplify two styles of visual self-representation that respond to and interact with the popular visual culture. Physicists draw on famous members of their discipline from the twentieth century and actively cultivate their images as popular icons of their discipline. Chemists rely on the strength of their emblematic objects and gestures, which have a deep historical root in popular visual culture. While the conservative strategy of chemists is still successful in terms of public visibility, the price is bland or comically depersonalized images and the adaptation to stereotypes that modern chemists might otherwise not always embrace.

The two different styles of self-representation by chemists and physicists are also seen in the way they represent their own disciplines. For many of the sciences the number of discipline images is much higher than the number of scientist images, and much more so for physics, which suggests that the abstract discipline is, in the self-representation, held more important than the people. In chemistry both image types differ primarily only in whether an anonymous chemist is in the foreground holding some emblematic glassware or in the background behind the glassware (see below). In physics, as mentioned above, many portraits of scientists depict theoretical physicists without emblematic objects. The discipline itself is represented as experimental by some apparatus on which several people may be working or, as a rule, standing behind. This latter image clearly opposes the popular characterization of physics as a brainy, hands-off science as encapsulated in the ubiquitous portraits of a wild-haired Einstein. Unlike the emblematic lexicon of glassware in chemistry, equipment, when foregrounded in physics images, is complex and electronically driven, frequently including oscilloscopes, lasers, mass spectrometers, and other electronic apparatus interconnected with wires and cables that fill whole rooms.[17] And unlike the socially isolated figure in typical portraits of both chemists and chemistry the majority of physics images show people in social interaction in both educational and research contexts.

Although there are exceptions, the visual self-representation of chemistry is very conservative, drawing on long established elements of popular visual culture. Whereas today's actual research includes complex instrumentation and teamwork in chemistry and physics alike, chemists tend to reiterate and reinforce the stereotype of the isolated researcher who employs historically outmoded methods and instruments. In contrast, physicists, who lack the heritage of visual stereotypes, have introduced a new lexicon of imagery that portrays their science as a modern instrument-driven and collaborative enterprise. Because that does not easily translate into simple emblems in the popular visual culture, they are more ready to develop a differentiated and up-to-date self-image of their discipline.

Different aspects of science highlighted by different institutions

Because our Internet search method is focused on image file names, the variation of search terms to explore broader visual associations with science is limited. Yet, at least for chemistry, a set of five search terms from the semantic field of chemistry, allows for the distinctions of five different aspects of the public self-image (Table 4.1).

Table 4.1 Internet image search results from the semantic field of chemistry.

Search term	Relative frequency %	Predominant image content
Chemical	30	Chemical plant
Chemicals	21	Industrial products
Chemist	10	Researcher
Chemists	2	Social context
Chemistry	37	Research apparatus

The term "chemical" is predominately associated with industrial sites (i.e., chemical plant exteriors and interiors) and the term "chemicals" with commercial chemical products (e.g., bottles of chemicals). The three terms "chemist," "chemists," and "chemistry" provide typical images each with scientists interacting with glassware, but in different manners: the first type of image features a person; the second indicates a social research context; whereas research instruments (glassware, apparatus, a complete laboratory) dominate the third. Since each of these image types focus on a different aspect of chemistry (chemical plant, industrial products, researcher, social context, and research apparatus), an analysis of their frequency and institutional contexts provides a differentiated view of the visual public self-image of chemistry.

Overall, our Internet image survey shows that both research instruments and chemical plants dominate the visual image of chemistry, and that the social context is almost absent (Table 4.1). Unlike the stereotypical image of chemistry in cartoons, where associations with chemical industry are extremely rare, half of the digital photographs present industrial plants or products. While this might be seen as an attempt to correct one stereotype, the relative lack of social contexts found in the sum of these photographs reinforces another one, that of the scientist working in isolation.

Based on a sample of 50 images for each image type, a closer look at the institutional context in which these images appear reveals how different institutions present different aspects of chemistry. Indeed, about 90 per cent of the images in our sample have been posted on the websites of institutions that are in various ways engaged in chemistry; that is, universities, schools, industry, and government (particularly governmental research institutes). These four institutions represent the breadth of chemistry's institutional reach, so their images represent the public self-images of chemistry.

Figure 4.9 shows the distribution of each of the five chemistry image types over the four chemistry institutions. Not surprisingly, industry focuses on chemical plants and products, whereas universities present chemistry primarily through research instruments and laboratories. What is striking, however, is that governmental institutions, including governmental

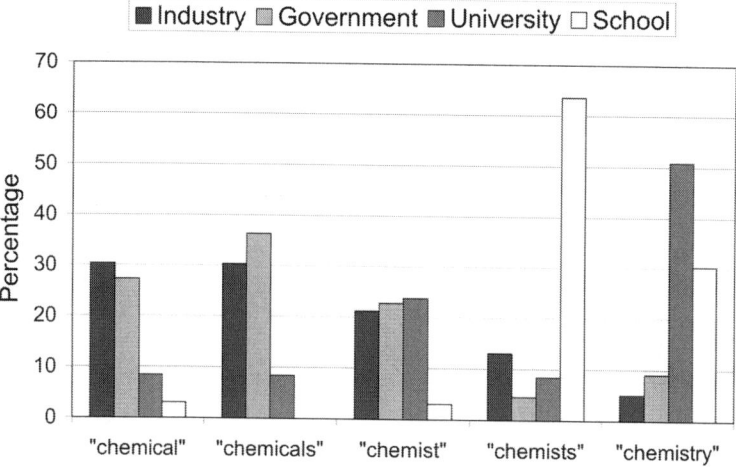

Figure 4.9 Distribution of chemistry image types over science related institutions on their websites.

research institutes, present chemistry in almost the same way as industry: they focus on industrial plants and products and almost ignore actual laboratory research and, particularly, its social contexts. On the other hand, for elementary and secondary schools, the social context (the interaction between students and between students and teachers) is the most important aspect of chemistry, which suggest that schools, and only schools, are strongly engaged in correcting the stereotype of the isolated researcher and thus in "humanizing" chemistry.

An analogous analysis for physics is faced with both the lack of correspondingly meaningful search terms and the fact that the self-image of physics is strongly dominated by universities. Yet, based on qualitative image analyses, some aspects of the public self-image of physics relative to that of chemistry are worth mentioning. First, images associated with the keyword "physicists" are extremely rare, because the social context is usually displayed in images associated with "physics"; this once more suggests that the social context is considered an integral part of the discipline of physics. Whereas universities highlight instrumentation and laboratories, government institutions clearly favor portraits of (theoretical) physicists over equipment and laboratories. This finding suggests that physicists in governmental research institutes, where cross-disciplinary departments of applied research are much more common than in universities, try to distinguish themselves from experimental and applied work. Furthermore, industrial aspects of physics are extremely scarce, despite its notorious industrial and government associations with weapons development, which physicists would presumably wish to de-emphasize in their public self-image.

The question of gender

The study of gender proportionality in science has had a strong tradition in feminist analyses of the disciplines for several decades, ranging from the dearth of female role models (associated with research primarily in the 1970s and 1980s) to the unequal allocation of male and female laboratory and office space at the most prestigious research institutions (Hopkins 2002). More recently, several studies have also analyzed the differential representation of female and male scientists in popular U.S. media. In 1990 LaFollette looked at the public image of science in popular U.S. magazines from the first half of the twentieth century finding that the illustrations in the articles "depicted women as minority characters in the drama of science," with women shown as technicians and assistants and men as supervisors (LaFollette 1988). In 2003 Flicker analyzed 60 feature films finding that "The role of the professional scientist is reserved for men," and that women were represented less than a fifth of the time as professional scientists (Flicker 2003; see also Pollak 2002). These studies suggest that the popular image of science, both visual and non-visual, is strongly dominated by male scientists, as it actually was, several decades ago. Today, the situation is much different in the United States (Table 4.2).[18] In chemistry, which so strongly dominates the popular image of science, women receive a third of all PhDs, while in biology the number of men and women receiving PhDs is almost equal. While most women work as professional scientists in industry, government, or universities, the number of women gradually drops with professorial rank in universities.

How then do scientists in their public self-image represent gender proportionality? Do they conservatively tender the outdated popular image, do they represent the actual gender ratio, or do they progressively represent themselves as gender balanced to overcome the bias? To answer this question, we have quantitatively examined images in our data set for chemistry and physics for gender ratios.

Overall, men dominate the public self-image of science, with physics presented as significantly more male dominated than chemistry (Figure 4.10). However, the male/female ratios are very close to those of the PhDs in both disciplines (Table 4.2), which suggests that the self-images depict the actual conditions and thus neither adapt to the popular image nor visually

Table 4.2 Male/female ratio of PhDs and faculty positions in U.S. universities.

Discipline	PhD	Asst. Prof.	Assoc. Prof.	Full Prof.
Biology	1.2	2.3	3.0	5.8
Chemistry	2.0	3.7	3.9	12.1
Physics	5.8	8.0	9.6	18.1

Adapted from Handelsman et al. 2005.

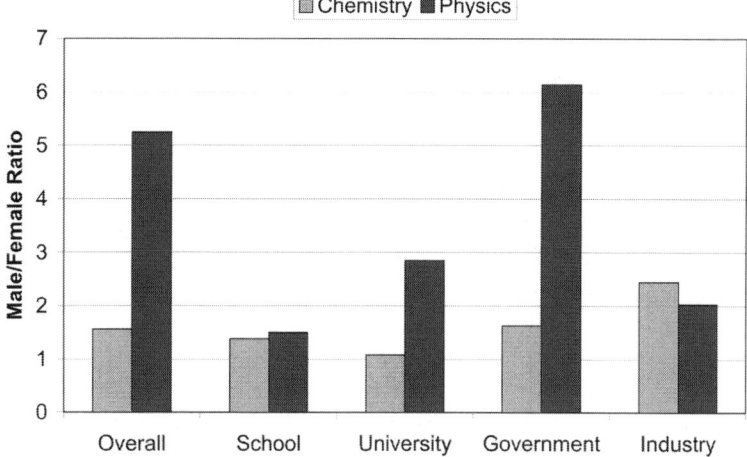

Figure 4.10 Male/female ratio in people-containing images of chemistry and physics by institutions.

encourage a gender balance. A breakdown by institutional context shows that for both chemistry and physics primary and secondary schools present a more gender balanced image, albeit not the actual gender ratio of their pupils which we assume to be balanced. The self-image that most closely displays a gender balance is that of chemistry at universities. This finding is surprising given the otherwise very conservative self-image of chemistry. The largest gender differentials for the institutional context of chemistry is found in industry, while for physics it is in government. If one takes only the images called "physicist," the male/female ratio rises to 19. This suggests that, unlike the chemists' preference for anonymous figures, the physicists' focus on personalities is particularly susceptible to the conservation of gender stereotypes.

A closer look at physics photographs shows that women are equally unassociated with the research aspects of physics, both experimental (indicated by complex apparatus) and theoretical (indicated by boards with equations and diagrams). Instead, when shown at all, they are predominately represented as secondary school students or lower division university students. A typical example might be the depiction of an older man dressed in a suit and tie observing a young woman in a lab coat interpreting x-ray data. Even though such pictures depict an educational context, the absence of images with reversed gender roles indicates the importance in the public self-image of physics of men remaining in charge.

In contrast, in images from chemistry, women appear to be almost equally as professionalized as male chemists. Both genders are shown in the professional scientists' gear (lab coats and goggles) and both interact with glassware and instruments in approximately the same ratio. Only industry, the most male dominated institutional context of chemistry, has highly differentiated images of men and women. In these, men are found in hard-hats in large scale industrial sites, while women are shown in their professional scientist gear in laboratories using the glassware and instruments found in the images associated with the other chemistry-related institutional contexts. This would seem to undermine an emerging progressive public self-image. Nevertheless, it may be interpreted quite differently as a gesture toward gender parity. Thus, these images project a conservative desire to retain gender distinctions while moving to a progressive public self-image of equal capability. They put forward the idea that both men and women are adept "in the field"—men at the industrial site and women in the lab (women are not under the auspices of an older male authority figure or in a teacher-student relationship but make distinct contributions in industrial chemistry). Ironically, of course the lab (feminized in these images) is the site normally associated with "brains" while the industrial site is associated with "brawn."

Much more so than the quantitative ratios, the different visual gender associations in chemistry and physics show how deeply seated our cultural narratives about the hierarchy of rigor in science remains in the public self-image of science—that physics is the hardest, most abstract science (i.e., masculine) while chemistry is the less mathematical and more life-related life science (i.e., feminine). The result of these narratives is clearly reflected in Table 4.2.

In sum, unlike the other aspects discussed in previous sections, the gender aspects of Internet photographs reveals the self-image of physics as extremely conservative as it continues to reinforce gender stereotypes of the popular image of science, while chemistry, despite its legacy of visual stereotypes, is socially progressive.

CONCLUSION

The public use of images is, despite its dramatic increase, still the least understood form of public communication, compared to communications in the written or spoken word that have been studies by rhetorics since antiquity. Although there were earlier attempts, like Roland Barthes's visual semiotics from the 1960s, it was not before the 1990s that scholars with such powerful catch-words as "pictorial turn" (Mitchell 1994) or "iconic turn" (Boehm 1994) tried to break open the limited scholarly focus on art history to establish a field called "visual studies" or *Bildwissenschaft* that broadly investigates the welter of images and their uses (Barthes 1964; Boehm 1994; Mitchell 1994). While this field is now flourishing and attracting scholars from various disciplines, its art historical origins still impacts the kind of questions that are posed and the methods used to study them. In this paper we have tried to broaden the focus by transferring questions from the field of public understanding of science to the visual field, by studying new types of images, and by using as far as possible established quantitative methods from empirical sociology, supplemented by qualitative and historical analysis.

Since the early 1990s, studies in the public understanding of science have grown at least as fast as visual studies and, although the public image of science has become an important topic, the public visual image of science has not, not even in terms of gender analysis. We assume that this is due to both methodological and conceptual barriers, and as a result we have put an emphasis in this paper on methods and conceptual clarification. Clearly, there is not one, but many public visual images of science, because there are various publics, each including image producers and consumers, various image types and visual media, and many different sciences. By focusing on two divergent public images of science, the popular visual image in clipart cartoons and the visual self-image of science in photographs, we have investigated the tensions that exist at the visual interface between science and the public.

Popular visual culture has preserved a visual image of science that, as a whole and in most details, dates back before the nineteenth century when science in today's sense hardly existed. This implies that there is a deeply seated level of the public understanding of science that has not been affected by the processes of the professionalization, diversification, instrumentalization, industrialization, commercialization, and the growth of science itself (by a factor in the order of 10^5) during the past two centuries. Because the past two centuries were also a period of improved public education and dramatically increased visual image creation and circulation by new media technologies, we may assume that such social and technological advances have had little impact on this level of visual understanding. From that we may conclude that it is unlikely that this popular image of science will easily change in the future.

In their self-images, scientific disciplines and institutions respond in different ways to popular visual culture. We have shown that chemistry and physics have opposing styles of self-representation. While chemistry dominates the popular image of science overall, such that stereotypes from premodern chemistry are the visual emblematics of today's science, modern physics, which emerged much later, has no clearly identifiable character in popular visual culture. In response, in their public self-image, chemists have largely adapted to the popular stereotypes in a conservative manner by featuring characterless and socially isolated chemists in stereotypical gestures with emblematic objects, which, although part of a rich historical tradition, represent pre-modern rather than modern chemistry. In contrast physicists, lacking such a heritage, have developed a progressive self-image dominated by electronic instrumentation, teamwork, and portraits of famous twentieth-century theoretical physicists. This progressive agenda, however, is undermined by gender representation, which reverses the ideology of progress, since self-images of physics tend to cultivate the popular stereotype that science is a male domain. Unexpectedly, chemists, despite conservative self-imaging, present themselves as approximately gender balanced.

With the exception of gender balance, chemists at universities do little to correct the popular clichés of their science. Only schools work hard to depict chemistry in its social context and therefore the more human side of science. The industrial side of science is only highlighted by government and industry and only for chemistry. As we have explored in another paper (Schummer and Spector 2006), these images are indebted to a number of iconological and aesthetic traditions including landscape, still-life, genre, and architectural painting and photography.

Because chemistry dominates the popular image of science, it requires particular attention in studies of the public image of science. In its conservative self-image, chemistry adapts to rather than corrects the popular visual image of science. Given the extremely conservative nature of the popular visual image of science as represented in clipart, one might argue that attempts to correct that by public self-images are ineffectual. However, within popular visual culture clipart represents probably the most conservative type of representation. Other types of popular visual culture like in magazines, movies, and TV would presumably be more amenable to new input from scientific self-images. Yet, if scientists in their public self-images, knowingly or not, reproduce the stereotypes of even the most conservative type, they reinforce clichés about science, which they otherwise complain about, in the entire popular visual culture.

NOTES

1. Earlier versions of this paper have been presented at the Commission of the History of Modern Chemistry Conference on "The Public Images of Chemistry in the Twentieth Century," Paris, France, 17–18 September 2004; in the Science Studies Seminar of the University of South Carolina, 16 March 2005; and at the Beckman Center for the History of Chemistry, Chemical Heritage Foundation, Philadelphia, PA, 23 March 2005.
2. We distinguish between the public self-image of science, which refers to the image scientists wish to project to the public, specifically on the Internet, from the self-image of science, which is private and idiosyncratic, dependant on the scientist or specific scientific discipline; the two, of course, overlap at points.
3. The actual search phrase has been "+scien* -(animal* flower* tree* herb technol* music* fiction)"; for the syntax of search phrases, see the website of http://www.clipart.com.
4. Search phrases need to be carefully selected so as to cover adjectives and nouns as well as singular and plural forms. For instance, for the biomedical sciences our search phrase is "+(biolog* medic*)."
5. In the early nineteenth century, "physics" was still the generic term for the natural sciences, almost similar to "natural philosophy." Modern physics emerged as its own discipline only in the second half of the nineteenth century by combining parts of applied mathematics with parts of what was

formerly called "experimental philosophy." For more details, see Stichweh (1984).

6. The percentage of chemistry images that contain glassware is actually much higher, since keywords usually do not note cases in which some glassware appears only in the background or is part of a larger experimental setting (see below).

7. To illustrate our keyword strategy, after sampling the images of research the best search phrase for capturing the field of reading and writing turned out to be "+(book* library document* text writ* read*)."

8. The gender ratio by keywords does not depict the actual gender ratio of the image contents, because gender specific keywords are more often used for images representing women than for images representing men. This illustrates a general shortcoming of our keyword analysis: elements or aspect of images that are taken for granted are not always indicated by keywords.

9. On the transformation of the "mad scientist" from literature to movies, see Toumey (1992).

10. More potently for clipart, "rocket science" probably has much less to do with a discipline than with the cold war space race that generated the much-loved term "it's not (ain't) rocket science." The sticking power of rocket science in the popular imagination is its association with a distinctively American ironic attitude toward intellectual labor that is encoded in this humorous phrase.

11. In terms of the number of publications indexed by their respective abstract journals, chemistry is as big as the total of all the other sciences; for some data, see Schummer (2004).

12. The history of the visual representation of glassware and other chemical instruments, which goes back via modern chemistry textbooks and early modern chemical craft textbooks to early alchemy, is well documented; see for instance, Weyer 1991; Knight 2003; Obrist 2003.

13. The visual history of the archetypical image of a chemist is presented in more detail in Schummer and Spector (2006, in preparation) as part of a special issue on the "Public Image of Chemistry." From our 2004 talk in Paris, Philip Ball has composed a brief picture story (Ball 2005).

14. This estimate is based on the latest available number of searches per day with Google, which was 250 million in February 2003 (Sullivan 2003; (http://searchenginewatch.com/reports/article.php/2156461).

15. There are some uncertainties about the image naming habits in different institutions due to different stages in the professionalization of website management. Because professional website management, for instance in online magazines, composes web page content out of text and image elements from databases, where image files are typically encoded by numbers, our approach focuses on less professional websites. This is desirable for our study, however, because we want to study the self-image of science rather than the image of science by web designers.

16. The actual number of images of famous physicists is much higher because these image files are named after the names of the physicists.

17. It is important to note, however, that these distinctions between the uses of instrumentation in chemistry and physics blur at their intersections in the fields of physical chemistry, chemical physics, material science, etc.

18. From Handelsman et al. (2005). The data is from 2001 to 2004 and based on the top 50 research departments of each discipline in U.S. universities according to the NSF ranking.

BIBLIOGRAPHY

The American Heritage Dictionary of the English Language (2000), 4th ed., Boston: Houghton Mifflin.

Ball, Philip (2005) "What is in the flask? The origin of the archetypical image of the chemist", *Nature*, 433 (6 January): 17.

Barthes, Roland (1964) "Rhétorique de l'image", *Communications*, 4: 40–51.

Boehm, Gottfried (1994) "Die Wiederkehr der Bilder", in G. Boehm (ed.), *Was ist ein Bild?*, München: Wilhelm Fink Verlag: 11–38.

Chambers, David W. (1983) "Stereotypic images of the scientist: The draw-a-scientist test", *Science Education*, 67: 255–265.

Flicker, Eva (2003) "Between brains and breasts—Women scientists in fiction film: On the marginalization and sexualization of competence", *Public Understanding of Science*, 12: 307–318.

Handelsman, Jo, Nancy Cantor, Molly Carnes, Denice Denton, Eve Fine, Barbara Grosz, Virginia Hinshaw, Cora Marrett, Sue Rosser, Donna Shalala, and Jennifer Sheridan (2005) "More women in science", *Science*, 309(5738): 1190–1191.

Hopkins, Nancy (2002) "Report of the school of science", in Reports of the Committees on the Status of Women Faculty, Massachusetts Institute of Technology, March 2002, http://web.mit.edu/faculty/reports/.

Knight, David (2003) "Exalting understanding without depressing imagination. Depicting chemical process", *Hyle: International Journal for Philosophy of Chemistry*, 9: 171–189.

LaFollette, Marcel C. (1988) "Eyes on the stars: Images of women scientists in popular magazines", *Science, Technology, & Human Values*, 13: 262–275.

—— (1990) *Making Science Our Own: Public Images of Science 1910–1955*, Chicago: University of Chicago Press.

Mead, Margaret and Rhoda Métraux (1957) "Image of the scientist among high-school students", *Science*, 126: 386–387.

Mitchell, W.J.T. (1994) *Picture Theory: Essays on Verbal and Visual Representation*, Chicago: University of Chicago Press.

Nye, Mary Jo (1996) *Before Big Science: The Pursuit of Modern Chemistry and Physics, 1800–1940*, Cambridge, MA: Harvard University Press.

Obrist, Barbara (2003) "Visualization in medieval alchemy", *Hyle: International Journal for Philosophy of Chemistry*, 9: 131–170.

Pollak, Melissa (2002) "Science and technology: Public attitudes and public understanding" *Science & Engineering Indicators* —2002, ch. 7: 1–6, http://www.nsf.gov/statistics/seind02/c7/c7h.htm.

Schummer, Joachim (2004) "Why do chemists perform experiments?", in D. Sobczynska, P. Zeidler, and E. Zielonacka-Lis (eds), *Chemistry in the Philosophical Melting Pot*, Frankfurt am Main: Peter Lang: 395–410.

—— (2006) "Historical roots of the 'Mad Scientist': Chemists in nineteenth-century literature", *Ambix* 53(2): forthcoming.

—— and Tami I. Spector (2006) "The visual image of chemistry", *Hyle: International Journal for Philosophy of Chemistry*, 12: in preparation.

Stichweh, Rudolf (1984) *Zur Entstehung des modernen Systems wissenschaftlicher Disziplinen: Physik in Deutschland; 1740–1890*, Frankfurt am Main: Suhrkamp.

Sullivan, Danny (2003) "Searches per day", SearchEngineWatch (February 23) http://searchenginewatch.com/reports/article.php/2156461.

Toumey, Christopher P. (1992) "The moral character of mad scientists: A cultural critique of science", *Science, Technology, and Human Values*, 17: 411–437.

Weingart, Peter (2002) "Of power maniacs and unethical geniuses: Science and scientists in fiction film", *Public Understanding of Science*, 12: 279–287.

Weyer, Jost (1991) "Chemie und Alchemie im 16. Jahrhundert und die chemische Fachliteratur jener Zeit", in *Von der Astronomie zur Alchemie—Bedeutende naturwissenschatliche Bestände des 15. und frühen 17. Jahrhunderts in der Historischen Bibliothek der Stadt Rastatt im Ludwig-Wilhelm-Gymnasium, Ausstellungskatalog*, Rastatt: 59–122.

Part III

Science Images

5   The Frog's Two Bodies
The Frog in Science Images

Bernd Hüppauf

TWO BODIES: LABORATORY IMAGES AND MYTHOLOGICAL IMAGES OF THE FROG

Among the animals endowed with a specific position in the history of the human self and of the relationship to nature, the frog needs to be discovered. In comparison to other animals with symbolic significance such as the lion or the eagle, the frog is an insignificant little creature. Certainly, it has gained a new prominence in recent decades as a public symbol, a heraldic emblem of the green movement. Taking the long history of frog imagery into consideration, it is surprising that the frog was chosen as one of the icons of a movement that supports a change of attitude towards nature. Its public image has completely changed from its earlier association with evil and the uncanny to become an object of love and identification. After hundreds of years as a popular symbol of fertility including sometimes terrifying implications, it has now come to signify fragility and serves as a reminder that nature's future is threatened. At a time when the body may disappear into virtual reality, the frog has come to symbolize resistance with its vulnerable body representing the weak and helpless, the victims of technological progress. It has become a medium for preserving the concreteness of a tangible nature and it is a signifier in a public program that aims to rediscover the body's indispensability for any definition of life.

A few noticeable exceptions aside, the frog has never been an object of high culture, and the exceptions, among them Aristophanes' *Frogs* and the anonymous *Batrachomyomachia*, *The Battle of Frogs and Mice*, are low genres, comedy, parody, and satire, and they associate the frog with negativity and mischief.[1] The rare references to the frog in modern literature and art are extensions of this pejorative image and the frog's marginality. La Fontaine' fable *La Grenouille* and Wilhelm Busch's poems on the finch and the frog, both satirizing the frog as an animal of vanity and stupidity are well known examples.[2] Yet from the first editions of the *Physiologus* on, the frog's position in European popular culture and mythology has been different. With the exception of extremely cold climate zones, frogs live all over the globe and there is hardly a culture without a frog mythology. Frogs exist, it can be argued, in three different varieties, as living animals, as a fetish, and as both real (miniature sculptures, drawings, amulets) and imagined images. Images of the frog have created a fetishized enigmatic animal empowered with magical qualities. The fetish is inconceivable without the living frog's body. Decisive for the production of the fetish is that subconsciously qualities are attached to this body, which transform it into a mental image that is endowed with energy and specific powers. Frog imagery associates the animal with sorcerers and witches and ascribes human qualities to this small creature that arouse deep emotions such as fear and, less commonly, also hope (Hirschberg 1988). It is no surprise that in systems of pre-modern cognition the three states of aggregation are inseparable. I will argue, however, that even under the conditions of the modern scientific laboratory, the frog's body is perceived as an image with qualities of a fetish.

In this essay, I will focus on frog imagery from the early period of scientific experimentation, the late eighteenth century, a time when fundamental changes in frog imagery can be observed. Compared to spectacular experiments upon animals, such as a Russian dog with two heads or the cloned sheep Dolly that have attracted wide media attention, images of the frog remain unnoticed and may seem banal. However, it is precisely the frog's unspectacular and unnoticed presence in the day-to-day routine of laboratories and in illustrated publications that make it exemplary. Indeed, it played a significant role in visualizing scientific experiments both in the sciences and in popular discourse. In comparison to an earlier and long period during which the frog was an animal in mythology and often weird imaginings, it had become a signifier of a scientific definition of the body.

Illustrations in nineteenth-century medical textbooks and encyclopaedias were designed to contribute to the process of disenchanting nature through science. The design and practices of laboratories in the mid-nineteenth century created the conditions that brought to perfection the mutual exchange between experimentation and imaging and as an implication a new imagery of the frog evolved that apparently severed its connections with a long tradition of myths and folklore. Illustrations in books and periodicals evolved in conjunction with scientific discourse and became the main media for a new image of the frog's body that followed the ideal of scientific objectivity. I would like to challenge this juxtaposition.

A scientific discipline's striving for theoretical generality by way of abstraction was no shield for protecting it from the impact of a derided past of imaginations, superstition, and the continued power of non-rational motivations. The positive belief in the power of verifiable knowledge and the scientists' ideal of a clear and simple image of life constituted not the only genealogy of almost two hundred years of experiments on frogs and their visual representation that was to become common with the expansion of the life sciences during the nineteenth century. Nineteenth-century laboratory experiments on frogs produced a visual discourse focused in the first instance on the production of objective knowledge but also unwittingly maintaining a connection to pre-modern and non-rational imaginings of animals. In physiological research, the frog was also associated with images of a hidden *negative memory*. For the experimenter the frog was at once both a mere object and an enigmatic animal, and the science illustrators extended this contradiction unwittingly to the image of the frog.

The clear separation of the body, defined as a mechanical structure, from the transcending dimension contained in myths and fantasies was illusory. The two bodies of the frog, one couched in myth and imaginary images and the other reduced to the diagrammatic patterns of a biologically defined morphology were never clearly separated. The supposedly meaningless laboratory pictures of the nineteenth and twentieth centuries were, in reality, metonymic replacements and transformations of deep seated and emotionally charged images of frogs and toads from myths and folklore. While serving as manifestations of the scientific ideal, these icons also made its fusion with an emotionally driven attitude to the animal's body tacitly visible. The production and interpretation of abstract images of laboratory frogs was grounded in a wider network of earlier images from fantasy and childhood that were transported through fairy tales, book illustrations, or narrated in myths and literature. In a cultural environment dominated by the sciences, which aggressively propagated the elimination of meaning and its delegation to a pre-scientific past, and which considered the contemporaneous fields of fantasy and imagination, literature, and the arts as dream —worlds incompatible with scientific knowledge, the frog imagery reveals an unwitting persistence of meaning.

Hundreds of thousands of living frogs were consumed and destroyed in laboratories. Their living bodies were connected to laboratory machines and prepared for the purposes of the laboratory by the application and intrusion of medical instruments and also, I want to argue, by the *techniques and instruments of iconic representation*. The animal body, thus prepared for its inclusion in the space of a modern laboratory, allegedly had no meaning. I would like to demonstrate, however, that the inseparable fusion of body and images of the body continued to convey meaning, through the new media, that was of a nature similar to the myths and imaginary images that predated the beginnings of serious experimentation in specific spaces of scientific purity. Frogs and frog images served scientific and didactic purposes to such an extent that the imagined frog became a furtive fetish of nineteenth-century laboratory experimentation.

The image of the sciences has always been created, and continues to be created, on the level of a subconscious communication of texts and their pictures with readers who may wish to perceive images as mere illustrations of the printed word but whose mind and psyche are also entangled in a whole web of non-linguistic information. There is evidence that popular images of science reflect deeply ingrained, durable, and stable stereotypes about science; the specific images of frogs in the sciences support this general observation. Obvious differences apart, the scientific image of the frog was never completely severed from its mythological origins. This continued fusion gave these images a specific power of persuasion beyond their intrinsic scientific functions.

In an essay on the visualization of Charles Darwin's theory of evolution, Stephen J. Gould introduces the term *canonical icon*. He refers to widespread and standard images that are associated with key concepts in science. "Nothing is more unconscious, and therefore more influential through its subliminal effect, than a standard and widely used picture for a subject that could, in theory, be rendered visually in a hundred different ways..." (Gould 1995: 41). The image of the frog was the canonical icon of the life sciences in the nineteenth and early twentieth centuries. It provided visual support for an unchallenged view of the body by serving as an unrecognized icon whose canonical status made the message appear beyond doubt. Perceived within the context of the nineteenth century experimental sciences, frog images are examples of *canonical icons* that follow the ideal of objectivity in methodically controlled experimentation and add a further dimension by providing visual legitimation.

In the late nineteenth century, frog images also played a significant role in the complex process of popularizing scientific research. Science images never appear in isolation. Each cluster activates memories of previous, similar images and together they form generalized images, often compressed to stereotypes and not dissimilar to archetypical images. They connect most of us by shaping common images and mapping imaginary spaces. The logic by which these imaginary spaces are formed, and their virtual topography, need to be explored. Laboratory experiments were carried out in a general framework that transcended the clearly defined space of the laboratory and included the rudimentary media networks of the time. Their images aimed to reach a wider audience of interested non-experts among the *Bildungsbürger* as well as entrepreneurs and technical and industrial organizations.

The new scientific practice of theory-driven laboratory experimentation required an epistemic frame of reference suited to the new scientific standards of systematic research but also one capable of expanding beyond the laboratory and into the media networks of this period. The available media were popular lectures, illustrated journals—first with copper or steel plate engravings and lithographs and, after ca. 1880, often with photographs—encyclopaedias, catalogues, and popular and illustrated handbooks.[3] Published images of laboratory practice presented the cutting up of a living body not only as a necessary means of scientific research but also as a media event of an exemplary nature that made the body disappear from sight. In both contexts, the scientific and the popular, images of laboratory animals were significant: they made them lose their bodies and turned them into abstract two-dimensional lines and shapes. These pictures were read as illustrations of scientific texts and they simultaneously contributed to producing a general image of science based upon a knowledge that was imaginary and at the same time conditioned by science images. The nineteenth-century practice of experimental physiology was linked to both a theoretical ideal and a media practice of popularization so that visual representations of animals became objects of scientific communication and at the same time contributed to shaping images of life—the life of animals and life in general, including human life—by super-imposing these new images on apparently incontestable and apodictic images of an older tradition. Images of frogs as canonical icons worked as a tacit invitation to transfer images from one context, the exterritorial space of the laboratory, to another context, popular culture and the life world. Scientific images of the frog, I want to argue, introduced spectators to a specific way of perceiving an animal's body and, furthermore, had an impact on general images of science and attitudes to life. This impact on the general perceptions of life was possible only because the images were not limited to their functions as science illustrations but maintained a connection to an age-old and nonscientific imagery, familiar to everyone.

FROGS AND TOADS IN MYTH, LITERATURE, AND THE ARTS—IMAGINED FROGS, THE FROG AND THE UTERUS, FROG AND/AS THE HUMAN

We know from archaeology that the legend of the frog as a magical animal goes back to pre-historical times (Hirschberg 1988: 77f.).[4] The frog was associated with secret knowledge. While ambivalence was characteristic of the frog's image, it was primarily grounded in negative emotions such as fear and repulsion. One of its characteristics was that it depicted uncanny strength in apparent weakness and was associated with life- giving water or with mud and disgusting slime. It signified fertility as demonstrated by clay figures of frogs with an exaggerated vulva on their back or the belly that have been found in various parts of Europe. Frogs are born in the water and live near it, so the fecundity of the wet but also the nauseating fluids of a dark world of fantasies of sin and evil are associated with their bodies. From numerous early sources we know that frogs and toads were used in magic and medicine (see articles on *Frosch* and *Kröte* in Bächtold-Stäubli 2005).[5] The powder of dried frogs and toads was believed to have healing qualities; frogs' entrails were used as charms and as medication in primitive medicine and by quacks in magic rituals. Frogs were never missing from a witch's brew. A recent example can be found in Grass's novel *Die Blechtrommel* where a group of mischievous children force dwarf Oskar to drink from a boiling brew made of repulsive ingredients including urine, snot, and a frog.

Frog charms were common from classical antiquity until well into the nineteenth century and were believed to protect against the powers of evil toads and frogs, the evil eye and various illnesses. Living frogs and images of frogs were attributed to gods and goddesses signifying specific faculties and powers. In certain cultures the frog was revered as a god or goddess and entire cults revolved around images or sculptures of frogs. In early Egyptian and Hebrew mythology, frogs belong to the realm of the gods and to magicians. The Egyptian gods Hah, Kek, Nau, and Amen were represented as frog-shaped (Germont 2001). "Later the frog came to mean both resurrection and the repulsiveness of sin, worldly pleasure, envy, greed and heretics" (Cooper 1992: 107).

This bronze sculpture from West Africa shows a two-faced head. In contrast to the two faced Janus-head of Roman mythology that symbolizes knowledge as it simultaneously looks into both the past and the future, this head has two faces in order to demonstrate that the evil of speaking

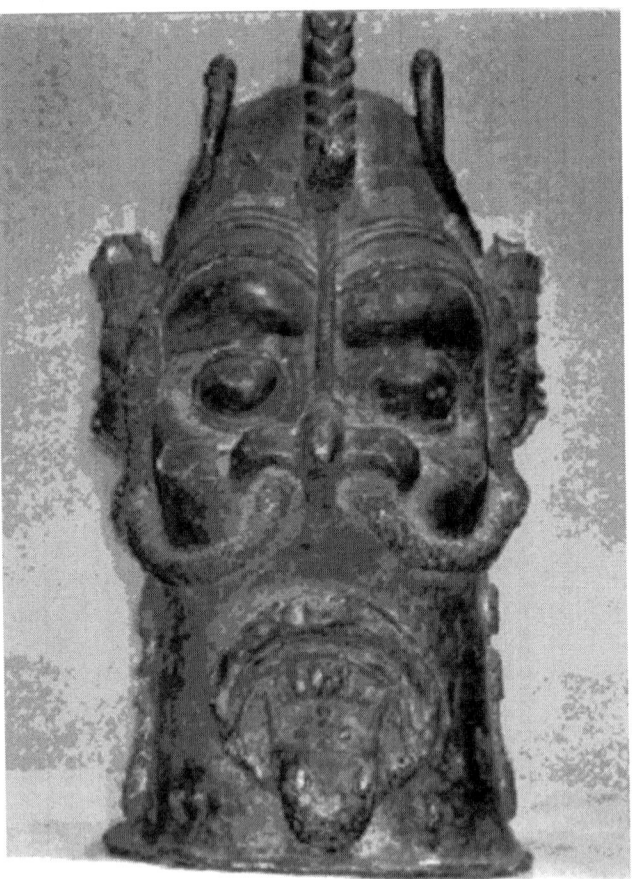

Figure 5.1 Head of an African sorcerer.

untruth occupies the entire person, front and back, and consumes all differences. The man is a vicious sorcerer who thinks and speaks evil and the evil is made visible through snakes curling around his head and a frog. What we see between his lips is not a mal-formed tongue but rather, what we see is untrue speech, visualized by a frog.

Notwithstanding the distance between the physiognomy of the frog and of primates in Linné's taxonomy, a close relationship between humans and frogs has always been suggested in popular culture. The frog was associated with human characteristics and has even been presented as a human in disguise, as the enchanted prince of fairy tales. In various cultures around the world it is depicted as surprisingly human and always ambivalent, the emblem of life and reproduction but also of deep-seated fear. It is surprising to note that—with the obvious exception of apes—there is no animal in the mythological tradition that is as human as the frog. The ugly little frog needs a small golden crown and an additional act of violence for its body to turn into that of a handsome young prince. Fables of Aesop and his successors use all sorts of animals to represent humans. Yet, few animals save the frog are imagined and imaged as a human being in disguise, enchanted and still human. The golden eyes of toads (*Kröten*) are windows on the soul, the croaks of toads (*Unken*) foretell a (bad) future, while swimming frogs have been associated with the fetus in amniotic fluid, and a frog lying on its back was commonly associated with a human baby's body.[6]

Particularly interesting is the connection of the uterus and the toad that exists in a range of variations. The uterus was seen as a toad-like organ that moved around in the female abdomen. This was of particular significance since the uterus was believed to be the organ of the female imagination. Until the eighteenth century medicine was convinced that the female imagination was capable of initiating pregnancy by imprinting the image of a male lover on the uterus. The image of a child or the image of disease was believed to be a cause for physical change, that is, pregnancy or illness (Fischer-Homberger 1979: 109). In traditional medicine the female *vis imaginativa* was believed to be located in the uterus or, as others observed, in the mind (Malebranche 1674/1980). In the eighteenth century an intensive debate on the physical power of imagined images evolved and until late in the century medicine charged the uterus with ill-directed imaginations resulting in a deformation of the fetus. Imagination had the power to deform the fetus, as it could be directed in the wrong direction, away from a form-giving male human being, resulting in the bad consequence of a disfigured child (Bundy 1927). The female imagination, not sufficiently controlled by reason, could go wild and it was believed that seeing images of deformed or monstrous bodies during conception or pregnancy would lead to deformations of the fetus. Frogs or toads were believed to be involved in the pregnancy *per imaginationem*. If a pregnant woman's gaze met that of a toad at a certain time of the year, she would give birth to a deformed child, since the perceived image would be turned into an image of the mind which, in turn, would be physically imprinted onto the fetus. "A remarkably persistent line of thought argued that monstrous progeny resulted from the disorder of the maternal imagination" (Huet 1993: 1). Ambroise Paré relays the following anecdote about a child born with the face of a frog. When the child's father was asked for an explanation, he explained that his wife was feverish and had been advised to hold a frog in her hand until it was dead. The woman was holding the frog while she had sex with her husband and this, he believed, was the cause for the deformation (Ambroise Paré: *Des monsters et prodiges*, quoted in Huet 1993: 16).

The *monstrous imagination* was a major topic of medical literature for two thousand years, until in the late eighteenth century a new image of the body emerged that separated the uterus from the imagination and prevented frogs or toads from contaminating female organs or even turning into one itself. This was both a turning point in the history of the body as well as the time in the history of medicine and zoology when experimentation on living frogs began. The shift separated the body from its environment and also led to a de-mystification of the frog. Yet frog mythology in a transformed fashion was carried over into the age of modern science to deeply affect its ideal of objectivity and, in particular, physiology.

THE FROG IN LABORATORY EXPERIMENTS—THE VANISHING OF THE BODY IN ABSTRACTNESS AND MECHANICAL OBJECTIVITY

One of the triumphs of the modern age was the elimination of fear-inspiring chimeras, of superstitions based on magical animals and the belief that images representing sacred or evil animals had power over men and women. Sacrificial animals, chimeras and monsters were subjected to scientific scrutiny and soon excluded from the taxonomy of zoology. As a rationally constructed system, nature offered no room for these animals and, consequently, their images were relegated to the sphere of the imagination and understood as non-real, the product of unchecked emotions, of fears and hopes that ought to be ruled over by reason.

Dissections, machines, and the disappearing body

The construction of life as the object of methodical research required the dismantling of a unity. The sciences abandoned concepts of purpose and meaning and defined life within the framework of methodical research. This definition required the separation from the sciences of the philosophical concept of life as an integrated unity, which abandoned the idea of a living organism as a whole and replaced it with experimental operations that require the cutting knife.[7] According to this theory, observation of parts and pieces of the disassembled body would reveal the mechanisms of life. The corresponding images are a radical example of an art that shuns the perception of the whole relegating it to myth or metaphysics.

This project produced remarkable results. It aimed at describing living organisms as a closed ensemble of distinct pieces and demonstrating how these pieces interact and move to make the body work efficiently. The living organism was no longer exposed to the *vis imaginativa* but interpreted as a system that could be disassembled into pieces that fit together in the way wheels and cogs, valves and transmission belts or bracelets and other elements of a machine interact and make it move and work. The best known elaboration of the analogy of body and machine was Descartes' *Discours de la methode pour bien conduire sa raison, & chercher la vérité dans les sciences* (Descartes 1637/1984–1991: part 5).[8] The Cartesian image of life became the dominant model during the two centuries that followed its invention. The mechanistic view of living organisms was extremely successful and made all other concepts appear obsolete. During the eighteenth century, the machine metaphor was radicalized through Lamettrie's *L'homme machine*. It now definitively excluded the psyche as an invisible condition of the whole. The machine, this theory asserts, functions equally well under every external circumstance, its functioning being determined by the quality of its elements and its intrinsic mechanism, independent of external factors. It might be worth mentioning that Lamettrie does not fail to refer to the frog. His frog no longer has the power to interfere with pregnancy and inflict harm on the human body. The frog now serves as the model for the analogy of a living organism and a machine.

The images discussed here are indicative of the self-image of the researcher as representative of the *avant-garde* of the age of science who adopts this theory of the body. Artists as science illustrators and scientists formed a close alliance that served the high ideal of explaining nature and visualizing this explanation in images. This ideal seemed beyond doubt and was the linchpin of the allegedly self-evident superiority of the modern culture built on a new concept of science as methodical research. This alliance led to the emergence of a new iconography in encyclopaedias, textbooks, and other scientific publications. It created an *abstract* and *spaceless* imagery that eliminated the body of the scientific object as well as all traces of the observers in an attempt to create images of abstract objectivity.

In sharp contrast to the laboratory situation, images from handbooks, encyclopaedias, and other works presented an empty space with no trace of actors and artists, not dissimilar to the contemporary theory of photography that interpreted photography as a means of nature to produce images of itself without interference from human subjectivity. This was only possible because the image presented the mutilated frog not as the work of the experimenter but as the result of an anonymous scientific process. In this image of the frog there could be no outside of science and of the laboratory. Indeed, for the theory of the time there was no outside of the scientific definition of the world. Magic was declared to be a phantom and non-existent.

The abstraction of the science image of the frog could be nothing other than the one and only true image.

The transformation of the frog into an abstract pattern was successful to the extent that its body began to disappear. A new medium replaced the imagined body with a new image, represented in abstract and diagrammatic pictures. They were based on a zoological image of the body and a definition of images as illustrations of scientific texts.[9] The images reduced animal bodies to an object of scientific investigation with no soul and life of their own, a mere means for attaining theoretical knowledge of biological organisms. They defined the body in terms of a complex machine that could be fully explained through methodical investigation and representation. Pictures served the purposes of a reductionist approach to animal experiments.

Images of the frog, more than images of other laboratory animals, were integral parts of a perceptual process that transformed living organisms into machines, which resulted ultimately in the disappearance of the body

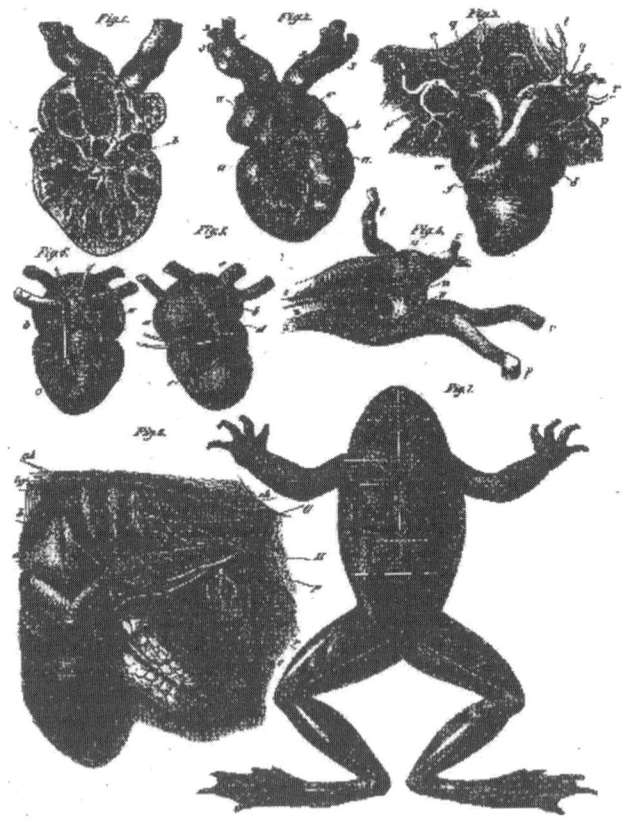

Figure 5.2 Frog disassembled for didactic purposes in medicine, printed from a copper plate (1878).

(Verdin 1882).[10] In science images frogs are shown in detail, but never as a body with emotions. The dissecting vision ensures that the frog of these images is neutralized, abstracted from reality, and demonstrably lacking the faculty to suffer. Its forced immobility is a constitutive part of the iconic message of the image. The analogy ensured that its body was independent of its environment and functioned under all external circumstances, even if it was reduced to single organs and limbs. Hence it could be visually represented as a store of parts providing the material for experiments. Science images of the frog closely followed this model of life and its imagery complied with the requirements of theory. They provided evidence for the theory by producing a visual field that governed the perception of the animal as a functioning system of discrete elements. They showed the frog's body in complete isolation, unrelated to concrete space and without motion, timeless and disassembled into discrete bits and pieces, limbs, organs, and nerve tissue, symmetrically arranged by the illustrator. We find arrangements in one picture frame of the body of a frog surrounded by isolated *interesting* parts or an assembly of body parts next to laboratory instruments and machines designed for frog experiments. The frog's body functioned as a machine, and, like every machine, it lacked the ability to suffer. Insofar as the ability to feel pain is a condition for the body's presence, these pictures rendered the frog's body absent. The frog's body, surrounded by laboratory instruments and cut open for the scientific gaze, is absorbed by, and acts as a dependent element of, the exterritorial space of the laboratory. Is this disappearance of the body on the science images different from the disappearance of the frog into the witch's bowl of a magical potion that was the consequence of a magical image of the animal?

The heart as a substitute

These images show the ultimate consequence of the tendency towards abstraction and anonymity. The body disappears and is being replaced with graphs, numbers, and imagined spaceless relations in a field of geometrical points. One can detect cynicism in these images. They are the result of a completely secularized view of life that has left behind any traces of the memory of a metaphysical image. Nor is it pagan. An experimental application of William Harvey's discovery of the circulation of the blood (*Exercitatio de motu cordis et sanguinis in animalibus*, 1628) is symptomatic. He explained the circulation of the blood in terms of a closed mechanical system, consisting of a pump connected to a system of pipes and valves. This explanation came close to the ideal of modern science developed in theoretical physics, which promised to explain everything, from the cosmos to the smallest unit, as examples of general laws which are simple and have the potential to reduce everything that exists to a few basic concepts and mathematical formulas. This fascination with physics was extended to the life sciences as the experiment described here demonstrates. Text

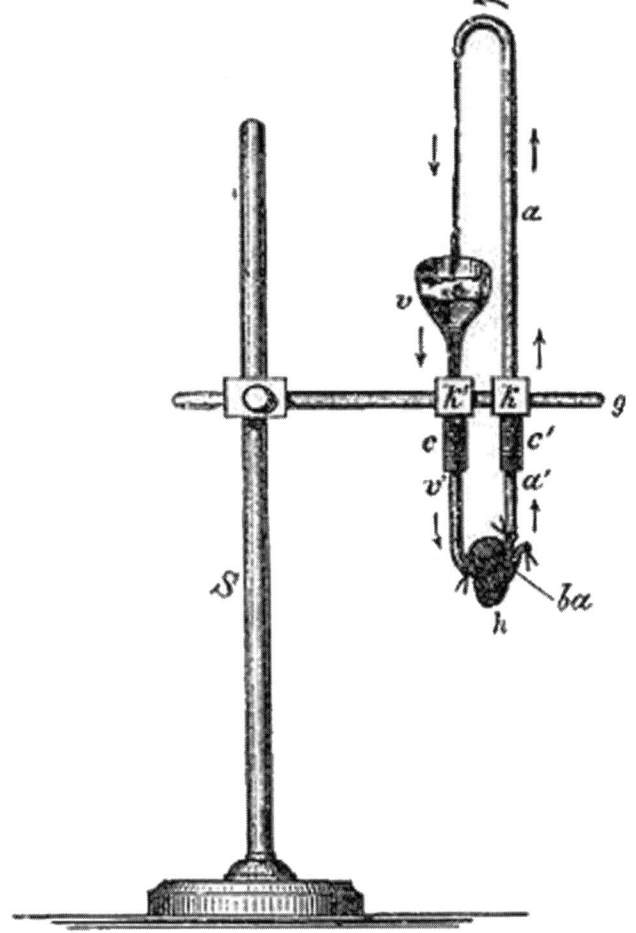

Figure 5.3 Artificial arrangement of blood circulation with a living motor (frog heart). Leipzig Univeristy.

and image introduce the reader to a simple artificial system made of metal, glass, rubber, and cork. Its operation requires the insertion of a pump. A pulsating frog heart is used as pump to make the simple machine work.

The description of this experiment for didactic purposes at the University of Leipzig is detailed and demonstrates "the fundamental fact of the blood circulation with such vivid clarity and elegance" that every teacher of physiology will benefit from it. The dry description refers to a *strongly pulsating heart* without further details. The demonstration required cutting open a frog's breast, extracting the heart, inserting two canulae and fitting the prepared heart into the system of pipes. Then it needed to be electrically stimulated to begin pulsating again and to set the system in motion by making the blood (or a physiological sodium chloride solution) move through the simple set of pipes.

In this image the frog is absent and present at the same time as it is represented by a frog heart. The body is reduced to its heart, the organ that is not only considered central to the body as a functioning machine, but also believed to be the location of emotions and the soul of humans.

Science image and cave art

Scientific images of frogs show no concrete and identifiable place. They reflect the concept of an empty space devoid of emotions and ethical imperatives. The images float on the page in a nowhere of abstraction. These images are designed with no regard to the material conditions of the laboratory and follow the principles of an abstract organization of the visual field. While they are reproduced on the pages of books and therefore are part of the drawing tradition which organizes its images within the frame of a well-defined and restricted space, science images do not comply with the requirements of framed pictures. Their margins are open, comparable to early paintings on cave walls. They do not follow requirements of mimesis. Science images of the frog are not representations but the product of the intention to construct an image independent of mimetic functions. They are characterized, as Sigmund Freud once said, by an over-emphasis on mental reality in contrast to material reality. They are, in his words, the product of a belief in the omnipotence of thoughts (*die Allmacht der Gedanken*) as shared by magical thinking (Freud 1919/1970: 267).

In a number of ways, these images have characteristics that make them comparable to the depiction of animals in the cave art of the Upper Paleolithic. They are disassociated from the three-dimensional world experienced outside the isolated space of a cave or a laboratory. There is no ground line in either the cave paintings or the science images. There is no hint of the environment in which these animals, imagined by Stone Age men inside a cave or by artists of the age of the scientific mind inside their studios, in front of a copper plate or drawing pad, live: no grass, no bush or tree, no water, no table, no dissecting hand, or rubbish bin. We see no trace of an originator or any subject involved in the construction and arrangement of the scene. It has been argued that with Paleolithic drawing, pictures came to express "man's thought about things rather than his sensations of them, or rather, when he tried to reproduce his sensations, his habits of thought intervened, and dictated to his hand orderly, lucid, but entirely non-naturalistic forms" (Fry 2003: 44). Fry argues that such "visual-conceptual habits of Neolithic man have sunk into our natures" and can be observed at present (Fry 2003: 45.) Authors who have written about connections between primitive art and modernity have supported this view. In an essay on African art, Elie Faure argues that "when modern science tries to express all its conquest in mathematical symbols, the mind is invariably brought back to primitive sources.... The result is always the impressive agreement between the most obscure and the highest form of reason" (Faure 2003: 55).

Upper Paleolithic parietal images have, what Halverson called their "own free-floating existence" (Halverson 1987: 66–67). Their forms are drawn "without regard to size or position relative to one another." These characteristics of early visual abstraction are shared by the abstract images of the scientific imagery of laboratory animals. They also float freely on the page, independent of any observed natural environment. This disconnection from an experienced environment is often reinforced by a further characteristic of parietal Paleolithic images. Animals have no hoofs, their legs ending in an undetermined airy fuzziness. If hoofs are shown, as in the Lascaux cave, they are drawn in such a way as to make visible their underside or hoof print.[11] This has led some pre-historians to suggest that these animals are depicted as dead. This seems an implausible reading of these cave paintings. Yet, the absence of anatomically correct legs is not a matter of chance nor is it a deficiency. The absence of legs to stand on is indicative of the "lifelessness" of abstract images. There is no mimetic realism implied. Legs have no function; this signifies a complete absence of activity including the activity of standing. This observation of abstraction and a lack of connections, shared by early cave paintings and primitive art and the scientific images of the nineteenth century, lead to a hypothesis. These pictures are the result of the laboratory as a non-territorial and *hyperreal* space. They are not connected to a social or natural environment, isolated in a space of the mind and, in contrast to their originators' intention, charged with qualities of magic.

FUSION OF MYTHOLOGICAL AND SCIENTIFIC IMAGES

The selection of the frog as the most popular animal for laboratory experiments needs to be seen in a wider context that transcends the frame of reference defined by the modern sciences. Science images of the frog also need to be perceived outside the context of a history of science as an institution driven by ideas of reason. They are symptomatic of a history in which the relationship between man and animal, frog and man, and in particular, frog and woman is far from rational, indeed, it is deeply emotional. The frog was chosen, I contend, because its image oscillated between a soulless primitive animal and the mysterious human and super-human qualities that were also ascribed to it. It is not unreasonable to associate these qualities with the female represented by the witch. The frog's body, operated upon and transformed into a science image, became an imaginary object inasmuch as the experimental process and the scientific gaze had prepared it. The transformation of an animal's body into science images introduced it into a web of signs that unwittingly incorporated memories of the scientists, who theorized about creating a new world of facts with no cognitive importance attached to images.

The crooked humanity that myths and magic had invested in the frog for many centuries continued to rule the perception in the age of rationalization, and scientists were no exception. It was the human being in the frog (*der Mensch im Frosch*) that stimulated the scientists' attention and laboratory practices for more than a century. Scientific experimentation on frogs was the extension of a long-term, twisted, and emotional relationship set up between humans and frogs. The frog was the ideal creature for scientific experiments because its image was invented to be both human and simultaneously it's opposite, a soulless reptile. It carried the signs of mysterious and incomprehensible qualities conjured up by the human imagination but could be just as easily equated with a machine. It lacked human faculties, in particular the faculty of suffering,[12] yet subconsciously it was also imagined as a metamorphic variant of the human with a body that incorporated the dark and evil dimensions of humanity. It was symptomatic of the ambiguous position occupied by the sciences in the late nineteenth century that traces of the mythological past were carefully eradicated and yet, they can also be seen in these images with no great interpretive effort.

Inquisition

These images leave no doubt about the status of the animal. It is a passive object in an empty space. There is no expectation that the object of the experiment will do anything, will ever speak and reveal the truth. Yet, the idea of extracting truth is the dominant motivation behind these images. Science images of the frog demonstrate that pain was not inflicted to make victims speak and that their truth was not linked to linguistic discourse. The body contained all the information and it needed to be made to perform in order to reveal it in a language of silence. Truth was equated with the information hidden in the mute flesh, in muscles, organs, and nerves and their silent movements. These parts of the body had to be isolated and their mute movements observed, but also physically interfered with in order to make them reveal their secret. They had to be connected to machines, measured and turned into quantifiable formulas in order to reveal information and make possible the recording of data. Recording the consequences of the interventions in the natural order of the body was considered the cardinal path to objective, scientific knowledge. The transformation of this process into images created a specific iconographic innovation, the visualization of a correlation between torture and knowledge.

The scalpel and various laboratory instruments and apparatus in conjunction with images served the purpose of endowing the mute animal body with a language comprehensible to human cognition. The animal body functioned as a medium to be wrought into new and anti-natural shapes that responded to the medical and physiological search for knowledge in a system of signs. The experimental physiologist cuts up the frog's body, stimulates nerves, and connects it with various instruments and machines in order to make it speak. These concrete operations on the frog's body are designed to extract knowledge from the body by making it perform and speak a silent language comprehensible to the expert. This code is then recorded through laboratory protocols, tracings, and diagrams, which make it possible to draw a stylized image of the body that makes the hidden secrets of its nature visibly comprehensible.

The space of the laboratory, equipped with experimental apparatus and medical instruments was not dissimilar to the chamber of the Inquisition and created similar conditions for inducing the frog's body to speak. It replaced the conventional meanings associated with the frog's body with its function in the process of producing abstract knowledge. The frog's body seemed to be particularly well suited for this type of performance since it was invisibly imbued with signs of human qualities from its mythical past. The "logic of torture" resurfaces in scientific images of frogs.[13] Both the Inquisition and modern science extract the truth from their victims, be they witches or, as with the images under scrutiny here, frogs. While science images of the frog carefully avoid all similarities with the human body, it is equally obvious that they reflect the convergence in this respect between the Inquisition and modern science. In every system of force and terror, it is the master of the rack and, ultimately, the executioner, the scientist in the modern laboratory, who makes the decision about right and wrong, life and death. The frog's judgement was death.

We are familiar with pictures of torture chambers from the twelfth and well into the nineteenth century when the Holy Inquisition tried to extract the truth from delinquents by making the body suffer damage as an object of cruelties and the soul suffer pain.[14] Who would not be reminded of these instruments and procedures of torture by many images of laboratory instruments? Perceived from this perspective, they begin to look remarkably similar. Images and descriptions inform us about precise procedures and the range of instruments, the tables and boards designed to tie down the accused and inflict all kinds of painful operations on the body of the delinquent. Included was pain inflicted on all parts of the body, occasionally the slow removal of limbs. Medical catalogues of the nineteenth and twentieth centuries show elaborate and diverse collections of instruments, tables, tilting boards, and grids that were designed to tie down frogs (and other animals) and immobilize their bodies with the aim of carrying out all desired experiments without any undesirable interference from the animal. It was an aim to keep the frog alive as long as possible after its body had been opened or limbs had been cut off. Laboratory protocols and images give the impression that they follow the same common iconic patterns established in the Inquisition's torture practices that aimed to make the accused speak.

E. ZIMMERMANN
LEIPZIG-BERLIN

Nr. 3565–3569. Froschbretter usw.

Frog Boards etc. — Planchettes pour grenouilles etc. — Porte-ranes.

Nr. 3565. Apparat zur Fixation der Wirbelsäule. An zwei prismatischen Stangen können je eine Klemme verstellt werden, welche in die Vertiefungen der Knochen eingreifen und die Wirbelsäule auf dem Brett festhalten.

Gewicht: netto 1,500 kg

Grundzahl: 170

Literatur: Cyon XII. 1

Nr. 3566. Einfaches Froschbrett aus Holz. An einem Stativstab befindet sich, verbunden mit einem Schwanier, das Froschbrett mit Löchern.

Größe der Nutzfläche: 160×90 mm

Gewicht: netto 1,500 kg

Grundzahl: 18

Nr. 3567. Froschbrett mit Knieklemme. Die letztere läßt sich an jeder Stelle der Hartgummiplatte, auf welche der Frosch in dem vorgesehenen Löchern festgebunden wird, befestigen und bietet dem Kalb sicheren Halt.

Größe der Nutzfläche: 170×95 mm

Gewicht: netto 1,500 kg

Grundzahl: 75

Nr. 3568. Froschpanzer nach Boehm.

Fixierungsapparat mit kegelig gebohlten Klemmen und Elektroden.

Der gesamte Apparat wird mit Stab St zur Befestigung an ein Stativ geliefert.

Gewicht: netto 0,760 kg

Grundzahl: 105

Literatur: Arch. f. exper. Path. and Pharm. Bd. 63

ca. ⅓ nat. Größe

Nr. 3569. Froschklemme zum Öffnen der Rückenwirbel. Die Ansätze der Rückenwirbel werden durch zwei verstellbare Klemmern geboten; der Kopf wird in einer davor angebrachten Klemme fixiert. Der gesamte Apparat kann in einer Tischklemme hoch und tief gestellt werden. Komplett inkl. Tischklemme.

Gewicht: netto 1,200 kg

Grundzahl: 93

Figure 5.4 A page from a sales catalogue of the E. Zimmermann Company advertises a variety of boards for fixing a frog for laboratory experiments.

However, a significant difference needs to be emphasized. Many Inquisition pictures show both the victim and the torturer. Clerics and judges were often arranged in a half circle around the torture table, dressed in full vestments and robes often with regalia. All were depicted with their eyes and ears fixed on the procedure and watching the victim with great attention. The combination in one frame of both the victim and the men responsible for her torture is indicative. The image leaves no doubt about the legitimacy, indeed of the inevitability of the procedure. The arrangement of the scene invests the observing men with full responsibility. They act in the name of a superior authority, the holy script, God. There was no need to hide the torturer or the torture chamber. On the contrary, the aim was to make everyone aware of the instruments of torture and the torment. An ideal of the Inquisition was not to use the instruments but merely to show them to the witch or the heretic with the intent of making them confess. We know, however, that there was no inhibition on the part of the church to applying the instruments physically, to make the accused speak. Inhibition could signify doubt and doubt was contrary to Christian dogma.

The images of scientific experiments do not show the men responsible for the arrangements. The new authority had no face and no ears. It is implicit, depersonalized, and anonymous. The cruel experiments are done in the name of science and its objects are depicted in complete spatial and temporal isolation. Whereas the torture of the Inquisition aimed at inflicting pain and causing horror, the scientific experiment's aim was different. Pain was immaterial, a mere side effect. Laboratory reports make it obvious that it was not noticed even if animals such as rabbits shrieked in pain. The frog was considered an animal incapable of suffering (Singer 1993).[15] The frog did not cry and showed no gestures of pain and therefore was an ideal object for revealing the hidden truth without producing undesirable side effects made by animals in pain. The silence of the frog precluded any emotions of inhibition. That the instruments were both shown to the distanced viewers of science images and put to use in the laboratories reveals a combination of unperceived anxiety and unwavering belief in the legitimacy of experimental practices.

THE LABORATORY AS A THEATRE OF SACRIFICE

Images of frogs contributed to a scientific practice that was at the same time an artistic practice. Laboratory reports often refer to experiments done on frogs in terms of a theatrical performance narrating event after event. The frog's body acts in a performance that provides answers to questions posed in the experiment. The choreography was a combination of scientific search and performance on the part of the animal whose body was prepared to form the center of all attention, exposed to the eyes of curious spectators.

A cynical joke could often be heard until recently: What is physiology? Physiology is the science of the dead frog (*die Lehre von toten Frosch*). This joke offers a revealing displacement in accordance with the metonymical operations that pertain to science images of the frog. Physiology is the discipline concerned with the chemical and physical functions of the living human body. The joke substitutes the dead frog for the living human body. This facetious displacement makes clear that the (dead) frog is the substitute for a human being in the same way as a sacrificial animal is a substitute for a spared human being. The secular imagery of the positive sciences, combined with elements of the iconic tradition of the martyr, the sacrificial lamb, or the torture victim, created the frog imagery. The frog was the frog but at the same time it was an emblem for the glorification of the sciences as a new religion. Frog iconography became the perfect medium for the *abstraction* and *objectification* of an originally religious dream. A critical reading of the impact of the Christian iconic tradition of pain and sacrifice on the images of the biological sciences could shed new light on the constitution of the sciences in the nineteenth century.

From early civilizations on, animals have served as sacred totems and as victims in ceremonial sacrifices. The fundamental idea of sacrificial substitution is built on animals and images of animals. We are familiar with rams and lambs, chickens, cocks and goats, or cattle as animals sacrificed in religious rituals. The frog had never been an animal apposite for sacrifice. The frog had to wait for the beginning of the scientific age and the emergence of the modern laboratory before being chosen as sacrificial animal. Physiologists furnished the spaces of their laboratories in such a way that it became possible to sacrifice this animal under the disguise of the rationality of scientific experiments. It is symptomatic of the transformation of ritual in the modern age that its crucial role in the creation of a self-image of the period of the positive sciences was un-ceremonial and its liminality went largely unnoticed.

No animal was sacrificed in the same way and in the same quantities for the sake of scientific knowledge as the frog was. Why the frog? That frogs had to be cut up and die in large numbers was considered an inevitable consequence of the definition of life by modern science. The scientific literature offered pragmatic explanations. It was argued that the frog's nervous system and its nerve fibers closely resembled those of human beings. It was therefore used as an ideal animal for experimental physiology's attempt to understand the human nervous system. Also, it was an easy animal to deal with: it existed in large numbers, was easily caught and handled, it exercised no resistance and—this was never said—it had no voice. It was therefore not uncommon to call the frog "dumb," unwittingly endowing it with a human characteristic. Some sources refer to their "painless death" and it was easy to discard them after their part in the experiment had come to an end. From the scientists' point of view, the frog was always well behaved. While these pragmatic arguments prevailed, there is evidence that they were little more than rationalizations for laboratory practices, which necessitated the science image to actively negate traces of the human in the frog iconology.

The scientific images served as a constitutive element in rationalizing the experiment in both meanings of the verb, they created a rational image in the sense of a Cartesian concept of nature and they also served as a means to cover non-rational motivations and darker affects and attach a pseudo-legitimacy to acts of sacrifice. It is my contention that it was precisely the human qualities attributed to the frog that made it so special and well suited as an object for laboratory experiments. The scientific images of frogs made visible an uneasy relationship of a rational individual who deals with frogs in the name of science and the subconscious desire to shed this self. Scientific images of the frog, it can be argued, were the result of a projection by the rational experimenter and his artist of an inverted image of their self. Guided by the scientist, the artist's focus was on hiding the tension between these two selves, one based on scientific theory and the other resulting from a mythology of a frog–man association. A result was the immobility and sterility of the pictures of frogs (and other animals) under laboratory conditions. Many scientific images display an impasse that results from invisible struggles between the universal ideal of rationalization and pre-modern myths, between hermetic exclusion and popularization, between spaceless abstraction and the concreteness of material laboratories, between purity and the necessity of bodies as mediums that inevitably contaminate the abstractions free from moral considerations. It can be added that a good part of the science images of the nineteenth and twentieth centuries were the product of such an impasse.

NATURALISM AND CONFLICTING THEORIES OF ART: THE CONSERVATIVE SCIENCE IMAGES AND WAYS OF SURMOUNTING IT

We know near nothing about the artists. Only a few of the images have a signature, which is always small and modestly placed not in the traditional place at the lower right side of a picture but almost invisibly hidden in or at the edge of the image.[16] It is as if the artists cannot resist the wish to be recognized as authors, yet they also want to remain an unrecognizable part of the anonymous process of modern science. The wish to hide and vanish from the surface of their own images could well suggest a bad artistic conscience. The artists may have had a hunch that their pictures imposed a constraint on the spectator's imagination and ways of seeing. The artists were supported by theories of art and literature of the late nineteenth century. Naturalism pursued the intention of removing the line that separated them from the sciences. They theorized about a structural homology between artistic and scientific experimentation as only different realizations of the same human mind. For naturalists the relationship between artistic techniques and instruments and laboratory techniques and instruments went beyond a mere analogy. Zola, referring to Claude Bernard, made it clear that the natural sciences provided the meta-discourse for the arts. He took his information on contemporary medical theory from Bernard's "An Introduction to the Study of Experimental Medicine" (1865). In Zola's treatise on the experimental novel (1880) medical experiment and artistic experience are put in close proximity. Through experimenting with imagined individuals the artist gains knowledge of nature in the same way as the scientist does by observing facts. In analogy, experiments carried out on the body of a frog would lead to linguistic and visual protocols of the world as it really is.

Supported by naturalistic theories of art and literature, the corresponding dominant ideology of the scientists of the time led the anonymous artists to use their means in order to visually mimic a mechanical concept of life. Artists who worked as illustrators acted in isolation from the world of the arts of their time and submitted their work to the ideology of mechanical objectivity. Given the power relationship between the sciences and the arts, they may have had no choice but to succumb to the domination of the sciences. In congruence with scientific theories of life, the iconographic tradition of the frog as a passive and motionless object of scientific observation persisted for over a century.[17]

The visual representation of the frog was as diagrammatical as possible without the object, however, disappearing into complete abstraction. The relationship of science images to life was predicated on the death of the observed object. The ensuing graphemes contained and also projected the object of experiments. Science images continued this heritage during a period when the arts experienced a revolution so that during the nineteenth century these artists worked in contradiction to images of life propagated in the arts. Impressionism, Expressionism, and Surrealism are labels referring to a later period in which the visual representation of men and women, nature and animals underwent fundamental changes. It was also the period when the new technologies of photography and film revolutionized ways of seeing and new knowledge about how bodies moved, and rendered visible a hitherto unknown reality. These innovations in the arts were inextricably intertwined with new philosophical theories of space and time and, what is particularly important in our context, of the concepts of the body and of life and death. In the face of these fundamental changes, the image of the scientists' frog remained unchanged and unchallenged. The potentially innovative power which the arts and literature after naturalism could have had on science images was lost. The separation between scientist and object, self and other maintained by science images with such rigor, was out of sync with simultaneous developments in the arts. A separation of the science images from the arts, it can be argued, contributed to the maintenance of a mechanical image of the body. It would be an untenable exaggeration to argue that orthodoxy in the sciences was based on the power of images. Yet the consistency of canonical icons acted as a strong agent against change and in favor of a Cartesian image of the body at a time when it had already become outdated. Artists, by adopting the scientists' naturalistic concept of life, created images that made a considerable contribution to the maintenance of this concept.

Science images did not passively represent scientific knowledge but created structures of seeing that organized entire perceptual fields. The artists' desire to produce images as an imitation of the precise linguistic code of the sciences corresponded to an *astigmatic vision* (see Trevor-Roper 1997: 49–63). Memories of images of a magical animal that had been identified with terrifying imaginations and evil disfigurations fell prey to the cold arrangements of the dissecting tables, the glass and metal of the laboratories. The artists aimed at the highest degree of objectivity and their strabismus prevented them from realizing the foggy magic that enveloped the objects of their images. This repression produced collective fallacies of vision. Memories filled with anxieties and hopes, laughter and terror remained absent from images in physiological handbooks and the readers' image of science. Into this void, perceptions of the body disappeared. The collective *deceptio visus* was neither an accident nor a simple mistake that could be rectified by diagnosis and correction. Collective scotomy was much more. It was the result of the desire not to see and to remain ignorant in the name of pure science.[18] An obliquity of vision was linked to the spectacular success of the experimental sciences. The inherent fallacies of perception remained hidden. Negativity and magic were and still are considered contrary to the frame of reference created by the rationality of modern science. To seize upon these qualities as a characteristics of science images is an attack on the integrity and self-understanding of the sciences. The unacknowledged relation to cave art and primitive art is but one of the non-rational elements in science images that need to be explored in more detail. Revealing a genealogy of the bad image of the frog in its scientific images is the same as revealing a cultural strabismus. The scientists' frog was immunized against negativity by the *mechanization* and *abstraction* of its image. Its image was rendered transparent and fully intelligible. Any attempt to correct this image involves a discomforting operation. To associate the canonical icon of experimental research with negativity and the violence of sacrifice inevitably raises resistance because seeing and producing science images as images of reason is central to understanding the self in the age of science. A new type of science image, emerging at present, may lead to new ways of seeing. These new images that do not pretend to represent a visual reality demand a new attitude towards images that is beginning to have effects also on our understanding of science images based on the ideal of objectivity. A new and close relationship of science images and contemporary art could open a new chapter in the evolution of science images and, as a consequence, of our image of the sciences.

NOTES

1. Until the late eighteenth century this parody on the war of Troy in hexameter was believed to have been written by Homer. An English translation under the title *Hesiod, the Homeric Hymns and Homerica: The Battle of Frogs and Mice* can be found at http://sunsite.berkely.edu/OMACL/Hesiod/frogmice.

html; an early annotated edition: [*Homer*], *Batrachomyomachia*, vorstellend die blutige und muthige Schlacht der Mäuse und Frösche. Mit Fleiss beschrieben und mit Anmerkungen ausgeschmückt, lustig und lieblich zu lesen. Von J(ohan) H. W(olsdorf). Hamburg 1780. Giacomo Leopardi's translations of the poem were followed by his *Paralipomeni della Batrachomyomachia* (posthumous 1842), a sequel to the antique poem.

2. An exception is Gertud Kolmar's poems on frogs, *Tierträume*, in Kolmar (1955: 90–99). Kermit the frog from the Muppet show may be the best known version of contemporary frog imagery.

3. Encyclopaedias and handbooks with frog illustrations were widely distributed in Europe. The *Atlas zur Methodik der physiologischen Experimente und Vivisectionen* (1876) by Elie de Cyon had a considerable influence on several disciplines of experimental science and is remarkable for its uncompromising plea for objectivity in the presentation of its materials and instructions on how to perform experiments on animals.

4. In 1962 Alois Gulder published the "Frauenkröte von Maissau" made of clay and believed to date back to the eleventh century BC.

5. A desirable distinction between frog and toad is beyond the scope of this essay.

6. *Grenouille* is the French word for frog and a baby's sleeping suit.

7. Husserl interpreted this split as the origin of the crisis of the sciences which, he argued, were completely dominated by the idea of rationality. Inevitably, this separation from the origin of all knowledge in the life world led to a change of meaning of the things themselves (*Sinnverwandlung der Dinge*). (Husserl 1936: 375f.).

8. Descartes's own experiments on living animals can be seen as an early model of nineteenth-century experiments. In his *Description of the Human Body* he depicts an experiment in which he "slice[d] off the pointed end of the heart in a live dog" and then touched it with his finger to feel the contractions (Descartes 1637: 317; see also the letter to Plempius dated 15 February 1638, p. 81f.).

9. The MPI for the History of Science, Berlin, designed a website "The Virtual Laboratory" with a frog as its logo. Available at: http://mpiwg-berlin.mpg.de/

10. The Max Planck Institut für Wissenschaftsgeschichte (Berlin) is doing innovative work in documenting the history of science images. My essay has particularly benefited from their project "Virtual Laboratory." Images reproduced on their website are representative of nineteenth-century frog imagery. I am particularly indebted to Drs Henning Schmidgen and Sven Dierig. Pertinent to the topic are also publications of the Berlin Helmholtz Center and, in particular, Bredekamp and Werner (2003).

11. For an interpretation of its significance for the merger of two-dimensional representation see Lewis-Williams (2002: 194–196).

12. Derrida, in a reference to Jeremy Bentham elaborates on the specific characteristics associated with suffering and the claim that animals differ from humans insofar as they lack the faculty to suffer: "Can they suffer?" The faculty to suffer is, he argues, different from the traditionally inferred faculties such as creating a language or thinking rationally because it implies *passivity* (Derrida 2003:. 191–208; see also Derrida 2002).

13. The term appears as early as 1632 in Friedrich von Spee's *Cautio Criminalis*.

14. It might be worth emphasizing that the aim of this torture is different from the oft quoted examples in Foucault's *Discipline and Punish: The Birth of the Prison*.

15. He traces the idea of a difference between man and animal to the erroneous belief in the animal's inability to suffer.
16. J. Glanadet or Perot are recurring names on the plates but often barely decipherable.
17. For a few years now computer programs have been used for teaching university students. They demonstrate the dissection of a virtual frog whose body pieces are re-assembled at the end of the demonstration so that the reborn animal can jump in a virtual pond saying farewell with a single croak.
18. Forgetting and neglect in the sciences are discussed in an insightful essay by Sacks (1995: 141–187).

BIBLIOGRAPHY

Bächtold-Stäubli, Hans (ed.) (2005) *Handwörterbuch des deutschen Aberglaubens*, vols 3 and 5. Herausgegeben von Hans Bächtold-Stäubli unter Mitwirkung von Eduard Hoffmann-Krayer. 9 Bände und ein Registerband. Verlagsgruppe Weltbild GmbH, Augsburg. (Original 1927 bis 1942, Walter de Gruyter, Berlin.)

Bredekamp, Horst and Gabriele Werner (eds) (2003), *Bildwelten des Wissens. Kunsthistorisches Jahrbuch für Bildkritik*, Berlin: Akademie Verlag.

Bundy, Murray Wright (1927) *The Theory of Imagination in Classical and Medieval Thought*, Urbana, IL: University of Illinois Press.

Cooper, Jean C. (1992) *Symbolic and Mythological Animals*, London: HarperCollins.

Derrida, Jacques (2002) "The animal that therefore I am (more to follow)", *Critical Inquiry*, 28 (2): 369–418.

Derrida, Jacques (2003) *Mensch und Tier. Eine paradoxe Beziehung*, Dresden: Stiftung Deutsches Hygiene-Museum.

Descartes, René (1637/1984–1991.) *Discours de la méthode pour bien conduir sa raison et chercher la vérité dans les sciences. Plus la dioptrique, les météores, et la géometrie, qui sont les essais de cette méthode (Discourse on the Method for Properly Conducting Reason and Searching for Truth in the Sciences, as well as the Dioptrics, the Meteors, and Geometry, which are essays in this method)*, in vol. 1 of *The Philosophical Writings of Descartes* , ed. and trans. J. Cottingham, R. Stoothoff, D. Murdoch, and A. Kenny, Cambridge, UK: Cambridge University Press.

Faure, Elie (2003) "The tropics", in Jack Flam and Miriam Deutch (eds), *Primitivism and Twentieth-century Art. A Documentary History*, Berkeley, Los Angeles, London: Berkeley University Press.

Fischer-Homberger, Esther (1979) "Aus der Medizingeschichte der Einbildungen", in E. Fischer-Homberger (ed.), *Krankheit Frau und andere Arbeiten zur Medizingeschichte der Frau*, Bern, Stuttgart: Verlag Hans Huber.

Freud, Sigmund (1919/1970) "Das Unheimliche", *Studienausgabe*, vol. 4, Frankfurt am Main: Fischer.

Fry, Roger (2003) "The art of bushmen", in Jack Flam and Miriam Deutch (eds), *Primitivism and Twentieth-century Art. A Documentary History*, Berkeley, CA: California University Press: 44–45.

Germont, Philippe (2001) *An Egyptian Bestiary*, London: Thames and Hudson.

Gould, Stephen J. (1995) "Ladders and cones", in Robert B. Silver (ed.), *Hidden Histories of Science*, New York: New York Review of Books: 401–412.

Halverson, John (1987) "Art for art's sake in the paleolithic", *Current Anthropology*, 28: 63–89.

Hirschberg, Walter (1988) *Frosch und Kröte in Mythos und Brauch*, Wien: Böhlau.

Huet, Marie-Hélène (1993) *Monstrous Imagination*, Cambridge, MA, London: Harvard University Press.

Husserl, Edmund (1936) *Die Krisis der europäischen Wissenschaften und die transzendentale Phänomenologie*, Husserliana vol. 4.

Kolmar, Gertrud (1955) *Das lyrische Werk*, Heidelberg: Lambert Schneider.

Lewis-Williams, David (2002) *The Mind in the Cave*, London: Thames and Hudson.

Malebranche, Nicolas (1674/1980) *De la recherche de la vérité*, English translation as *The Search after the Truth* by Thomas M. Lesson and Paul J. Olscamp, Columbus, OH: Ohio State University Press.

Sacks, Oliver (1995) "Scotoma and neglect in science", in Robert B. Silver (ed.), *Hidden Histories of Science*, New York: New York Review of Books: 141–187.

Singer, Peter (1975/2002) *Animal Liberation*, New York: Ecco.

Singer, Peter (1993) *Praktische Ethik*, Stuttgart: Reclam.

Trevor-Roper, Patrick (1970) *The World through Blunted Sight. An Inquiry into the Influence of Defective Vision on Art and Character*, London: Thames and Hudson.

Verdin, Charles (1882) *Catalogue des Instruments de Precision Servant en Physiologie et en Medicine*, Chateauroux: Collection Rand B. Evans.

6 Science from Hell
Jack the Ripper and Victorian Vivisection

Colin Milburn

1888...if any fluids typified that year then blood and ink would surely be the main contenders.

(Alan Moore and Eddie Campbell, *From Hell*, 1999)

Everybody is a book of blood; Wherever we're opened, we're red.

(Clive Barker, *Books of Blood*, 1984)

DEAD LETTERS

The killer known to history as Jack the Ripper, who murdered at least five and perhaps as many as ten women in the Whitechapel district of London during the autumn of 1888, was also, it would seem, an inveterate writer of letters. Between 1888 and 1891, upwards of 300 letters, postcards, notes, scraps of graffiti, telegrams, and other correspondences purporting to be written by Jack the Ripper were received by Scotland Yard, the City of London Police, various newspaper offices, prominent public figures, and private homes all over London and beyond (Evans and Skinner 2001). While the vast majority of these communications were obvious hoaxes, police investigators nevertheless passed several of the more credible letters to the newspapers for facsimile publication, in hopes that some witness might recognize circumstantial details or even features of the handwriting that would disclose the murderer's identity. The deluge of "Ripper letters" circulating in the postal system and reproduced by the press—both those purporting to be written by the killer and those from concerned citizens hoping to offer insight on the case—joined the frenzy of sensationalistic newspaper reportage of the "Whitechapel horrors," the several coroner reports on the victims' mangled bodies, and the police drawings and pho-tographs of the crime scenes to form a media mosaic clustered around a phantasmagoric criminal, a figment of the gothic imagination. While the brutal Whitechapel murders in 1888 of Polly Nichols (31 August), Annie Chapman (8 September), Catherine Eddowes and Elizabeth Stride (both

125

30 September), and Mary Kelly (9 November) are tragic facts of history, the figure of Jack the Ripper was entirely a composite construct of the Victorian media ecology (Curtis 2001). Taking a cross-section of this media ecology, I aim to show how circulating images of illicit writing, diabolical murder, and scientific research triangulated a Ripper mythos for Victorian culture, a nightmarish fable of outcast London that implicated experimental physiology and medicine as accessories to serial atrocity.

I begin with two of the "Ripper letters." The first, received by the London Central News Agency on 27 September 1888, three weeks after the gruesome murder of Annie Chapman and only three days before the "double event" murder of Elizabeth Stride and Catherine Eddowes, was written in red ink and adorned with bloody fingerprints (Figure 6.1). Coining the "trade name" of "Jack the Ripper," the letter reads as follows:

Figure 6.1 "Dear Boss" letter (front and back), 27 September 1888 (UK National Archives ref. MEPO 3/3153. Courtesy of the National Archives. This material is in the copyright of the Metropolitan Police and is reproduced here by the permission of the Metropolitan Police Authority). The fading postscript ("They think I'm a doctor," etc.) is now almost indiscernible; it is located on the back of the letter in the bottom left corner, perpendicular to the body of the text.

Dear Boss,

I keep on hearing the police have caught me but they wont fix me just yet. I have laughed when they look so clever and talk about being on the right track. That joke about Leather Apron [an early suspect in the case] gave me real fits. I am down on whores and I shant quit ripping them till I do get buckled. Grand work the last job was. I gave the lady no time to squeal. How can they catch me now. I love my work and want to start again. You will soon hear of me with my funny little games. I saved some of the proper <u>red</u> stuff in a ginger beer bottle over the last job to write with but it went thick like glue and I cant use it. Red ink is fit enough I hope <u>ha. ha</u>. The next job I do I shall clip the ladys ears off and send to the police officers just for jolly wouldn't you. Keep this letter back till I do a bit more work, then give it out straight. My knife's so nice and sharp I want to get to work right away if I get a chance. Good Luck.

Yours truly

Jack the Ripper

Dont mind me giving the trade name

[PS] Wasnt good enough to post this before I got all the red ink off my hands curse it. No luck yet. They say I'm a doctor now. <u>ha ha</u>

The second letter was received on 16 October 1888 by George Lusk, president of the Whitechapel Vigilance Committee (Figure 6.2). The Vigilance Committee had recently organized to hunt down the person (or persons) responsible for the escalating mutilation-murders of Whitechapel prostitutes, from whose flayed bodies select vital organs had been removed. The letter arrived in a small cardboard box along with a portion of human kidney. As one of Catherine Eddowes's kidneys had been stolen by her murderer, this ghastly missive was given particular attention by Scotland Yard. The letter reads:

From hell

Mr Lusk

Sor

I send you half the Kidne I took from one woman and prasarved it for you tother piece I fried and ate it was very nise. I may send you the bloody knif that took it out if you only wate a whil longer

signed

Catch me when you can Mishter Lusk

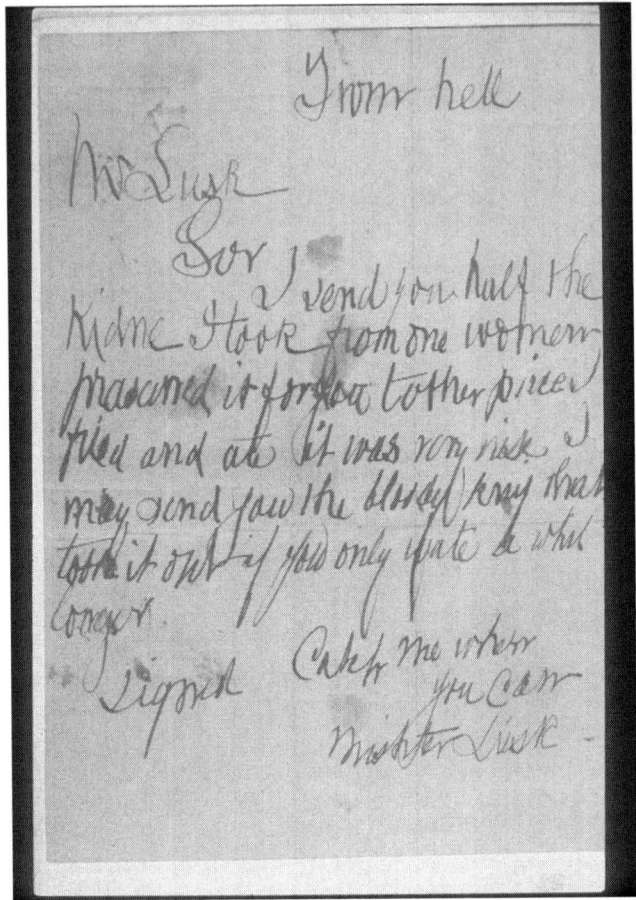

Figure 6.2 "From Hell" letter, 16 October 1888 (Royal London Hospital Archives, ref. GB 0387 LH/X/97. © Royal London Hospital Archives. Courtesy of Royal London Hospital Archives). This image is reproduced from an original photograph of the letter; the letter itself went missing from City Police archives sometime in the 1960s.

These two letters, among the most widely publicized of the voluminous Ripper correspondence, register three aspects of the killer's "profile" commonly accentuated by the Victorian newspapers and by hundreds of other Ripper letters. These were the "writerly" and "inscriptional" obsessions of the murderer as a user of media networks; the "diabolical" and "gothic" qualities of the murders; and the patterns of "scientific experimentation" seemingly discerned in this series of crimes; that is, the persistent question of physiological or anatomical "knowledge" that haunted the coroner's inquests, police reports, and newspaper accounts of the crimes.

The very existence of these Ripper letters cemented the killer's persona as a writer and a consummate navigator of the postal system, a logorrheic composer of tantalizing documents directed to the forensic gaze of the police, the press, and various authority figures around the city. Signed by "Jack the Ripper" and often following the style and diction of the "Dear Boss" and "From Hell" letters, these texts emanated an imaginary "author-function" that appeared to unify and resolve them en masse, neutralizing their contradictions as the fantasized origin and organizational principle of their circulation (Foucault 1977). Certainly, the letters as a whole are deeply inconsistent in terms of their content, and their attempts to enact various faux dialects are radically unconvincing. Scotland Yard immediately dismissed almost all of them as attention-seeking counterfeits (although a few were less obviously so, especially one postcard that seemed to accurately predict the "double event" murder of Stride and Eddowes in the same night). Despite their overtly fraudulent nature, however, the letters signed by "Jack the Ripper" contributed to the murderer's reputation as a heckler of the police who used the media channels of Victorian England to send various textual clues, scripts, tracks and traces of his criminal autobiography, playing "funny little games" and boldly taunting: "Catch me when you can." A repetitive, serial author of grisly letters who writes, after all, from the depths of hell.

Jack the Ripper's epistolary self-fashioning as a demonic force "from hell" in the letter to George Lusk provoked numerous copycat Ripper letters that referenced satanic and infernal motivations. The occultist Dr. Roslyn D'Onston (alias of Robert Donston Stephenson) further stoked such associations in promoting his theory that the Whitechapel killings were really the "unholy rites" of an elaborate necromantic ceremony, figuring Jack the Ripper—"The Whitechapel Demon"—as a black magician guided by "agencies of evil spirits and demons" (D'Onston 1888: 5). (A few of Stephenson's contemporaries even suspected him of being the satanically inspired Ripper; see Whittington-Egan 1975: 68–98; Harris 1994). An extensive diabolic vocabulary equally fueled the Victorian journalistic conventions of "sensation-horror" reportage that bodied forth a gothic idiom in narrating the murders (Curtis 2001). Typically describing Jack the Ripper as the "demon of Whitechapel," a "ghoul" or "vampire" stalking the streets to fulfill his "fiendish lust," a "crazed devil" and a "cruel monster" acting out of motives "beyond human imagination," the London newspapers shrouded the killer in a semiotic cloak of sadistic perversion, monstrous fury, and unspeakable appetites (Curtis 2001: 109–163; Walkowitz 1992: 192–201).

Such "appetites" are displayed in the "From Hell" letter when the author implies that he cooked and devoured a portion of Eddowes's kidney ("I send you half the Kidne I took from one woman and prasarved it for you tother piece I fried and ate it was very nise"). The syntactical overlapping here of the "raw" and the "cooked," the shared textual space of the "abattoir" and the "kitchen," or the "savage" and the "civilized," would itself seem to correspond to the gothic models of "split personality" offered up by several contemporary commentators (including famous journalist W.T. Stead) to explain the killer's psychology as a "savage-savant." Journalistic accounts of the killer as a subject divided between fierce intelligence and raging violence frequently relied on direct allusions to Robert Louis Stevenson's *The Strange Case of Dr. Jekyll and Mr. Hyde* (1886) (Smith 2004: 76–81; Curtis 2001: 32–119). In Stevenson's gothic fable, the war between the civilized and the atavistic self is a physiological reality, and this horror-fantasy image provided an immediately accessible means for the London press to comprehend the mind of the Whitechapel killer (Frayling 1986; Walkowitz 1992: 206–211). Moreover, frequent citations of Stevenson's quintessential narrative of the "mad scientist" in the context of the murders imaginatively supplemented theories that the Whitechapel killer himself belonged to the scientific professions.

Indeed, the "From Hell" letter notes those ubiquitous hypotheses circulating in the Victorian media ecology that Jack the Ripper might be a physiologist or medical experimenter: "They think I'm a doctor now, ha ha." This particular representation of the Ripper entered public discourse with the assertion, raised repeatedly during the official inquests into the victims' deaths, that the killer displayed a certain scientific "knowledge" in performing these eviscerations and removing select organs. The specter of human vivisection was first suggested by police surgeon Dr. George Bagster Phillips on 19 September 1888 during the Annie Chapman inquest. Phillips described the "deliberate, successful, and apparently scientific manner in which the poor woman had been mutilated," and he speculated that the "mode in which the knife had been used seemed to indicate great anatomical knowledge" ("The Whitechapel Murder" 1888: 98). The question of medical or scientific motives became an early focus of the investigation, and Coroner Wynne Baxter later publicly announced that the uterus of Annie Chapman

> had been taken by one who knew where to find it, what difficulties he would have to contend against, and how he should use his knife so as to abstract the organ without injury to it. No unskilled person could have known where to find it or have recognized it when it was found. For instance, no mere slaughterer of animals could have carried out these operations. ("The Whitechapel Murder" 1888: 105)

The implication that a sadistic medical scientist could be conducting hideous vivisection experiments or "operations" on the women of Whitechapel spread almost instantly throughout the Victorian media and the popular imagination.

"Medical knowledge" became a major motif of reportage on Jack the Ripper and of police inquests as the slayings increased from two to four during the month of September. One letter written to the *Evening News* on 17 September 1888 by an "Ex-Medico's Daughter" hypothesized that the Whitechapel killer was a "medical maniac," a "human vivisectionist" whose crimes were committed "in the cause of science" for gynecological purposes, to reveal "the mysterious changes that take place in the female sex at about the age of these poor women" (quoted in Walkowitz 1992: 209). This writer's portrait of Jack the Ripper as a maniacal hybrid of gynecologist and vivisectionist echoed the antivivisection propaganda produced at the *fin de siècle* that often strategically linked women and experimental animals together as victims of medical violence (Elston 1987). As Coral Lansbury has argued, "women were the most fervent supporters of antivivisection, not simply for reasons of humanity, but because the vivisected animal stood for vivisected woman: the woman strapped to the gynaecologist's table, the woman strapped and bound in the pornographic fiction of the period" (Lansbury 1985). Such fears found a visible exemplar in a real-life killer who seemingly had transferred his scientific attention from tortured animals to tortured women.

I am less concerned with the "real" identity of Jack the Ripper than with the cultural logics that were animated to make sense of the "Whitechapel horrors": what Elana Gomel has called the "bloodscripts" that emerge at sites of excessive violence to narrativize and account for violent subjectivities (Gomel 2003); or what Mark Seltzer has called the "splatter codes" of the pathological public sphere linking word counts and body counts, relaying wounded bodies through public media (Seltzer 1998). For Jack the Ripper, never identified and never caught, became a cipher, a blank space into which numerous social anxieties could be channeled. Among these contemporary anxieties were longstanding class conflicts (between the west of London and the crime-ridden East End) and anti-Semitism (John Pizer, a Jewish butcher allegedly nicknamed "Leather Apron," was an early Ripper suspect, and ambiguous graffiti found at the scene of Catherine Eddowes's murder—"The Juwes are the men that will not be blamed for nothing"—threatened to inflame public hostilities), as well as concerns about the effectiveness of police in preventing crime (Walkowitz 1992). But perhaps most prominently the killings came to instantiate an entrenched iconography of male violence against women that eerily echoed images of the opening, dissection, and mutilation of women prevalent in Victorian medical science (Walkowitz 1992; Showalter 1992; Jordanova 1989).

Though there were several criminal "types" offered as possible culprits—including the degenerate aristocrat, the working-class "Jack," and the sinister foreigner—the mad medical scientist was perhaps the most compelling and enduring for the Victorians (Frayling 1986). Indeed, nearly half of over 300 Ripper-related letters addressed to the City of London police from 1888 to 1889 speculated that the killer was a human vivisectionist (Curtis 2002). One letter signed by "Jack the Ripper" sent to Dr. Thomas Openshaw (the medical examiner of the "Lusk kidney") seems to endorse this prominent theory by ending with the following rhyme: "O have you seen the devle/ with his mikerscope and scalpul/ a looking at a Kidney/ with a slide cocked up" (Letter reprinted in Evans and Skinner 2001: 67). This ditty fashions the killer as a diabolical (if surprisingly illiterate) physiologist crudely poeticizing his own specular exploits. Altogether, the media confluence of newspapers, police reports, autopsies, and the flood of Ripper correspondence produced a resonant configuration of the Whitechapel killer as gothic vivisector and literary experimenter: a satanic scientist who *writes*.

Indeed, Jack the Ripper writes in blood: "I saved some of the proper red stuff in a ginger beer bottle over the last job to write with." And even as the killer appeared to document his crimes by dispatching hundreds of macabre memos, Victorian culture imagined the murders themselves as acts of bloody writing. Coroner Baxter, summing up the Chapman inquest, stated: "The body [of Annie Chapman] had not been dissected, but the injuries had been made by some one who had considerable anatomical skill and knowledge. There were no meaningless cuts" ("The Whitechapel Murder" 1888: 105).

The cuts are now *meaningful* marks. The bleeding female corpses literally become semiotic, bearing legible inscriptions—even "writing" as such (Derrida 1974). Under the gaze of medical examiners and police investigators, the ghastly wounds appear as material signifiers of great complexity, suggesting not only a motive—"The conclusion that the desire was to possess the missing abdominal organ [the uterus] seemed overwhelming" ("The Whitechapel Murder" 1888: 105)—but also a suspect profile. As Dr. Frederick Gordon Brown, surgeon of the City of London Police Force, remarked during the subsequent inquest into Catherine Eddowes's murder: "The way in which the mutilation had been effected showed that the perpetrator of the crime possessed some anatomical knowledge...[and the pilfering of Eddowes's kidney in particular required] a good deal of knowledge as to the position of the organs in the abdominal cavity and the way of removing them" (Brown quoted in "The East-End Murders" 1888: 223). The kind of knowledge potentially gained, according to Brown's official written testimony, by a skilled professional "in the habit of cutting up animals":

> I believe the perpetrator of the act must have had considerable knowledge of the position of the organs in the abdominal cavity and the way of removing them.... It required a great deal of ["medical" —*deleted*] knowledge to have removed the kidney and to know where it was placed, such a knowledge might be possessed by some one in the habit of cutting up animals.... (Coroner's Inquest 1888: 207)

While belated deletion of the word "medical" from the archived inquest record here seemingly leaves open possibilities that other professionals accustomed to "cutting up animals"—such as butchers, hunters, or cooks—might possess sufficient "anatomical knowledge" to have carried out these murders, the lingering trace of "medical" experimentalism could not so easily be effaced from public media. One Ripper letter even makes the equation directly in doggerel poetry, associating the killer with a man "Mad on Vivisection (the cutting up of animals).... Who wrote essays on women bad" (Letter reproduced in Evans and Skinner 2001: 288–289). In this letter, the "cutting up of animals" means vivisection, and its practice links to the sadistic scriptor in Whitechapel who "wrote essays" on the bodies of wayward women.

The mutilated bodies become "essays" documenting their own status as subjects of experimental science gone mad. The thoroughly "meaningful" cuts announce the skillful method through which they were performed and the "scientific" goal for which they were intended. They are read through a certain interpretive lens as the signature of the vivisectionist, making visible the motive and occupation of their author. Frances Power Cobbe, outspoken Victorian feminist and antivivisection crusader, seeing signs of physiological knowledge in the crimes, announced her support of police bloodhounds for tracking down the killer: "Should it so fall out that the demon of Whitechapel prove really to be...a physiologist delirious with cruelty, and should the hounds be the means of his capture, poetic justice will be complete" (Cobbe 1888).

The horrid events in Whitechapel seemed to fulfill Cobbe's worst nightmares. After all, the possible existence of such a demon physiologist recalled assertions in her Dante-esque tract *The Nine Circles; or The Torture of the Innocent* (1893) that vivisection would inevitably turn respectable scientists into skilled torturers filled with inhuman lusts. We see dire pronouncements throughout the writings of Cobbe and other antivivisectionists, such as G.M. Rhodes's *The Nine Circles of the Hell of the Innocents* (1892), prefaced by Cobbe, that the science of vivisection unleashes base and devilish forces lurking in human beings. Likewise, in H.G. Wells's 1896 novel of wild vivisection, *The Island of Dr. Moreau*, the shipwrecked Prendick observes the outcome of vivisection and feels "a nasty little sensation, a tightening of my muscles.... It's a touch—of the diabolical, in fact" (Wells 1896/1993: 23). The vivisector Dr. Benjulia in Wilkie Collin's 1883 novel *Heart and Science* confronts an eyewitness to his harsh treatment of a monkey directly on this point: "Do you think I'm the Devil?" (Collins 1883/1996: 109). Benjulia's own brother later echoes this intimation when, watching the physiologist erupt into a vivisectional fervor, whispers to himself, "I begin to believe in the devil" (191).

Even professional physiologists tended towards a lexicon of diablerie when writing on the topic of vivisection. In 1882, Ronald Ross—fifteen years prior to his celebrated work on malaria that would be recognized by the Nobel Prize in 1902—composed a gothic science-fiction story called "The Vivisector Vivisected." In this satirical tale, two daring vivisectors use an artificial heart-machine to resurrect a deceased fellow physiologist, making him over as a test organism. The victim-scientist perceives this fate as his own personal inferno, seeing signs of hellfire all around as he says to his experimental tormentors:

> "I went in for physiology, and so got ruined. I say! [W]on't the devil give me hot for my vivisecting—for the cutting—eh? for the fastening up—eh?...[W]hat is this? what is this? I am dead! Begorra, I died just now—I died of a cut on the head, and drank a bottle o' whiskey upon it to die drunk! Oh! Lord I see it—ochone! I am in hell, and I am drunk still!" He wrenched again at his wrists, screaming.... "Ah! Krimy!" groaned the man, his eye wandering down to the instrument stuck in his chest, the stitches in his skin and the tubes leading to the pumps: "[O]ch! St. Pathrick, I see it! And my punishment is, to be done to as I have been done by. And you are a couple of devils, and I a vivisection; and I shall be vivisected for iver and iver, wurrld without end—oh! Lord—damn—damn—damn—." (Ross 1882/1988: 347)

Ross's vivisectors appear to each other as devils lurking in stygian pits of the experimental lab. A similar tableau of the pandemonic laboratory was evoked by Cambridge physiologist Michael Foster for his vehement 1874 defense of vivisection in *Macmillan's Magazine*. Upholding the necessary value of vivisection for the advance of medical knowledge, Foster neverthe-less allows that, in terms of actual practice, "a physiologist...might be an angel in the bosom of his family, but a demon in the laboratory" (Foster 1874: 368). In late Victorian England, then, vivisection was frequently con-figured as a science from hell—and in the autumn of 1888, Jack the Ripper became its poster boy.

Rhetorical critiques of experimental physiology from the antivivisec-tionist camp undoubtedly bolstered the rumor the Jack the Ripper was a crazed vivisectionist practicing his nefarious science in the back alleys of the East End. Antivivisectionists had been responsible for arousing public suspicion that the practice of vivisection would dehumanize students of science, turning them into heartless sadists filled with uncontrollable bar-baric desires, or vicious monsters abandoned of all morality (French 1975). Lewis Carroll suggested in 1875 that the vivisection of animal victims would inevitably drive scientists to vivisect human victims, and "successive generations of students, trained from their earliest years to the repression of all human sympathies, shall have developed a new and more hideous Frankenstein—a soulless being to whom science shall be all in all" (Car-roll 1875: 854). In Carroll's invocation of a Frankensteinian monstrosity inextricable from the practice of science, we see mirrored a similar claim by H.G. Wells, who writes in a postscript to *The Island of Dr. Moreau* that, "[s]trange as it may seem to the unscientific reader, there can be no denying that...the manufacture of monsters—and perhaps even of *quasi*-human monsters—is within the possibilities of vivisection" (Wells 1896/1993: 88). Wells's phrase *"quasi*-human monsters" would seem to refer not only to the constructed Beast-Folk of his gothic scientific romance, but also to their half-monstrous creator, the vivisectionist Moreau.

But as much as these critics contributed to the idea that vivisection produces monsters and pointed to Jack the Ripper as exemplifying their warnings, there is something more epistemic in the way that the authorial hand of vivisection was read through the ghastly wounds of eviscerated prostitutes during the autumn of 1888. When police surgeons, coroners, reporters, and even medical practitioners and scientific researchers took seriously the signifying wounds of the female bodies as leading inexorably back to the inscribing hand of some "half-mad physiologist," as the *East London Observer* put it on 22 September, working to secure "living tissues or organs from a healthy subject for experiments" (quoted in Curtis 2001: 227), we see more than the vitriol of antivivisectionists—who were fairly marginal, anyway, even among animal welfare groups (Ritvo 1987: 157–166)—but a cultural discourse in which "mad science" is the logical context of serial slaughter (Turney 1998: 43–63). The successive iteration of mangled bodies in Whitechapel immediately appeared to be a form of communicative inscription—"There were no meaningless cuts"—and this was a particular form of inscription that experimental physiology had already claimed as its own.

I want to argue that nineteenth-century vivisection—by which I mean the set of technical and discursive operations that made vivisection available as an experimental method for Victorian physiology—produced an interpretive community conditioned to observe "meaning" in wounds. That is to say, the practice of vivisection constructed an epistemic frame of reference, a specific hermeneutic horizon, expanding beyond the physiological laboratory and into the media networks of Victorian culture, which primed readers to "see" vivisection within serial mutilations of bodies. In this frame of reference, the methodically wounded body *signifies*: it gives forth biological secrets from within its organic depths while simultaneously announcing on its bleeding surface its transformation into an object of scientific inquiry. Enmeshed in the space of the experimental laboratory, manipulated and recreated by the scalpel and the graphical recording instrument, the vivisected body conveys meanings unavailable to the closed and contained body. In other words, the Victorian discourse on experimental physiology rendered vivisection as a media practice, and the vivisected body thereby became a vehicle for scientific communication, a book of blood. In doing so, it created the conditions of possibility wherein the carnal signatures "From Hell" and "Yours Truly, Jack the Ripper" could be read as the infernal marks of a vivisection event.

MEANINGFUL CUTS

The notion that Jack the Ripper might be a crazed doctor or physiologist had become so widespread in the aftermath of the first two Whitechapel murders that the scientific community moved to defend their profession. *The Lancet* ran an article cautiously admitting "the improbability of any-one but an expert performing the mutilations described in so apparently skilful a manner," especially considering that Annie Chapman's uterus had been removed by someone seemingly possessed of "such knowledge of ana-tomical or pathological examinations as to be enabled to secure the pelvic organs with one sweep of a knife" ("The Whitechapel Murders" 1888). The article nevertheless stated that the theory of a medical scientist turning to murder was "highly improbable, although it may have a small basis of fact, which will require exercise of much common sense to separate from the sensational fiction that surrounds it." In any case, *The Lancet* chastised Coroner Baxter for publicizing his speculations about medical organ har-vesting without regard to potential public reactions:

> The public mind—ever too ready to cast mud at legitimate research— will hardly fail to be excited to a pitch of animosity against anatomists and curators, which may take a long while to subside. And, what is equally deplorable, the revelation thus made by the coroner, which so dramatically startled the public last Wednesday evening, may prob-ably lead to a diversion from the real track of the murderer, and thus defeat rather than serve the ends of justice. ("The Whitechapel Mur-ders" 1888)

Scientists and physicians as far away as Australia also became concerned about the public "pitch of animosity" arising from unchecked international rumors that a medical maniac was vivisecting prostitutes in London. Dr. Andrew Wilson wrote in the *Port Philip Herald*:

> The medical evidence given at the inquest on the last victim [Annie Chapman] bore out that the manner in which the woman's body was mutilated gave evidence of some acquaintance with anatomy. Upon this declaration a writer in a London evening paper had the effrontery to suggest that the murders may have been the work of some physiolo-gist who desired to gain possession of human organs for purposes of science! Such an outrageous proposition, it is to be presumed, only requires to be mentioned to be reprobated as itself the product of a dis-eased mind. Anatomists do not require to kill subjects to obtain human organs for investigation, and it is not possible to name any "vivisec-tion" experiment in the same breath with such a brutal and unmeaning crime. (Wilson 1888)

Defending experimental physiology by mocking the mad scientist rumors, Wilson insists that "vivisection" cannot be associated with "brutal and unmeaning crime." Experimental physiology is not the product of a "diseased mind" but of a *self-evidently* sane mind, and far from "unmeaning crime," vivisection instead would be, according to Wilson's logic, deeply "meaningful." While defending experimental physiology against Jack the Ripper, Wilson here echoes a motif of Victorian scientific discourse that vivisection produces meaning, that it is a method for generating signification: a scientific merger of the pen and the knife.

Vivisection was regularly figured as the conjoined action of dissection and writing, a mode of producing graphical meaning across the organic body. Michael Foster, for example, writes of vivisecting a rabbit:

> Did the reader ever see a rabbit completely under the influence of Chloral?... You prick with a needle the exquisitely sensitive cornea of its eye; it makes no sign, save only perhaps a wink. You make a great cut through its skin with a sharp knife; it does not wince.... Yet it is full of action. To the physiologist its body, though poor in what the vulgar call life, is still the stage of manifold events, and each event a problem with a crowd of still harder problems at its back. He therefore brings to bear on this breathing, pulsating, but otherwise quiescent frame, the instruments which are the tools of his research. He takes deft tracings of the ebb and flow of blood in the widening and narrowing vessels; he measures the time and the force of each throb of the heart, while by light galvanic touches he stirs this part or quiets that; he takes note of the rise and fall of the chest-walls...he divides this nerve, he stimulates that, and marks the result of each...and having done what he wished to do, having obtained, in the shape of careful notes or delicate tracings, answers to the questions he wished to put, he finishes a painless death by the removal of all the blood from the body, or by any other means that best suit him at the time. (Foster 1874: 370–371)

Foster describes the opened animal body as a medial surface, a "stage of manifold events," that in itself "makes no sign"—it does not produce its own meanings—but upon which meanings and "answers" are made to appear through the work of vivisection. The "instruments which are the tools of his research" are used to take "deft tracings of the ebb and flow of blood," to measure, to record, and to "take note": "he...marks the result of each." The theatrical performance of vivisection on the "stage" of animal form involves an intimate choreography where cutting and inscribing, excavating and marking, exposing and signing intertwine and become indistinguishable:

> The chemistry of living beings, one would imagine at first thoughts, might be investigated without distressing the organisms...[but the]

shifts and changes of the elements within our bodies are too subtle and complex to be divined from the results of the chemical laboratory; the physiologist has to search for them within the body, and to mark the compounds changing in the very spot where they change; otherwise all is guess-work. (Foster 1874: 373)

The vivisector digs into the animal body to precisely "mark" the "very spot" of buried chemical compounds, a scientific treasure hunt that eliminates fruitless speculation by drawing a piratical map into the flesh of the ruined organism.

In the same vein, several Victorian novels featuring vivisection emphasize the hybrid nature of instruments of writing and instruments of cutting within the space of the physiological laboratory: the pen and the knife serving the same purpose of making meaningful the mute animal structure. For instance, in H.G. Wells's *The Island of Dr. Moreau*, the clinical and calculating Moreau explains his program of vivisection research to the shipwrecked Prendick through recourse to inscriptional metaphors and graphic demonstration. Exploring "the extreme limit of plasticity in a living shape" (Wells 1896/1993: 48), transforming animal bodies into ersatz humanoids, Moreau has come to see pain as a durable trace of animality: "pain, Prendick, is the mark of the beast upon them, the mark of the beast from which they came!" (48). Moreau, with an "artistic turn of mind" (47), uses pain like a writer or painter would use ink or paint, coating over the "mark of the beast" by "dip[ping] a living creature into the bath of burning pain" (51) and then engraving the tortured organism with humanity: "Moreau took them [infant animals] and stamped the human form upon them" (53). For Moreau, the body is a medium to be "carven and wrought into new shapes" (46), "stamped," or written upon by instruments of pain: "As he spoke he drew a little penknife from his pocket, opened the smaller blade, and moved his chair so that I could see his thigh. Then, choosing the place deliberately, he drove the blade into his leg and withdrew it" (48). This evocative "penknife" metonymizes vivisection as a media practice, as a science of drawing and withdrawing meaningful cuts from malleable tissues.

Similarly, in Collins's *Heart and Science*, the authorial hands and cruel instruments of the vivisector Dr. Benjulia leave behind trails of "horrid stains, silently telling their tale of torture" (Collins 1883/1996: 185). Vivisection becomes a form of storytelling, writing Grand Guignol fables with the body's spilt fluids. Certainly, for some Victorian intellectuals working across the fields of literature and science, like H.G. Lewes and George Eliot, the relationship between the pen and the knife went beyond analogy, standing rather for the potential of novelistic experimentation to parallel physiological experimentation in exposing the workings of human consciousness: a "recognition of the potential affinity between the literary and scientific imaginations, between the real investigations of the scalpel and the imaginary invasions of the pen, between vivisection and fiction" (Menke 2000: 647).

Considering this frequent conjunction of the pen and the knife in Victorian discourse on vivisection—biological science imagined as both serial killing and serial writing—the forensic reading of the autopsied victims and the autopsied letters in the Jack the Ripper case predictably focused on questions of "signature" evidence, and therein, among other tantalizing clues and possible leads, discovered signs of physiological knowledge and its disciplinary tools. Chief Inspector Swanson's report to the Home Office on the murder of Annie Chapman states the opinion "that the murderer was possessed of anatomical knowledge from the manner of removal of viscera, & that the knife used was not an ordinary knife, but such as a small amputating knife" (Swanson 1888: 67). Several Ripper letters captured the overlay of violence and writing in the killer's signature by focusing on this murder weapon as a pen as much as an "amputating knife," a stylus as much as an experimental instrument. One letter sent to Chief Commissioner Charles Warren of Scotland Yard on 3 November 1888, signed "J. T. Ripper" and announcing more vicious murders to come ("one or two more will feel my knife"), ends with the postscript: "next one I copp I'll send the toes and earoles to you for supper." The postscript is followed by the Ripper's signature, and below this, a drawing of a man's hand holding a narrow knife upright by the fingertips, as one would hold a pen. The tip of this knife rests just under the signature, as if the knife has been the instrument used to write this letter, to sign off the bloody deed (Letter reproduced in Evans and Skinner 2001: 107).

In this missive, there is little difference between a letter and a dismembered body (both here imagined as items that travel through the postal system, "earoles" alongside text, like the kidney and note sent to Lusk), between a crime scene and a kitchen (both here imagined as places where one might prepare a grisly "supper"), or between a epistolary pen and an amputating knife (both here imagined as instruments for signing murder). All of these features—evoked repeatedly in the innumerable hoax letters, medical autopsies, police reports, and newspaper stories constituting the Victorian media mosaic of Jack the Ripper—lent themselves towards an already emergent bloodscript of the experimental vivisector. For Victorian culture had been taught to perceive the authorial signature of vivisection in the mangled body, to understand vivisection as a media practice working to draw meaning from living flesh. In essence, through the didactic efforts of both professional physiologists and their antivivisectionist opponents, the Victorian public had been encouraged to consider vivisection—as Thomas De Quincy had suggested of murder—as one of the fine arts.

The revolution in English physiology that began during the 1870s under the influence of Michael Foster, John Scott Burdon Sanderson, and Edward A. Schäfer received considerable public attention both in Britain and abroad. Importing techniques and laboratory protocols from the physiological schools of Germany and France—notably refined practices of experimental vivisection—and renovating the institutional framework of physiological education in England between 1870 and 1900, the scientific successes of the new English physiology soon achieved international prominence (Geison 1978; Romano 2002). The publication of Foster's *Textbook of Physiology* and his foundation of the British Physiological Society in 1876, the inauguration of the *Journal of Physiology* in 1878, along with the publication of Burdon Sanderson's notorious two-volume *Handbook for the Physiological Laboratory* in 1873, gave exposure to the new techniques and provided a strong image of experimental physiology to the Victorian public—and in the process provided tangible targets for the newly organized antivivisectionist movement (French 1975). The contents of certain continental physiological textbooks that informed the reformation of English physiology became widely known through their citation in both professional scientific publications and antivivisection treatises. (Ironically, English physiologists sometimes themselves gained an impression of continental technique through the translations and quotations contained in antivivisection publications; physiologist Gerald F. Yeo once wrote that "my knowledge of [Italian physiologist Paolo Mantegazza's] work was derived solely from Miss Cobbe's writings, and may be quite incorrect" [Yeo 1882]). These texts, along with graphic diagrams of the physiological laboratory, its instrumentation, and its experimental organisms disseminated by English scientists, shaped the public image of the vivisected body as a communicative medium.

The *Atlas zur Methodik der physiologischen Experimente und Vivisectionen* (1876) by Russian-born physiologist Elie de Cyon was not only highly influential in the disciplinary world of experimental vivisection, but also something of a *bête noire* among Victorian antivivisectionists for its unabashed celebration of scientific violence towards "animal material." De Cyon defended himself in the English press, sarcastically counterattacking those critics who stood aghast at his depiction of the work of the vivisector as the work of the artist, the belletrist, the sculptor (de Cyon 1883). In a passage from the *Methodik* frequently quoted by the Victorian media, de Cyon writes:

> The true vivisector...must approach a difficult vivisection with the same joyful excitement, with the same delight, as the surgeon when he approaches a difficult operation from which he anticipates extraordinary consequences. He who shrinks from the section of a living animal, he who approaches a vivisection as an unpleasant necessity, may perhaps be able to repeat one or two particular vivisections, but will never become an artist in vivisection.... The sensation of the physiologist when, from a gruesome wound, full of blood and mangled tissue, draws forth some delicate nerve thread . . . has much in common with

that of a sculptor. (De Cyon quoted in Cobbe 1882: 611; Coleridge 1882: 227)

The "artist in vivisection," like a "sculptor," working within the mess of meaningless material, the unformed medium of the "gruesome wound, full of blood and mangled tissue," pursues a line of significance, shaping the meaning of his material at the moment when he "draws forth some delicate nerve thread," drawing out the hidden value from the otherwise senseless mass. The vivisector literally and figurally "draws forth" the vital thread of physiological meaning from out of the organic medium; he discovers and makes visible the truth of the organism that had been concealed by the shapeless chaos of the bleeding body. Vivisection gives meaning to wounds, it finds the "delicate" within the "gruesome," and it exposes the work of fine art that lies dormant within the block of animal matter.

We find similar accounts of the vivisector as media artist in the work of Claude Bernard, who personally trained several leading Victorian physiologists—including Burdon Sanderson—in vivisection techniques, and whose experimental protocols profoundly influenced development of the new physiology in England (Richards 1986). Bernard's most famous text, *An Introduction to the Study of Experimental Medicine* (1865), construes the physiological experimenter as producing meaning at the surface of the animal body. Bernard writes, "In a general and abstract sense, an experimenter, then, is a man who produces or induces, in definite conditions, observed facts, to derive from them the instruction which he wishes,—that is, experience" (Bernard 1865/1927: 21). The vivisectionist produces or induces "observed facts" by cutting into the organism, and the meanings of these material signs are then "derived" hermeneutically, interpreted as the experimenter "wishes." The "experimenter," as the French language would here suggest, "experiences" the meanings of those facts that he causes to take place on the body of the organism: he is simultaneously a producer and an interpreter, a writer and a reader. For the observed "facts" are read, "instruction" is derived. Yet as we see throughout Bernard's text, these physiological "facts" are literally the *wounds* as such, and the instructive value of these meaningful wounds is to expose not only the truth of the organism, but more specifically, the truth of the experimenter and his desire, the truth of vivisection itself:

Facts materially alike may have opposite scientific meanings, according to the ideas with which they are connected. A cowardly assassin, a hero and a warrior each plunges a dagger into the breast of his fellow. What differentiates them, unless it be the ideas which guide their hands? A surgeon, a physiologist and Nero give themselves up alike to mutilation of living beings. What differentiates them also, if not ideas? (Bernard 1865/1927: 103)

Bernard suggests that the same "fact," the same act of physical "mutilation," conveys different "meanings" relative to the intentions and the "ideas" with which it was informed. Ideas would thus be intimately "connected" to the material facts of wounds, whose significance would be determined by the authorial intentions guiding their differential production.

The material facts of the vivisected body are therefore signifiers of both physiology and character, namely, the character of the scientist and his ideas. For Bernard immediately continues:

> A physiologist is not a man of fashion, he is a man of science, absorbed by the scientific idea which he pursues: he no longer hears the cry of animals, he no longer sees the blood that flows, he sees only his idea and perceives only organisms concealing problems which he intends to solve. (103)

Under the physiological gaze, the animal body is rendered a pure medium. Blood is no longer visible as blood, cries are no longer audible as cries: these expressions of the body have become transparent to the scientist who sees the organism only as a vehicle for communicating occulted biological "problems." But these problems are overdetermined by the volition of the scientist and his acts of inscription, for the vivisectionist "sees only his idea" within the bleeding wounds he produces. The cuts on the body are full of meaning, but the scientist reads therein only his own *intentionality*. For the "scientific idea" embedded in the signifying wounds on the animal body contains the man of science: he is "absorbed by the scientific idea which he pursues." Within the wounds, the vivisectionist discovers *himself*. In other words, the wounds appear as the scientific *signature* of the vivisectionist.

But this signature of the vivisectionist is legible only through the lens of conditioned interpretation: "It is impossible for men, judging facts by such different ideas, to ever agree: and as it is impossible to satisfy everybody, a man of science should only attend to the opinion of men of science who understand him, and should derive rules of conduct only from his own conscience" (103). Bernard here argues that because the "facts" of wounds can be judged in such different ways, depending on the ideas with which they were *conceived* and the ideas through which they are *received*, only a common horizon of expectation will allow scientifically generated wounds to escape the restrictive realm of social mores and thereby achieve their fullest experimental value. That is, in order to "agree" on the scientific meanings of the wounds generated by vivisection, and to distinguish this practice from common murder—from the crimes of a "cowardly assassin" or the perversities of a "Nero"—one must be disciplined to see like a vivisectionist. Only "men of science who understand" will be able to discern true vivisectional intent in the bleeding configurations of violated organisms. The properly trained physiological gaze, then, will see "cutting up animals" as experiment rather than butchery, will see "mutilations" as authentic signs of "scientific ideas" rather than as "meaningless cuts." Men of science, according to Bernard, would be conditioned to see vivisection itself as the efficient meaning of eviscerated and ripped apart bodies.

For the new experimental physiology in Victorian England, then, establishing the legitimacy of vivisection practices would necessarily derive not only from a rhetoric of future clinical applications—perhaps the main argumentative strategy used by defenders of vivisection to make their science palatable to the Victorian public (Richards 1986, 1987; Rupke 1987)—but also, following Bernard's logic, from the heuristic construction of a kind of general observer trained to read wounds in the correct way, just like a vivisectionist. Proper differentiation of "vivisection experiment" from "unmeaning crime" demands "men who understand," observers who see "ideas" embodied within the gashes of tortured animals. And indeed, in the process of publicizing and defending their scientific enterprise, Victorian physiologists made the public capable of seeing both the future benefits of vivisection research and the honorable "ideas" of vivisection *itself* within the bleeding body. As Michael Foster quoted in his 1881 Inaugural Address to the Physiology Section of the International Medical Congress: "For either in this way, namely through death and wounds, through dissection and, as it were, by a Caesarean operation, will truth be brought to light or otherwise it will lie for ever hid" (quoted in Richards 1992: 166).

Experimental physiology created a way of seeing the bleeding body as a medium, whereupon every cut becomes an extension of a scientific system of inscription, a seriation of legible meanings. Even critics of vivisection in Victorian England came to understand the physiological researcher as uniquely capable of "understanding...the full meaning and extent of the waves and spasms of agony he deliberately creates" (Cobbe 1894: 666–677). Vivisection makes meaning take place on the wounded body: pain becomes legible, torture signifies, and wounds are writing.

Within Victorian culture, then, scientific discourse constructed an interpretive community able to understand the scientific meaning of wounds, to see vivisection within images of flayed and rended bodies (Figure 6.3). So it is not really surprising that this interpretive community would then also see vivisection in the image of Catherine Eddowes at the site of her murder in Mitre Square, Aldgate (Figure 6.4). In both of these images, the wounded body literally gives forth writing: there are no meaningless cuts.

SCENES OF THE CRIME

The laboratory and the crime scene both emerge as spaces of writing, enclosures in which wounded bodies are made to signify. Where the Ripper strikes, letters and more letters proliferate, spinning a literary cocoon around the corpses. Near Annie Chapman's head, investigators find a torn

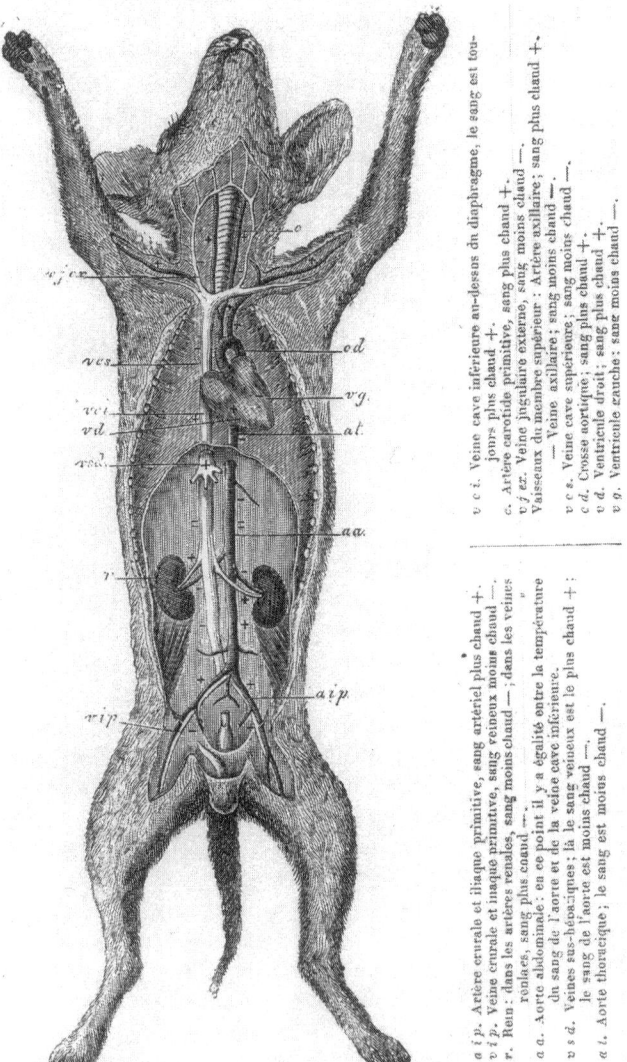

FIG. 3. — Représentation du système circulatoire du chien.

Figure 6.3 Textbook image of vivisection: "Représentation du système circulatoire du chien" (Bernard 1871) (Courtesy of Widener Library, Harvard University).

envelope bearing fragments of addresses ("Sp" [Spitalfields?] and "Sussex Regiment") and, on the back, the lone letter "M." In Goulston Street, investigators find a blood-soaked shred of Catherine Eddowes's apron, discarded by the killer directly below a brick archway adorned with mysterious chalk graffiti: "The Juwes are the men who will not be blamed

Figure 6.4 Inquest sketch: Catherine Eddowes crime scene, 30 September 1888. Sketch made by Frederick William Foster, the City of London Architect and Surveyor (Royal London Hospital Archives, ref. GB 0387 LH/X/97. © Royal London Hospital Archives. Courtesy of Royal London Hospital Archives).

for nothing." No less mysterious are the two deep chevrons sliced into Eddowes's face—noted in Dr. Frederick Brown's postmortem examination as "on each side of cheek a cut which peeled up the skin, forming a triangular flap about an inch and a half" (Coroner's Inquest 1888: 205)—which some mythographers of the case have seen as alphabetic carvings, doubled Vs or perhaps an M, personal signs of the Ripper (Fido 1987: 75; Harrison 1993: 170; Feldman 2002: 56-7; also Didbin 1978). At the murder scenes, cuts become telltale clues in a broader syntax of criminal marks, inscrutable communiqués made by chalk, pen, and knife.

Likewise in the physiological laboratory, the wounded body communicates through its articulation with tools of inscription: the scalpel, the kymograph, and the myograph. An illustration plate from Burdon

Sanderson's 1873 *Handbook for the Physiological Laboratory*—reproducing images from the works of German physiologist Eduard Pflüger and French physiologist Etienne-Jules Marey—shows the living wounds and severed muscles of frogs hooked up to myographic instruments (Figure 6.5). Nerve-muscle extracted from the sliced skin, strung between electrodes and recording levers of the instruments, becomes medial connective tissue in linking marks of the scalpel to marks of the stylus, knife to pen. The ripped amphibian body disperses into the lines of the laboratory. Physiology transfers itself into the graphical curve, a calligraphic expression of the meaning of wounds which Marey, in *La méthode graphique en sciences experimentales* (1878), appositely called the "language of the phenomena themselves" (quoted in Brain 2002: 166). In these instrumental images from the physiological laboratory, the vivisector is invisible, the organism as such disappears, and all that remains are written traces of occulted biology: the irrefutable truth of the organism recorded as a machinic line of automatic writing, the graphical relay of natural processes (Brain 2002). Vivisection initiates a chain of signification, the "observable facts" of wounds spewing forth the "language of the phenomena themselves," material signifiers drawing out other material signifiers, and everything is red.

The image of the nineteenth-century physiological laboratory as the scene of writing, a spatialized chain-reaction of material signifiers, prefigures the modern lab as a vast "system of literary inscription" orchestrated by scientists who appear as "compulsive and almost manic writers" (Latour and Woolgar 1986: 52, 48). In this active site of systemic inscriptions, the experimental arrangement of the vivisected body with graphical recording instruments enfolds the real into discursive networks of interlocked graphemes, constituting what Hans-Jörg Rheinberger has called the "graphematic space" of science:

> The whole experimental arrangement...has to be taken as a graphematic articulation. Written tables, printed curves, and diagrams are further transformations of a graphematic disposition of pieces of matter, a disposition that is embodied in the design of the experiment itself.... It is not simply the measuring devices that produce the inscriptions. The scientific object itself is shaped and manipulated 'as' a traceable conformation. Temporally and spatially, the object *is* a bundle of inscriptions. It displays only what can be handled in this way. (Rheinberger 1997: 111)

The vivisected body, traced by the scalpel and relayed by the instrumentation of the laboratory, becomes itself a literary object precisely because the laboratory apparatus and the physiological gaze has prepared it for presentation in this manner. The metamorphosis of bleeding wounds into scientific inscriptions drawing forth other scientific inscriptions within the boundaries of the laboratory spins a material web of signs that even incor-

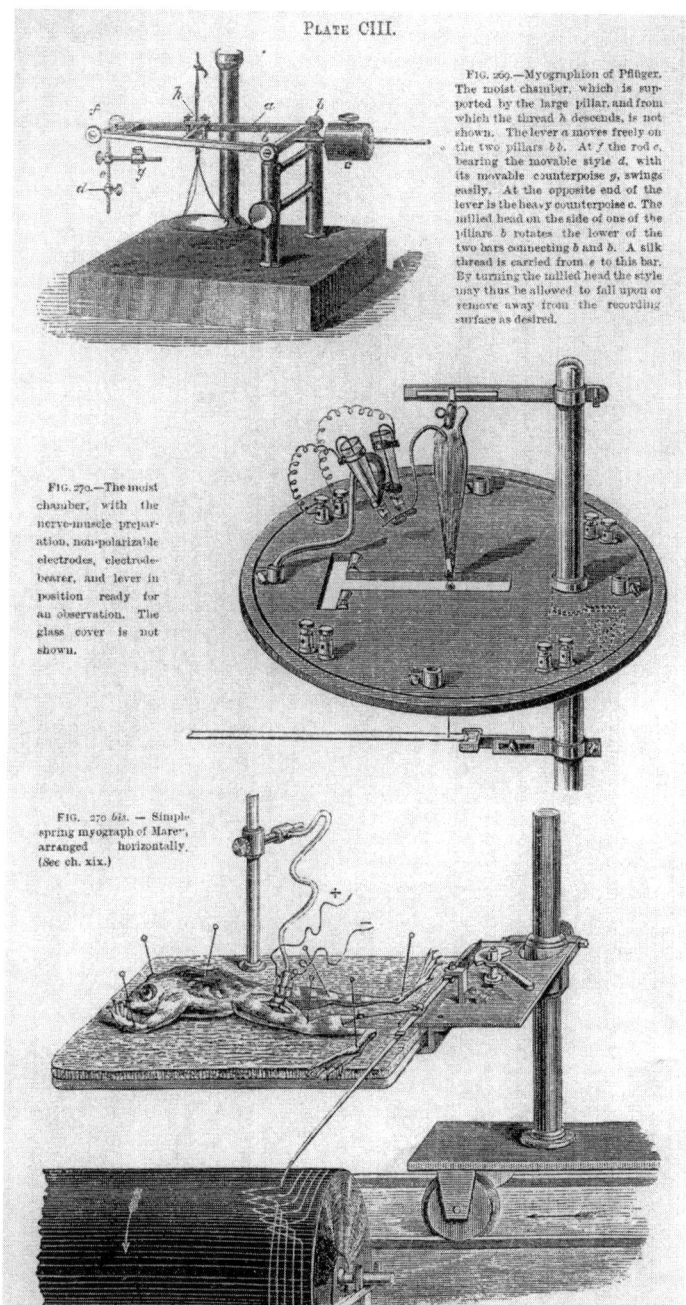

PLATE CIII.

FIG. 269.—Myographion of Pflüger. The moist chamber, which is supported by the large pillar, and from which the thread *h* descends, is not shown. The lever *a* moves freely on the two pillars *b b*. At *f* the rod *c*, bearing the movable style *d*, with its movable counterpoise *g*, swings easily. At the opposite end of the lever is the heavy counterpoise *c*. The milled head on the side of one of the pillars *b* rotates the lower of the two bars connecting *b* and *b*. A silk thread is carried from *e* to this bar. By turning the milled head the style may thus be allowed to fall upon or remove away from the recording surface as desired.

FIG. 270.—The moist chamber, with the nerve-muscle preparation, non-polarizable electrodes, electrode-bearer, and lever in position ready for an observation. The glass cover is not shown.

FIG. 270 *bis.* — Simple spring myograph of Marey, arranged horizontally. (*See* ch. xix.)

Figure 6.5 Handbook for the Physiological Laboratory: Myographic apparatuses of Pflüger and Marey (Burdon Sanderson 1873) (Courtesy of Widener Library, Harvard University).

porates the scientific author presiding "manically" over the operation. If these reduplicated images from *The Handbook for the Physiological Laboratory* present the space of vivisection as authorless, as a space of automatic writing where phenomena record themselves directly into the graphical line, this is only because the hand of the vivisector has already been traced by the wounds themselves. The vivisectionist has become the absent subject of the scene of vivisection: as Claude Bernard writes, "The physiologist is...absorbed by the scientific idea which he pursues" in the laboratory to such an extent that, indeed, "We cannot imagine [him] without his laboratory.... Without it [the lab], neither experimenters nor experimental science can exist" (Bernard 1865/1927: 148).

For the scientist, there is no outside the laboratory, no "without" the laboratory: the scientist is "absorbed" by the experimental ideas that are given material form inside the space of physiological inscription, the space of wounding. The scientist dissolves into this space, discernable only as a ghost, a signature.

The enchained graphemes of the vivisection laboratory appear to contain and reciprocally project the subject of vivisection; that is, the subject position of the vivisector as "scientific author" (see Biagioli and Galison 2002)—which is also to say, the locus of "physiological knowledge" as such, the putative "signified" or the "meaning" of those laboratory marks made by pen, knife, and recording instrument. Such marks signify physiological knowledge, which is precisely why the crime scenes of Jack the Ripper, also spaces of graphematic violence, so vividly animated the subject of vivisection, exhibiting marks of a skilled hand ostensibly disciplined by the dissection room, a hand accustomed to sampling from what Bernard called the "ghastly kitchen."

Indeed, Bernard's culinary metaphor itself conceptualizes the vivisection lab as a medial space outwardly projecting the authorial position of scientific knowledge. Narrating the pilgrim's progress of physiology, Bernard describes the experimental laboratory as an enclosed room of horror beyond which lies a realm of profound meaning, the dazzling knowledge of "life" as such: "If a comparison were required to express my idea of the science of life, I should say that it is a superb and dazzling lighted hall which may be reached only by passing through a long and ghastly kitchen" (Bernard 1865/1927: 15). The laboratory, while leading the scientist towards revelation, towards a transcendental signified, is itself deeply material; it is the space of the signifier, the carnal surface of incision. Like the vivisected organism, the laboratory operates as a book of blood, promising meaning and the light of truth on the other side, a "passing through" towards rare knowledge, but appearing on its crimson surface as nothing so much as a butcher's block, a wet and grisly place where the "cutting up of animals" might also entail cooking and eating them. This ghastly kitchen, a domestic room of unmentionable "appetites," opens up the "science of life." The professional progress of the physiologist towards knowledge occurs then within a moist training ground of both medical slaughter and medical cuisine. Or, as Jack the Ripper wrote of his own ghastly experiments: "I send you half the Kidne I took from one woman and prasarved it for you tother piece I fried and ate it was very nise."

On 10 November 1888, at the scene of Mary Jane Kelly's death—the last and most gruesome of the five so-called "canonical" Ripper murders—the *Daily Telegraph* reported:

> A most horrifying spectacle was presented to the officers' gaze, exceeding in ghastliness anything which the imagination can picture. The body of the woman was stretched out on the bed, fearfully mutilated. Nose and ears had been cut off, and although there had been no dismemberment, the flesh had been stripped off, leaving the skeleton. The nature of the other injuries was of a character to indicate that they had been perpetrated by the author of the antecedent crimes in the same district; and it is believed that once more there are portions of the organs missing. That the miscreant must have been some time at his work was shown by the deliberate manner in which he had excised parts, and placed them upon the table purposely...Dr. J.R. Gabe, who viewed the body, said he had seen a great deal in dissecting rooms, but he had never witnessed such a horrible sight as the murdered woman presented. Before anything was disturbed a photograph was taken of the interior of the room.... It was evident that a large keen knife had been used by a hand possessed of some knowledge and practice. ("The East End Tragedies" 1888: 337–338)

The enclosed space of the Kelly murder, inside her room at 13 Miller's Court, 26 Dorset Street, Spitalfields—the only Ripper murder to take place completely indoors—appears to the forensic "gaze" as a scene of violent writing. For each injury translates a "character" in the style of "the author of the antecedent crimes," an author with a trained "hand possessed of knowledge and practice." This hand has left a signature in its "deliberate" and "purposeful" manner of cutting and removing organs: this is no "unmeaning crime." And at this scene of violent authorship, with its residues of deliberation and purpose, knowledge and practice, the medical witness Dr. J.R. Gabe can only summon an inadequate comparison to the horrors of the "dissecting room."

This crime scene, "exceeding in ghastliness anything which the mind can picture," nevertheless conjures pictures of clinical dissections, the deliberate and purposeful "ghastliness" of the physiological laboratory. This domestic room of vicious slaughter is haunted by Bernard's "ghastly kitchen." For once again, organs are missing, and in their worrisome absence appear signs of an appetite for both cutting and cooking—the same conflicted appetites given life in the vivisection lab, which Wilkie Collins called "the atrocities of the Savage Science" (Collins 1883/1996: 136). The forensic gaze upon

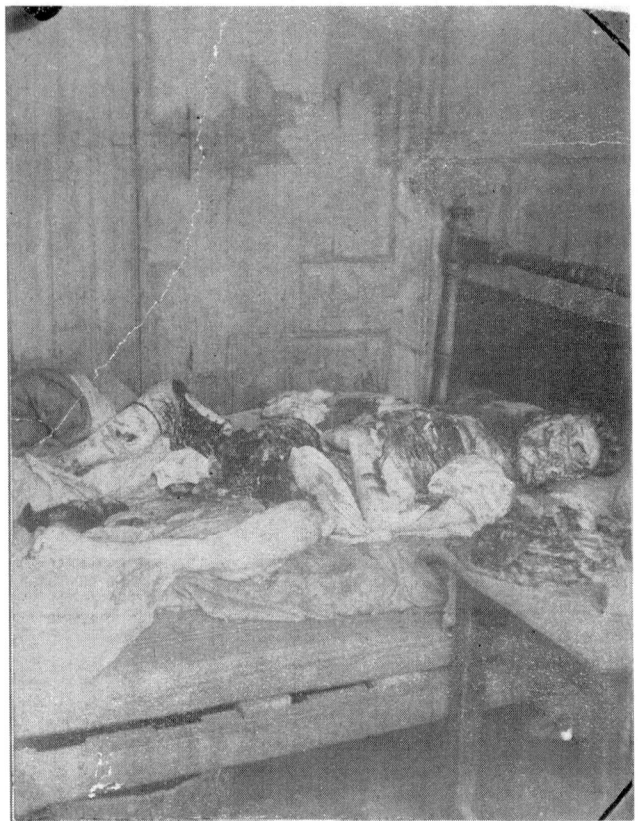

Figure 6.6 Crime scene photograph: Mary Jane Kelly, 9 November 1888. (UK National Archives ref. MEPO 3/3155. Courtesy of the National Archives. This material is in the copyright of the Metropolitan Police and is reproduced here by the permission of the Metropolitan Police Authority). At least two photographs were taken of Kelly's body at the crime scene; both are currently held at the National Archives. In this photograph, the dubious letters "FM" are said to be seen on the wall immediately above Kelly's left hand, among other fugitive and fleeting signs.

the Kelley scene thus connects the dots between serial murder, experimental biology, and incisive writing, drawing conclusions even as it produces more writing—the reportage of the crime scene—and even as it produces other forms of graphical representation: "Before anything was disturbed a photograph was taken of the interior of the room" (Figure 6.6).

This photograph extends and retraces, as another iteration, the enchained characters, records, graphemes, and graphics framed by the room at 13 Miller's Court. Certainly, Victorian photography was readily pictured as an inscriptional, even "autographic" practice—as the titular metaphor of William Henry Fox Talbot's pioneering photography book, *The Pencil of Nature* (1844), might suggest—whose value as "eyewitness" documentation for both forensic and natural science could be secured only within a broader interpretive architecture of drawings, diagrams, and textual rede-scriptions (Tucker 2005). But more particularly, this photograph of Mary Kelly has continued to disclose traces of writing long after its original cre-ation, revealing uncanny afterimages of letters etched in the sanguinary walls of Kelly's room. Although no investigator on the scene made note of any mural messages, the very fact that a photograph had been taken trig-gered contemporary rumors that hitherto obscured writing might yet be revealed. As one newspaper reported:

> Profiting by previous blunders the police called a photographer to take a picture of the room before the body was removed from it. This gives rise to reports that there is more handwriting on the wall, though three or four people who were allowed into the room say they did not observe it; but possibly they were too excited to notice such a detail. ("Fiendish Atrocity" 1888)

While such "handwriting on the wall" was not to be forthcoming for sleuths in 1888, latter-day Ripperologists studying this photograph have reputedly discovered significant scrawls in the blood spray just above Kelly's body: confessional graffiti, satanic sketches, the initials "FM," and other fanciful spoor of the Ripper (e.g., Begg et al. 1996: 487 ["Wood, Simon"]; Harrison 1993: 100–101; Feldman 1998: 71–72). Others have even seen words, numerals, and elaborate designs printed on Kelly's body itself (e.g., Derrico 2005; Keating 2005; Ryan 2005). These phantom etchings have materialized only belatedly, inadvertently, as artifacts of this photograph's reproduction and circulation (the image has appeared in dozens of publica-tions: first in Lacassagne 1899; famously in Rumbelow 1975; with greater clarity of the supposed graffiti in Knight 1976; etc.). The photo seems to carry along its own contextual mythos of criminal logorrhea and autho-rial purpose, inducing pareidolia and ensuring that all marks are remark-able, recoverable as signs, productive of new meanings. Jacques Derrida has written: "By all rights, it belongs to the sign to be legible, even if the moment of its production is irremediably lost, and even if I do not know what its alleged author-scriptor meant consciously and intentionally at the moment he wrote it, that is, abandoned it to its essential drifting" (Derrida 1982: 317). Hence, in the drifting of a police photograph, accidental sprays of arterial blood become decipherable as writing: at the graphematic scene of the crime, as in the vivisection laboratory, there can be no meaningless cuts.

By the end of November 1888, the medico-scientific community had formed a cohesive front on the implausibility of the "scientific Ripper" the-ory. Where articles in *The Lancet* cautiously entertained the topic through September and October, by December medical writers raised the idea only for ridicule. Even Dr. Thomas Bond's report to the Home Office following the postmortem examination of Mary Jane Kelly actively directs attention away from notions of physiological knowledge and scientific sadism. Despite expert testimonies in the previous cases and opinions of eyewitnesses at the crime scenes, Bonds writes that the entire series of murders was, after all, simply unmeaning crime: "In each case the mutilation was inflicted by a person who had no scientific or anatomical knowledge. In my opinion he does not even possess the technical knowledge of a butcher or horse slaughterer or any person accustomed to cut up dead animals" (report reproduced in Evans and Skinner 2000: 361). Bond instead shifts the motivational profile of the killer away from "scientific sadism" towards either "erotic mania" or "religious mania." With nary a contrary voice in their ranks by the end of 1888, medical researchers adopted strategies for crushing lingering rumors about the Whitechapel killer's ties to scientific medicine similar to their strategies for dealing with antivivisectionist rhetoric broadly, publicly disputing the very idea as nothing but ignorant folly while upholding the inherent morality of trained scientists (see Bryant 1977). Thus would physiology seemingly save itself from association with its most damaging exponent, Jack the Ripper.

But as I have attempted to suggest, it was not simply antiscientific critiques of vivisection that led to widespread imaginary connections between the Ripper's crimes and the experiments of the physiological laboratory. It was also the responsibility of physiologists themselves, who had deployed a certain vivisectional gaze, a certain readerly frame of reference, in remediating the practices of experimental physiology in England. After all, the scientific community's galvanized refutations of the mad vivisector theory did not dispel a cultural desire to see scientific meaning in the victims' wounds. Even years after the last murder, the vivisection theory retained potency. As late as 1895, the *Chicago Times-Herald* alleged that Scotland Yard detectives, guided by the clairvoyant Robert James Lees, had long ago apprehended Jack the Ripper but concealed the investigation, precisely because the killer turned out to be a respected medical researcher and practitioner of vivisection:

> When "Jack the Ripper" was finally run to earth it was discovered that he was a physician in good standing, with an extensive practice. He had been ever since he was a student at Guy's Hospital, an ardent and enthusiastic vivisectionist. Through some extraordinary natural contradiction, instead of the sight of pain softening him, as is the case with most devotees of scientific experiments, it had an opposite effect. This so grew upon him that he experienced the keenest delight in inflicting tortures upon defenceless animals.... [A]n exhaustive enquiry before a commission in lunacy developed the fact that while in one mood the doctor was a most worthy man, in the other he was a terrible monster. He was at once removed to a private insane asylum in Islington, and he

is now the most intractable and dangerous madman confined in that establishment. ("The Capture of Jack the Ripper" 1895)

While this newspaper *exposé* of the secret capture of a lunatic vivisection-ist and subsequent elite cover-up was in all likelihood a hoax (Harris 1987, 1994), it nevertheless reflects the tenacity of the "mad scientist" as a preva-lent and compelling profile of the Ripper, one that has endured in popular media throughout the last century. (For example, Sir William Withey Gull, Royal Physician and arch vivisectionist, has been fingered as the Ripper by several modern accounts that connect mad science to elaborate conspiracy theories of Freemasonry and royal scandal; see Knight 1976; Moore and Campbell 1999; see Odell 2006: 125–181). The survival of the sadistic scientist theory notwithstanding, there is some figural sense in which the Whitechapel killer, regardless of his "real" identity, was indeed intimately involved with the arts of Victorian vivisection. Because as a product of the Victorian media ecology, the Ripper inhabited the same frame of refer-ence, the same cultural logic, as those laboratory physiologists who were understood to render the body legible as a text of experimental operations, an inscriptional surface for the vehiculation of scientific meanings. With horrifying literalness, Jack the Ripper exposed the Victorian image of the vivisected body as a grisly epistle addressed to the scientific gaze, a letter signed in ink and blood, a postcard, as it were, from hell.

ACKNOWLEDGMENTS

I would like to thank Luis Campos, Seeta Chaganti, Mark Jerng, Matthew Jones, Elizabeth Lee, Sharrona Pearl, Robert J. Richards, and Laura Thie-mann Scales for offering extremely insightful suggestions during develop-ment of this essay. Acquisition and reproduction of archived images was graciously assisted by Jonathan Evans and Kate Richardson at the Royal London Hospital Archives and Museum, Paul Johnson and Nicholas Coney at the UK National Archives, and David Capus at the Metropolitan Police Service. For their help in tracking the Ripper, a special thanks goes to my fellow sleuths Alisa Braithwaite, Nadine Knight, and Amanda Teo.

BIBLIOGRAPHY

Begg, Paul, Martin Fido, and Keith Skinner (1996) *The Jack the Ripper A-Z*, revised ed., London: Headline.
Bernard, Claude (1876) *Leçons sur la chaleur animale, sur les effets de la chaleur, et sur la fiévre*, Paris: J.-B. Baillière.
——— (1865/1927) *An Introduction to the Study of Experimental Medicine*, trans. Henry Copley Greene, New York: Macmillan.

Biagioli, Mario and Peter Galison (eds) (2002) *Scientific Authorship: Credit and Intellectual Property in Science*, London: Routledge.

Brain, Robert (2002) "Representation on the line: Graphic recording instruments and scientific modernism", in Bruce Clarke and Linda Dalrymple Henderson (eds.), *From Energy to Information: Representation in Science and Technology, Art, and Literature*, Stanford, CA: Stanford University Press: 151–177.

Bryant, Ian (1977) "Vivisection: A chapter in the sociology of Victorian science", *Ethics in Science and Medicine*, 4: 75–86.

Burdon Sanderson, John Scott (ed.) (1873) *Handbook for the Physiological Laboratory*, 2 vols., London: J. & A. Churchill.

"The Capture of Jack the Ripper" (1895) *Chicago Times-Herald*, 28 April 1895; reprinted in Melvin Harris (1987) *Jack the Ripper: The Bloody Truth*, London: Columbus Books: 84–92.

Carroll, Lewis (1875) "Some popular fallacies about vivisection", *Fortnightly Review*, 23: 847–854.

Cobbe, Frances Power (1882) "Vivisection and its two-faced advocates", *Contemporary Review*, 41: 610–626.

—— (1888) "Detectives: To the editor of the Times", *The Times*, 11 October 1888; reprinted in Stephen P. Ryder (ed.), *Casebook: Jack the Ripper* [Accessed: 24 January 2006], http://www.casebook.org/press_reports/times/18881011.html.

—— (1894) *The Life of Frances Power Cobbe*, vol. 2, Boston: Houghton Mifflin.

Coleridge, Lord Chief Justice [Stephen] (1882) "The nineteenth century defenders of vivisection", *Fortnightly Review*, 38: 225–236.

Collins, Wilkie (1883/1996) *Heart and Science: A Story of the Present Time*, ed. Steve Farmer, Ontario, Canada: Broadview.

Coroner's Inquest (1888) "No. 135, Catherine Eddowes Inquest, 1888", Corporation of London Record Office; reprinted in Stewart P. Evans and Keith Skinner (eds) (2000), *The Ultimate Jack the Ripper Companion: An Illustrated Encyclopedia*, New York: Carroll & Graf: 199–215.

Curtis, L. Perry (2001) *Jack the Ripper and the London Press*, New Haven, CT, and London: Yale University Press.

—— (2002) "Responses to the Ripper murders: Letters to Old Jewry", *Ripperologist*, 40: 3–7.

de Cyon, Elie (1883) "The anti-vivisectionist agitation", *Contemporary Review*, 43: 498–516.

Derrico, Eddie (2005) "Letters on Kelly's Leg", *Casebook: Jack the Ripper—Message Boards*, Stephen P. Ryder (ed.) [Accessed 27 June 2006], http://www.casebook.org/forum/messages/4921/20457.html.

Derrida, Jacques (1974) *Of Grammatology*, trans. Gayatri Chakravorty Spivak, Baltimore, MD: Johns Hopkins University Press.

—— (1982) "Signature event context", in *Margins of Philosophy*, trans. Alan Bass, Chicago: University of Chicago Press: 307–330.

Dibdin, Michael (1978) *The Last Sherlock Holmes Story*, New York: Random House.

D'Onston, Roslyn [Robert Donston Stephenson] (1888) "The Whitechapel demon's nationality: And why he committed the murders, by one who thinks he knows", *Pall Mall Gazette*, 1 December 1888; reprinted in Jarett Kobek (ed.) (2006), *Crowley's Ripper: The Collected Work of Roslyn D'Onston*, revised ed., Kobek.com Klassic Reprint: 3–7; available at http://www.kobek.com/crowleyripper.pdf.

"The East End tragedies. A seventh murder. Another case of horrible mutilation" (1888) *Daily Telegraph*, 10 November 1888; reprinted in Stewart P. Evans and Keith Skinner (eds) (2000), *The Ultimate Jack the Ripper Companion: An Illustrated Encyclopedia*, New York: Carroll & Graf: 336–45.

"The East-End Murders" (1888) *The Times*, 5 October 1888; reprinted in Stewart P. Evans and Keith Skinner (eds) (2000), *The Ultimate Jack the Ripper Companion: An Illustrated Encyclopedia*, New York: Carroll & Graf: 215–25.

Elston, Mary Ann (1987) "Women and anti-vivisection in Victorian England, 1870–1900", in Nicolaas A. Rupke (ed.), *Vivisection in Historical Perspective*, London: Croom Helm: 259–294.

Evans, Stewart P. and Keith Skinner (2000) *The Ultimate Jack the Ripper Companion: An Illustrated Encyclopedia*, New York: Carroll & Graf.

—— (2001) *Jack the Ripper: Letters from Hell*, Phoenix Mill, Stroud, UK: Sutton.

Feldman, Paul H. (2002) *Jack the Ripper: The Final Chapter*, reissue ed., London: Virgin Books.

Fido, Martin (1987) *The Crimes, Detection and Death of Jack the Ripper*, London: Weidenfeld and Nicolson.

"Fiendish atrocity: The details of another Whitechapel murder" (1888) *Atlanta Constitution*, 12 November 1888; reprinted in Stephen P. Ryder (ed.), *Casebook: Jack the Ripper* [Accessed: 15 June 2006], http://www.casebook.org/press_reports/atlanta_constitution/881112.html.

Foster, Michael (1874) "Vivisection", *Macmillan's Magazine*, 29: 367–376.

Foucault, Michel (1977) "What is an author?" in Donald F. Bouchard (ed.) (trans. Donald F. Bouchard and Sherry Simon), *Language, Counter-Memory, Practice*, New York: Cornell University Press: 113–138.

Frayling, Christopher (1986) "The house that Jack built: Some stereotypes of the rapist in the history of popular culture", in Sylvana Tomaselli and Roy Porter (eds), *Rape: An Historical and Cultural Enquiry*, Oxford: Blackwell: 174–215.

French, Richard D. (1975) *Antivivisection and Medical Science in Victorian Society*, Princeton, NJ: Princeton University Press.

Geison, Gerald L. (1978) *Michael Foster and the Cambridge School of Physiology: The Scientific Enterprise in Late Victorian Society*, Princeton, NJ: Princeton University Press.

Gomel, Elana (2003) *Bloodscripts: Writing the Violent Subject*, Columbus, OH: Ohio State University Press.

Harris, Melvin (1987) *Jack the Ripper: The Bloody Truth*, London: Columbus Books.

—— (1994) *The True Face of Jack the Ripper*, London: Michael O'Mara Books.

Harrison, Shirley (1993) *The Diary of Jack the Ripper*, New York: Hyperion.

Jordanova, Ludmilla (1989) *Sexual Visions: Images of Gender in Science and Medicine between the Eighteenth and Twentieth Centuries*, Madison, WI: University of Wisconsin Press.

Keating, David (2005) "What do the numbers 28:4 mean on Kelly's leg?", *Casebook: Jack the Ripper— Message Boards*, Stephen P. Ryder (ed.) [Accessed 27 June 2006], http://www.casebook.org/forum/messages/4921/17193.html.

Knight, Stephen (1976) *Jack the Ripper: The Final Solution*, London: Harrap.

Lacassagne, Alexandre (1899) *Vacher l'eventreur et les crimes sadiques*, Lyon: A. Storck; Paris: Masson.

Lansbury, Coral (1985) *The Old Brown Dog: Women, Workers and Vivisection in Edwardian England*, Madison, WI: University of Wisconsin Press.

Latour, Bruno, and Steve Woolgar (1986) *Laboratory Life: The Construction of Scientific Facts*, Princeton, NJ: Princeton University Press.

Menke, Richard (2000) "Fiction as vivisection: G. H. Lewes and George Eliot", *ELH*, 67: 617–653.

Moore, Alan, and Eddie Campbell (1999) *From Hell*, Paddington: Eddie Campbell Comics.

Odell, Robin (2006) *Ripperology: A Study of the World's First Serial Killer and a Literary Phenomenon*, Kent, OH: Kent State University Press.

Rheinberger, Hans-Jörg (1997) *Toward a History of Epistemic Things: Synthesizing Proteins in the Test Tube*, Stanford, CA: Stanford University Press.

Richards, Stewart (1986) "Drawing the life-blood of physiology: Vivisection and the physiologists' dilemma, 1870–1900", *Annals of Science*, 43: 27–56.

—— (1987) "Vicarious suffering, necessary pain: Physiological method in late nineteenth-century Britain", in Nicolaas A. Rupke (ed.), *Vivisection in Historical Perspective*, London: Croom Helm: 125–148.

—— (1992) "Anaesthetics, ethics and aesthetics: Vivisection in the late nineteenth-century British laboratory", in Andrew Cunningham and Perry Williams (eds), *The Laboratory Revolution in Medicine*, Cambridge, UK: Cambridge University Press: 142–169.

Ritvo, Harriet (1987) *The Animal Estate: The English and Other Creatures in the Victorian Age*, Cambridge, MA: Harvard University Press.

Romano, Terrie M. (2002) *Making Medicine Scientific: John Burdon Sanderson and the Culture of Victorian Science*, Baltimore, MD: Johns Hopkins University Press.

Ross, Ronald (1882/1988) "The vivisector vivisected," in Eli Chernin (ed.) (1988), "An artificial heart revives a corpse: Sir Ronald Ross's unpublished story of 1882, 'The Vivisector Vivisected'", *Perspectives in Biology and Medicine*, 31: 341–352.

Rumbelow, Donald (1975) *The Complete Jack the Ripper*, London: W. H. Allen.

Rupke, Nicolaas A. (1987) "Pro-vivisection in England in the early 1880s: Arguments and motives", in Nicolaas A. Rupke (ed.), *Vivisection in Historical Perspective*, London: Croom Helm: 188–209.

Ryan, Stuart (2005) "The mystery image", *Casebook: Jack the Ripper—Message Boards*, Stephen P. Ryder (ed.) [Accessed 27 June 2006], http://www.casebook.org/forum/messages/4921/20907.html.

Seltzer, Mark (1998) *Serial Killers: Death and Life in America's Wound Culture*, New York: Routledge.

Showalter, Elaine (1992) *Sexual Anarchy: Gender and Culture at the Fin de Siècle*, London: Virago.

Smith, Andrew (2004) *Victorian Demons: Medicine, Masculinity, and the Gothic at the Fin de Siècle*, Manchester: Manchester University Press.

Swanson, Donald (1888) "Murder of Annie Chapman", Metropolitan Police Criminal Investigation Department, Scotland Yard; reprinted in Stewart P. Evans and Keith Skinner (eds) (2000), *The Ultimate Jack the Ripper Companion: An Illustrated Encyclopedia*, New York: Carroll & Graf: 66–69.

Tucker, Jennifer (2005) *Nature Exposed: Photography as Eyewitness in Victorian Science*, Baltimore: Johns Hopkins University Press.

Turney, Jon (1998) *Frankenstein's Footsteps: Science, Genetics, and Popular Culture*, New Haven, CT: Yale University Press.

Walkowitz, Judith R. (1992) *City of Dreadful Delight: Narratives of Sexual Danger in Late-Victorian London*, Chicago: University of Chicago Press.

Wells, H. G. (1896/1993) *The Island of Dr. Moreau: A Variorum Text*, ed. Robert M. Philmus, Athens, GA: University of Georgia Press.

"The Whitechapel Murder" (1888) *The Times,* 27 September 1888; reprinted in Stewart P. Evans and Keith Skinner (eds) (2000), *The Ultimate Jack the Ripper Companion: An Illustrated Encyclopedia,* New York: Carroll & Graf: 102–107.

"The Whitechapel Murders" (1888) *The Lancet,* 29 September 1888: 637; reprinted in Stephen P. Ryder (ed.), *Casebook: Jack the Ripper* [Accessed: 8 January 2006], http://www.casebook.org/press_reports/lancet/lancet880929.html.

Whittington-Egan, Richard (1975) *A Casebook on Jack the Ripper,* London: Wiley.

Wilson, Dr. Andrew (1888) "The Whitechapel murders—Old and new," *Port Philip Herald,* 1 November 1888, sec. Health; reprinted in Stephen P. Ryder (ed.), *Casebook: Jack the Ripper* [Accessed: 4 January 2006], http://www.casebook.org/press_reports/port_philip_herald/881101.html.

Yeo, Gerald F. (1882) "Letter to the editor", *Contemporary Review,* 41: 897–898.

7 The Scientist as Personality
Elaborating a Science of Intimacy in the Nadar/ Chevreul Interview (1886)

Charlotte Bigg

I

On 31 August 1886 the French chemist Michel Eugène Chevreul turned 100 years old. The government deemed the event a suitable occasion for a national festival. No expenses were spared to honor the pioneer of organic chemistry, the director of Gobelins' manufacturing for over forty years, the author of the celebrated theory of the contrast of colors, the patriot who had twice refused the Tsar's offer to leave his country for Russia, and the benefactor of mankind whose improved soap and candles were used in every home in the nation. All in all fifty-two speeches were made, a medal was struck, a statue was unveiled, the major Parisian theatres ran plays written for the occasion, delegations of foreign scientists and the President of the Republic assembled, the Paris town hall hosted a banquet for a thousand guests, a festival of poetry and music was held in Chevreul's honor, and a candle-lit procession illuminated the streets of Paris on that evening.[1]

The Third Republic was notoriously prone to such great public celebrations, from the burial of Victor Hugo in the previous year, to the French revolution's centenary in 1889. These popular ceremonies were orchestrated by the government with a characteristic display of the young Republic's symbols, busts of Marianne, elected representatives and ubiquitous flags, arguably substitutes for the religious festivals favored by the Second Empire. The distinctive values of the Third Republic, Secularity, democracy, and progress were to take the place of religion and political and social conservatism (Nelms 1987).

In the ideological battle waged by the new regime, science and technology played a central role. Against the Empire's transcendent values, the republicans believed science would bring economic progress, social welfare, and, most importantly, the recovery of national prestige following defeat at the hands of the Germans in 1871. Scientists were hailed as heroes and were literally made to take the place of religious icons as objects of public worship. The republic's secular sanctification of Pasteur is one notorious instance (e.g., Geison 1995: 259–274). On a more modest scale, Chevreul's centenary fulfilled the same function.

The representations of scientists on such occasions were primarily conventional, in the sense that they were portrayed in their public, official capacities rather than as private persons: as icons to be celebrated, heroes to be emulated, public figures to be respected. And when their private selves were mentioned, scientists appeared as hard-working, self-sacrificing, honest, modest individuals devoted to the greater good. The iconography produced by these occasions is unmistakable, as Figure 7.1 reminds us, taken from the special issue of the nationally-distributed journal *Le Courrier*

M. Chevreul.

Figure 7.1 Chevreul centenary issue in *Le Courrier Français* 3: 35 (31 August 1886), 5.

Français devoted to Chevreul's centenary. The press played an important role in consolidating and spreading such visual and literary conventions of representing scientists.

The printed media could, however, sometimes function as a locus for experimentation, creativity, and renewal in matters of representation: not only materially, for instance with the introduction in the later decades of the nineteenth century of a number of techniques for transferring photographs to the print medium, but more fundamentally in fostering new cultural representations of science and scientists. One such example is the unorthodox contribution to the celebration of Chevreul's centenary published in the *Journal Illustré* on 5 September 1886. This journal published for the occasion an interview between the photographer Nadar (Félix Tournachon) and Chevreul. It consists of a transcript of the conversation, questions by Nadar, and answers by Chevreul on a variety of topics, accompanied by a series of photographs of the two men chatting, taken by the interviewer's son, Paul Nadar (see Figures 7.2–7.5). This is a remarkable document in many respects: part showmanship, part advertising stunt, part technological prowess, and even part science fiction. All these elements were brought together by Nadar to produce a self-consciously novel manner of representing the scientist that focused on his private, intimate self. The Nadar/

Figures 7.2, 7.3 The first and penultimate pages of the *Journal Illustré* issue dedicated to the Nadar-Chevreul interview (*Journal Illustré* 23:36 (5 September 1886): 281, 287) (Bibliothèque du Pays Châtelleraudais).

Chevreul interview arguably constitutes a milestone in the emergence of a new convention of portraying scientists as personalities.

II

Picking up the classic analysis running from Edgar Allan Poe (*The Man of the Crowd*) through Charles Baudelaire to Walter Benjamin, Richard Sennett has argued in his *Fall of Public Man* that the nineteenth century saw profound social changes in the conduct of public life. He contends that in its later decades "personality entered the public realm," one manifestation of the increasingly pervasive secular worldview of European societies. Immediacy of sensation and perception prevailed in all spheres, echoed for instance in Baudelaire's "Modernity is the transitory, the fugitive, the contingent" (cited in Nochlin 1972/1990: 28). In science, it meant a move away from metaphysics and towards phenomena and experimentation, while impressionism and naturalism were its embodiments in literature and the arts. In society also, immanence became more valued than transcendence, and the emphasis placed on appearance: Behavior, dress, the immediate environment were now conceived as reflections of the inner self. People in public ceased to be represented by rank or titles, and instead were differentiated according to personal characteristics. In short, "beliefs in society [...] centered on the immediate life of man himself and his experiences as a definition of all that he can believe in" (Sennett 1986: 150–152).

Even social relationships were redefined, in particular behavior in public. Since external details were taken to reveal inner character, people in public sought to shield themselves from the scrutiny of others. In theaters, the public increasingly policed itself into silence. On the street, one attempted to be unremarkable and to control one's feelings. In contrast, public figures were expected to display the character and personality which the spectators themselves repressed. It was in these conditions that the cult of the personality could come into existence: Politicians displayed their character more than their political beliefs and actors' opinions were echoed throughout the press.

The most cursory survey of today's media coverage of public figures shows the endurance of this trope into the twenty-first century; and that it is by no means restricted to politicians and actors, but also includes scientists. Focusing on the case of the Nadar/Chevreul interview, this paper examines the contexts of the elaboration of a new type of scientific persona in the printed media and the resources Nadar mobilized to this end; and how different material, visual, and narrative techniques, in particular scientific techniques, were brought together into a "science of intimacy" whose first subject was Chevreul. My analysis relies primarily on the published version of the interview that appeared in the *Journal Illustré*, but I have also made use of unpublished materials pertaining to the interview.

 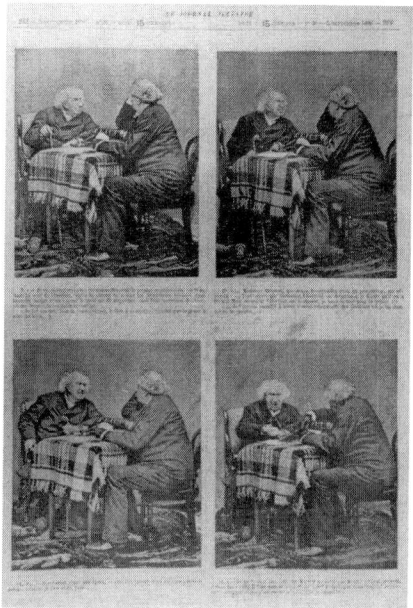

Figures 7.4, 7.5 The second and last pages of the Nadar-Chevreul interview. The whole issue consists of alternating pages of text, as in Figure 7.3 (transcript of the conversation), and pages of photographs with captions as above (*Journal Illustré* 23:36 (5 September 1886): 284, 288) (Bibliothèque du Pays Châtelleraudais).

The *Journal Illustré*'s was actually a much-reduced version of the original project, which featured a great many more photographs taken during the *entretien* and included the planned publication of a book by Nadar, *L'Art de Vivre Cent Ans*.[2]

III

The *Journal Illustré*'s coverage of Chevreul's centenary contrast with contemporary visual representations in several crucial respects. To begin with, Chevreul is portrayed in a very plain setting, unlike the glorifying pedestals of the official celebrations. The photographs show him sitting opposite a slouching Nadar in front of a hastily drawn screen. The table is covered with a woolen rug and a crumpled carpet shows between its unadorned legs. While in several of the unpublished photographs, Chevreul is dressed up and posing for a formal portrait, he seems during the interview to have made himself comfortable, shedding his top hat, tie, and even his shoes, swapping them for slippers, as shown in Figure 7.5. The recorded conversation between the two men deals with such mundane topics as Chevreul's eating and drinking habits and the physiological causes of his longevity; anecdotes about Nicephore Niepce are told, and the respective merits of balloons and airplanes are weighed. In every respect, the interview seems to privilege the trivial, the quotidian about Chevreul, and while it is acknowledged that he is a great scientist, little mention is made of his scientific achievements.

Perhaps the most remarkable aspect of this interview is the way in which it is introduced and set up by the "editors" of the *Journal Illustré* (to all evidence Nadar himself). This, it was claimed, was a new, revolutionary type of journalism, only made possible by technological innovation. Nadar, famous photographer and balloonist, socialite, and caricaturist, a passionate advocate of the *Plus Lourd que l'Air* had, the "editors" contended, now invented a method which "will mark the annals of the human mind."[3] Acknowledging the recent fashion for what was becoming known as "interwiews," a form of journalistic writing borrowed from the United States, the "editors" pointed out that this kind of reporting could easily misrepresent the thought and words of the person interviewed:

> When one of our peers in journalism [...] makes an "interwiew," that is, goes to ask questions to this or that celebrity of the day, he can only offer his easily mistaken ear and his treacherous memory to the crowd avid for his stories. Yet the sole purpose of such a piece of information resides in its absolute mathematical precision [...]

> Only science could provide a solution to this problem of inaccuracy [...] and give the reader the "proof" of the veracity of the reproduced conversation. (*Le Journal Illustré* 1886: 282)

To guarantee the accurate rendering of Chevreul's utterings, Nadar had resorted to photography which had arguably so far improved since Daguerre that it had become truly instantaneous. Further, Nadar had proposed to record the conversation by means of the phonograph, which he claimed to have had first imagined in the 1850s, describing it as a "daguerreotype of sound." Unfortunately, the new device did not yet allow the recording of spontaneous speech, so it was replaced by what was referred to as "the official, if not automatic stenographer" (Nadar 1886: "Avis des éditeurs"). Co-ordination between the recording of sound and image was effected by the calling of numbers, making it possible to match photographs closely with the words spoken at the time of exposure, as the captions on Figure 7.5 demonstrate.

In this way, "Nature caught by surprise" could be displayed. "For the first time the reader becomes the spectator [...]. For the first time, he does not need anyone to listen and to see for him: it is he, he himself, who sees, who listens." From wherever he was, the "editors" contended,

> From the back of his flat, sitting at the fire, in his sick-bed, from the lost extremity of the most distant solitudes, onboard a ship beyond the sea

horizon, he is present always and everywhere, always where he wishes to be, penetrating to places no one has ever been to, in the most secret, most remote intimacy of the greatest as well as of the lowest men. [...] He sees, he *hears* the first and the last actors of all the dramas and comedies of our daily life. (Nadar 1886)

The reader was thereby informed, entertained, and educated, the explicit aims of the whole enterprise. The first application of this new method was to Chevreul, from whose lips the reader would gather the secret of his longevity and his exact opinions about science and method, thus making these pages a "genuine encyclopedic breviary of the nineteenth century" (Nadar 1886).

Mechanical and (quasi-)automatic recording, it was therefore argued, guaranteed authenticity in capturing and rendering the visual and auditory components of the conversation, such that the reader became a virtual witness of the exchange between Nadar and Chevreul. Instead of admiring him from a distance, as conventional reporting and public celebrations required, the reader was now in a position to experience the great scientist for him- or herself.

IV

This document has often been referred to as the first photographic interview. It evidently belongs to this period of widespread invention and experimentation with sound and image recording devices which led to cinematographic cameras and phonographs in the 1890s, from Thomas Edison's kinetoscope and phonograph, Edweard Muybridge's zoopraxiscope, to Jules Janssen's photographic revolver and Etienne-Jules Marey's chronophotography. Nadar was well acquainted with these developments, having attended, on Marey's invitation, the first European demonstration of Muybridge's technique in 1881. In fact, he had been friends with Marey since the 1860s, sharing with him a fascination for new technologies, including airplanes, photography, and telegraphy (Braun 1992: 35, 37, 52).

Nadar's set-up indeed bears more than a passing resemblance to some of the proto-cinematographic devices, especially those combining sound and image. The phonoscope, for instance, developed a few years later by Marey's assistant Georges Demenÿ, was an apparatus for visualizing speech (originally devised for teaching speech to the deaf) through the projection in rapid succession of a series of photographs of a person articulating a sentence—one of the Marey laboratory's earliest attempts at synthesizing movement. Demenÿ described his invention in the popular science journal *La Nature* as "speaking photographs" that "preserved the physiognomy [of speech] just like the phonograph preserved voice. By adjoining the latter to the *phonoscope* the illusion will be complete" (see Figure 7.6) (Demenÿ 1892: 515).

Figure 7.6 Georges Demenÿ's articulating 'vive la France!' (detail). The original caption reads 'Spécimen des photographies parlantes. Photographie des mots *vive la France!* (Demenÿ 1892).

Nadar drew an explicit analogy between the interview set-up and Marey's chronophotographic method: just as "doctor Marey was able to capture and preserve all the successive moments of the flight of birds," Paul Nadar had captured "the attitudes and successive expressions of the venerable centenarian whose face underwent transformations following each evolution of the dialogue, while each play of his physiognomy was recorded by the operator" (Nadar 1886: "Avis des éditeurs"). Nadar had recorded the successive moments of the motion of Chevreul's *thoughts*, as one contemporary commentator put it.[4] Although the twelve photographs that appeared in the *Journal Illustré* only give an imperfect impression of the intended effect, the technical set-up and the resulting images imitate the chronophotographic method. And indeed, the initial project had been to publish over one hundred pictures, a task which proved technically (and most likely financially) impossible (see Figures 7.7, 7.8, 7.9).[5]

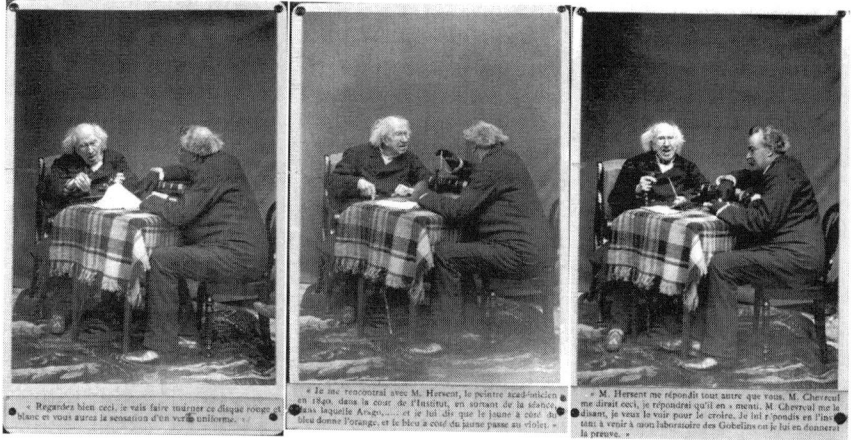

Figures 7.7, 7.8, 7.9 Examples of photographs that were not used in the *Journal Illustré* issue. Chevreul is shown explaining his theory of the contrast of colors. The text is reproduced in the *Journal Illustré*, see Figure 7.3 (Bibliothèque Nationale de France).

The Nadar/Chevreul interview, then, was put forward as a transposition of chronophotography to the realm of personality. Mechanical recording techniques were extended beyond the physiological characteristics of animal and human motion to encompass the psychology, the thoughts, and character of the subject under investigation. If scientists had often enough involved their own bodies in chronophotographic experiments as the phonoscope example shows, Nadar went a step further and subjected a scientist's *mind* to Marey's technique.

More generally, Nadar drew heavily on scientific rhetoric and emphasized scientific method. He contended that the interview embodied "the conquest of the document by science" (*Le Journal Illustré* 1886: 282). For instance, he repeatedly quoted Chevreul on a central component of the scientific method, observation: "scientific observation solves all material and moral problems," "in all matters, one must prove, one must show,"[6] "I want to show, because it is when I see that I believe"![7] Chevreul had summarized his *Methode à posteriori expérimentale* as follows: "1. observation of a phenomenon; 2. reasoning, whose aim is to discover the immediate cause of the phenomenon; 3. experiment to control the outcome of the reasoning" (Chevreul 1866). Nadar made the principles guiding the interview accord with those guiding Chevreul's work, considering Marey as the finest proponent of Chevreul's *méthode*:

Marey was destined to become the future head of the *a posteriori* school, the foundation of the science of the old Chevreul. Like the stubborn centenarian, when Marey has seen, he will want to see again. He will not rely on his eye, his hand, and his ear. Legitimately cautious

against the illusions of sight and hearing, he only trusts in the automatic, the incontrovertible sincerity of the tool which he has trusted to
see, touch and hear for him. (Nadar 1899: 301).

Observation required a careful disciplining of the scientist in order to
ensure reliable results, and scientists worked to become as transparent as
possible, notably by adopting mechanical recording technologies such as
photography. Although the text of the interview places much emphasis
on recording technology, it could in this way be presented as the means
of providing an unmediated, true to nature, transparent reproduction of
the conversation. Deploying the techniques of automatic recording and the
rhetoric of mechanical objectivity, Nadar claimed to give a "mathematically exact," "authentic," and truthful portrayal of Chevreul unmediated
by journalistic subjectivity.

In a potentially ironic reversal, Nadar thus applied in the interview this
experimental method to the chemist himself. Nadar took on the pose of the
self-effacing scientist observing and recording the external characteristics
of his subject of investigation, while Chevreul was turned into the object of
investigation—though Nadar's presence in the text and in the photographs
made him simultaneously an object of his own observation.

Clearly, though, this was not quite science. Instantaneous photography
made visible the otherwise imperceptible succession of bodily motions.[8]
But what was the added value of Nadar's set-up for studying a conversation
or Chevreul himself? The nature of the journal, Nadar's own background,
the emphatic prose suggested that if this document was an experiment, it
was one of a particular, media-oriented kind—not that such hybrids were
particularly unusual. A common property of the proto-cinematographic
devices was their easy transit between scientific laboratories, research institutes, exhibitions and fun fairs. Many of the optical devices for researching
movement rapidly made their way to fun-fairs, where they enjoyed a brief
life as entertaining novelties. Muybridge had first become involved with
successive photography to investigate the sequence of motion of the legs of
running horses, but he later tried to convince Edison to commercialize his
instrument for entertainment purposes. Nadar insisted, however, that his
set-up had lost none of the original scientificity of Marey's method.

V

Of course the alleged transparency of Nadar's set-up was carefully constructed. Not only was the electric phonograph replaced by a perhaps automatic but first and foremost human stenographer, the text of the interview
itself was heavily edited, as shown by the variations between the different
versions that survive. Indeed, we know that the interview took place over
several days. The choice of the photographs and slices of conversation to be

reproduced added of course another element of subjective interpretation. The *Journal Illustré* featured for instance a long section on one of Nadar's obsessions, hot-air balloons, while Chevreul's account of his theory of colors was largely omitted.

So while Nadar explicitly disrupted conventional representations and rules of reading and replaced them with "Nature itself," he was in fact engaged in the elaboration of new conventions for representing spontaneity and authenticity. This is the essence of the realist project in art and literature in the nineteenth century, and in particular of the Naturalistic movement in French literature, a movement which itself claimed to be inspired by scientific method. The most prominent of the naturalists, Emile Zola thus wrote in *Le Roman Expérimental*, first published in 1880:

> If it has been possible to transfer the experimental method from chemistry and physics to physiology and medicine, as Claude Bernard had, it can be transferred from physiology to the naturalist novel. From this day, science thus enters this realm of ours, us the novelists, analysts of man in his social and individual action. (Zola 1890: 6)

In mobilizing scientific techniques and rhetoric so explicitly in portraying Chevreul, Nadar was in line with the naturalistic discourse. Observation, attention to detail, authenticity in the portrayal of the social and material realities of the times were the main features of naturalism, as displayed throughout the volumes of Zola's *Rougon-Macquart* series and which we find expressed in similar terms in the interview. In 1885 naturalism also made its first appearance on stage, with the setting-up of André Antoine's *Theatre Libre*, an experimental theatre supported by Zola himself, and which started out by staging some of his plays. The Théâtre Libre caused a scandal throughout the late 1880s in Paris for Antoine's hyperrealist staging. In particular, Antoine asked his actors to speak normally, rather than declaiming as was the norm; he replaced footlights with natural lighting, and encouraged actors to turn their back to the stage when the plot required it, just like Nadar on the interview photographs. The analogy between the interview and theatrical play was furthered by the introduction of sub-text: Chevreul's "Permettez," "Alors," "Voyons," are reproduced in the text; a comment by Nadar at one point is qualified by the word "whispering" in brackets. Such interjections, just like Chevreul's slipper are prime examples of "superfluous details" to the narration that, according to Roland Barthes, through their very redundancy create an *"effet de réel"* and are accordingly a hallmark of realist literature (Barthes 1969/1984: 179–187).

For all its claims of novelty, the Nadar/Chevreul interview in fact imitated the naturalistic project to apply scientific method to the representation of individuals and societies, and it borrowed its conventions to convey the impression of unmediated reality. Nadar appears in this perspective as the director of a naturalistic drama with Chevreul as its main character.

VI

In order to understand not only in what way Nadar mobilized scientific techniques and rhetoric in constructing this representation of the scientist, but also the rationale for elaborating this kind of representation, we need to take into account the immediate purposes the interview was made to serve and the wider context in which it was elaborated.

One answer is that the interview also served the prosaic purpose of advertisement. Already Chevreul's birthday had given rise to several initiatives: newspaper advertisements claimed that the lungs of the illustrious centenarian had been preserved thanks to the *Gerandel* throat drops; a Bordeaux pharmacist had likewise suggested that Chevreul would become immortal if he regularly took his special concoction.[9] The *Journal Illustré* interview was in part an advertisement for Eastman products. Not only was Eastman mentioned several times in the course of the conversation; his products also figured as an advertisement at the bottom of one of the pages: "The images in this issue were obtained with Eastman film which produces an exposure in one two-thousandth of a second" (*Le Journal Illustré* 1886: 287).

Since the move from wet to dry photographic plates in the late 1870s, a whole business of processed, ready-made dry plates had grown, especially in the United States, with George Eastman becoming one of the major manufacturers. In late 1885, Eastman had begun commercializing stripping film as a substitute for glass plates. Simultaneously he started investing in the European market, setting up a central bureau for European sales in London and franchising other establishments throughout the continent (Jenkins 1975: 66–121). In France, the *Office Général de Photographie* on 53 rue des Mathurins became Eastman's privileged partner, as the advertisement in the *Journal Illustré* indicates.[10] This was Paul Nadar's atelier, that a couple of years later also helped commercialize Eastman's new camera aimed at a mass market, the Kodak, as Figure 7.10 suggests.

The interview is then what we would nowadays call an infomercial. The emphasis on spontaneity, on absolute precision in the recording, and truth in the rendering were in part attributed to the new Eastman film, which, it was claimed, enabled photographs to become truly "instantaneous" for the first time. In practice of course the difference between a photograph made in a hundredth of a second and one made in one two-thousandth of a second, while certainly crucial for projects such as Marey's, was insignificant for the recording of a conversation. Further, the introduction's deliberate emphasis on this new interviewing method's ability to reach each reader (readers became spectators, they each attended the event, each of them in the first row), corresponded to Eastman's project to transform photography into an amateur pursuit, to simplify its practice by making training superfluous, and thereby to conquer a large market. The next logical step was that each of the readers might have photographed the conversation them-

Figure 7.10 Kodak-Nadar advertisement, 1888 (*Revue Bleue,* 15 December 1888).

selves as the Kodak advert suggested two years later: as Figure 7.8 implies: "photography by all and for all." The exclusivity and uniqueness of the interview was made possible, it was argued, by the new Eastman film; but it was implied by the Eastman advertisement on the same page that each purchaser of photographic film would be able to reproduce such a feat. Selling the exclusive to a large public was already then a fine sales argument.

The rise of a large market for new technologies was not confined to photography: the press was another case in point. After the freedom of the press law was passed in 1881, the press grew exponentially in France: more than a million papers were sold daily in Paris alone in the late 1880s. More than anything else, this contributed to turning journalism into a commercial venture, which implied the industrialization of newspaper production as well as a profound change in the content, style, and appearance of newspapers. Photolithographical advances, e.g., the replacement of photoengraving by the half-tone process for instance made possible the inclusion for the first time of a large number of illustrations. This development, which in Britain was dubbed "new journalism" privileged reporting and news, that became the "heart of the newspaper" as *Bel Ami* stated in Maupassant's eponymous novel published in 1885. News, novelties, sensationalism, and commercialism were the hallmarks of the new journalism, in which Nadar, stuntman and acknowledged self-publicist, was at home (Wiener 1988: esp. 1–90).

When Nadar promised that the reader, from wherever he was, could from now on be himself in the presence of Chevreul, he was supplying an exaggerated vision of the transformed perceptions of space and time which telegraphy, photography, and railways had initiated, and which his predictions about airplanes and phonographs furthered (no wonder Jules Verne had taken Nadar as his model for Michel Ardan, the brash hero of his early science-fiction novel *De la terre à la Lune*). Instantaneity captured mechanically or electrically and reproduced at a great distance seemed to abolish time and space, and blurred the boundary between presence and absence, in contrasting it to the biological time represented by Chevreul's great age.

This made for a dramatic account suited to the emerging new journalism, intended to create a sense of wonder akin to the visit of the great exhibitions so prized by late nineteenth-century Europeans, with its typical display of futuristic, amazing technology. Just like exhibitions, the interview testified to a continuum from science to technology to entertainment, a system which relied on the creation of a large market for innovative and entertaining technology, a growing urban public with a modicum of spare income to spend on entertainment and consumer goods, whether newspapers or photographic cameras.

Adopting Walter Benjamin's pessimistic take on this development, in this economy, the scientist appears to be reduced to a mere *actualité* destined to disappear from the next issue, leaving space for the next amazing novelty. Nadar took on here the role of the all powerful reporter delivering stories (and products) to an avid public. Science, via the techniques and rhetoric of precision and objectivity and the scientist Chevreul, served at best as a pretext, at worst as legitimating agents for a new journalism that was in fact driven by commercial interest. The insistence on authenticity and immediacy only sought to dissimulate the fact that this system destroyed its very authenticity through endless mechanical reproduction (Benjamin 1936/1996: esp. section II).

VII

What position did the reader occupy in this configuration? He or she was drawn into this techno-commercial system, and not just as a customer. If

Nadar's set-up and the newspaper were presented as transparent media, it was up to the reader to observe Chevreul's behavior and interpret the evidence supplied. In Benjamin's words, every spectator became an expert of sorts. This in principle empowered the reader, but only on the condition, of course, that he or she possesses the requisite skill to make sense of "Nature caught by surprise" (Nadar 1886). For this a science of a less mechanical and of a more intuitive kind was required for reading Chevreul's personality from the evidence supplied; a hermetic science of correspondence between appearance and essence, physiognomy. This is the final component of Nadar's science of intimacy I want to consider.

Félix Nadar first became involved with photography in the 1850s when helping his brother Adrien make a series of photographs for the physiologist Guillaume-Benjamin Duchenne de Boulogne. Based in the Salpêtrière Hospital in Paris, Duchenne had set out in the 1840s to "discover the laws which guide the expression of human physiology," as he put it, "provoking, by means of electrical current, the contraction of the muscles of the face, to make them speak the language of passions and of feelings, and starting from the expressive muscle to reach the soul which sets it into motion." In 1854, Duchenne asked Adrien Nadar to assist in the production of a photographic taxonomy of facial expressions for publication in his volume *Mécanisme de la Physionomie humaine, ou analyse électro-physiologique de l'expression des passions applicable à la pratique des arts plastiques* (Duchenne de Boulogne 1862: quotations on pages xi–xii).

The following year Adrien and Félix Nadar realized the *Têtes d'expression de Pierrot*, a series of photographs of the famous mime Debureau acting out attitudes such as "laziness," "surprise," or "attention," echoing Duchenne's project on a theatrical mode (see Figures 7.11 and 7.12). Nadar was well acquainted with Duchenne, sharing with him a profound admiration for the work of Rodolphe Töpffer, the inventor of the comic strip (Jammes 1978: 215–220; for a detailed comparison of Töpffer and Duchenne's physiognomical methods see Dupouy 2005: 24–60).

For the *Têtes d'expression*, the Nadar brothers were awarded the gold medal of the 1855 Paris Exhibition but they soon fell out over the nature and extent of their respective contributions to Duchenne's project, and over the use of the name Nadar. In the ensuing legal battle, Félix Nadar asserted his views on photography:

> What cannot be learnt is the moral intelligence of your subject—it is this rapid tact that puts you in communication with the model, and leads you to assess and guide him towards his own habits and ideas and following his character, and enables you to give, not trivially and by chance, an indifferent plastic reproduction within the reach of any laboratory assistant, but the most familiar and the most favorable likeness, the intimate likeness [*la ressemblance intime*]. This is the psychological side of photography, the term does not seem over-ambitious to me.[11]

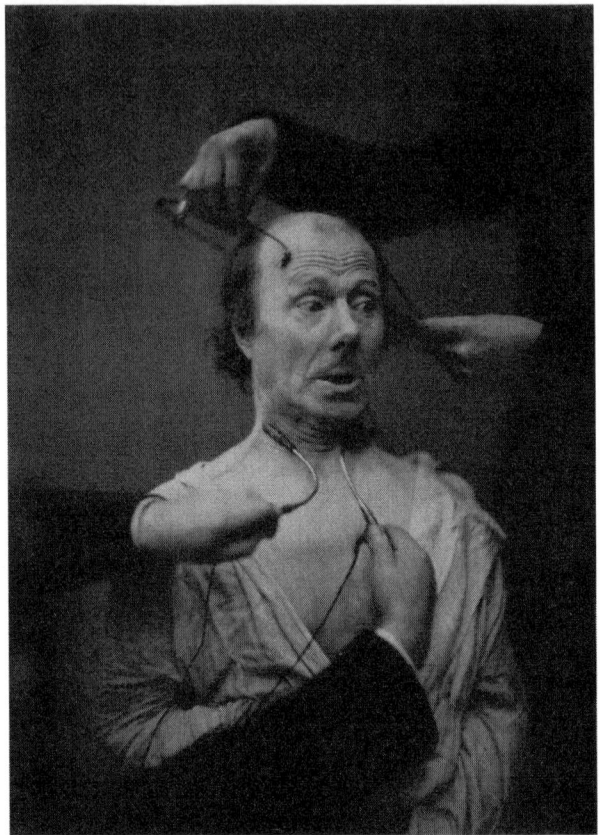

Figure 7.11 Guillaume-Benjamin Duchenne de Boulogne; with Adrien Tourna-chon, "Frayeur" (fear), illustration for *Mécanisme de la Physionomie Humaine*, 1854 (École Nationale Supérieure des Beaux-Arts, Paris).

This intimate likeness Nadar appeals to is fundamentally physiognomic. According to the founding father of the science of physiognomy, Johann Casper Lavater, the role of the portraitist is to capture the features of the person and reveal its inner truth rather than to represent or enhance their beauty (Lavater 1775–1778/1969). Lavater advocated realism in representation, but a realism that revealed a deeper truth. This notion was central to Balzac, for whom the insignificant details, the missing button, the untied laces, revealed social, economic, and personal characteristics of the persons he described in his novel, notably in the opening scene of *Le Père Goriot*, in which each aspect of its inhabitant's dwellings is minutely described, providing so many clues as to his psychological state.

The physiognomic taxonomies which run from Lavater to Duchenne and Nadar's Pierrot series provide all the clues necessary for such an interpretation to those wishing to really know who people were. The interview

Figure 7.12 Adrien Tournachon, Gaspard Félix Tournachon (Félix Nadar), Le mime Debureau: Pierrot plaidant (Pierrot pleading) (1854–1855) (Bibliothèque Nationale de France).

could similarly function as an exemplary case for the readers to practice with, courtesy of Nadar.

VIII

In this final *coup d'eclat* before his retirement, Nadar returned to his early experiments with physiognomic photography, applying his consummate skill to capture exactly these details important to interpretation, offering to the cued reader all the material necessary for deciphering Chevreul's inner self through his appearance and utterances. Chronophotography, naturalism, and physiognomy were combined in Nadar's science of intimacy, an attempt to recreate *presence* and to capture Chevreul's personality in the context of the commercialized new journalism. Throughout the interview, Nadar's insistence on authenticity implied that the scientist's private self was more real than the public figure on which the official celebrations focused; and that personal acquaintance was a better source of knowledge than official representations could ever give.

The Nadar/Chevreul interview exemplifies the emergence of the public figure of the *personality* discussed by Sennett (1976/1986). As one of Nadar's correspondents pointed out, Nadar had "perfected the very contemporary art of undressing people in public."[12] Chevreul's science was not so much the focus of the article, but rather his personality: his (and Nadar's) ideas about current affairs primed over his scientific achievements. But science was nonetheless present throughout, put forward as the means of creating this immediacy and intimacy, of making this type of representation possible. The readers were thereby shown what it was like to *be* an eminent scientist.

While the public celebrations of Chevreul imposed an interpretation of his work to the public and gave a standard picture of the personality of the great scientist, the readers of the *Journal Illustré* were given the impression that they could make their own opinion of the scientist. By removing the pedestal on which the republic sat Chevreul, Nadar claimed to bring the public figure closer to the readers. But the very realism and vividness of the scientist's portrayal were the result of careful staging. However intimate Chevreul might appear, he was on a stage, displaying a personality mediated by material and discursive technologies. Nonetheless, the reader was given the impression that the distance between the spectator and the scientist was no longer created by ritual, as in the official demonstrations, nor by the dissimulated technologies but it was implied by an intrinsic difference in character. In this respect, Nadar's article is a forerunner of the modern celebrity interview: a star (or Nobel Prize winner) will pose for photographs in his or her own home, and answer questions totally unrelated to his or her work. It is the very fact that the public figure has a personality which makes them interesting and worth listening to. In a sense, when Nadar advised the readers to learn from the old man's wisdom, he was telling them that this was the source of his greatness, and what distinguished him from them. Chevreul's scientific achievements were to be taken as the expression of such a remarkable personality.

The public's role differed markedly in both types of representations: in the official ceremonies the spectator was kept at a distance from the stages, while in the interview the reader "is present always and everywhere, always where he wishes to be, penetrating to places no one has even been to, in the most secret, most remote intimacy of the greatest as well as the lowest men" (Nadar 1886: "Avis des éditeurs"). On the other hand in the latter, the public was ubiquitous but physically absent, while in public demonstrations the onlooker could at least have a collective sense of being. Actual presence at a public event implied that the spectator could to some extent interact with it, while the anonymous reader of a paper was made wholly passive.

Opposed, but also complementary to the official heavy machinery of glorification of Chevreul, Nadar's portrayal provided in effect a novel means of sanctification of the scientist. Nadar assembled and concentrated in the interview all the elements instrumental to the rise of the *personnalité* at the very moment when the traditional symbolism and separation between public and private beings was being reaffirmed in the commemoration: photography, an impetus to naturalism in art and literature, experimental science with its concern for immediacy and truthfulness, and journalism, the place where the new styles of representation of public figures were fostered and commercialized.

ACKNOWLEDGMENTS

For their valuable comments and suggestions I am indebted to Simon Schaffer, Annik Pietsch, Bernd Hüppauf, Cornelius Borck, Peter Geimer, Frédéric Graber, as well as the audience at the University of Cambridge's History and Philosophy of Science Departmental Seminar.

NOTES

1. See, e.g., Célébration du centenaire de M. Chevreul. 31 août 1786–31 août 1886, Rouen: Impr. de J. Lecerf, 1886; de Font-Reaulx 1888; Hommage à M. Chevreul, à l'occasion de son centenaire, 31 août 1886 par MM. M. Berthelot, E. Demarçay, Dujardin-Beaumetz, E. Gautier, Ed. Grimaux, G. Pouchet, Ch. Richet et F. Alcan, Paris: F. Alcan, 1886; Guiard 1886.
2. Nadar, "L'Art de Vivre Cent Ans" is an undated, unpaginated, hand-written manuscript, kept at the Bibliothèque Nationale de France: mss naf 13828 [copie 258 ff] (in the following cited as "Nadar 1886"). On the basis of this manuscript, another version of the interview has been reconstituted and published in Reynes (1981). In addition, over 80 photographs taken of and around the interview exist, e.g., at the BnF and at the Médiathèque de l'Architecture et du Patrimoine in Paris (those kept in the latter can be viewed online: http://www.mediatheque-patrimoine.culture.gouv.fr)
3. Nadar 1886: "Avis des éditeurs." All translations are mine unless otherwise indicated. On Nadar's life and work see Gosling (1976).
4. "Lettre critique de M. Gustave Grignan," quoted in Nadar (1886).
5. According to G. Reynes, Nadar initially proposed the interview to the Figaro, but was turned down on account of the technical difficulty of printing so many photographs. A partial version of the interview featuring only 12 photographs subsequently appeared in the Journal Illustré (Reynes 1886: 156).
6. "En toutes choses, il faut prouver, faire voir," cited in Nadar (1886).
7. "Je veux faire voir, parce que c'est quand je vois que je crois," cited in Reynes (1981: 169).
8. On instantaneous photography in science and art in the period see Geimer (2004).

9. As reported in "Les réclames Chevreul," *Le Courrier Français* 1886: 14.
10. "les pellicules EASTMAN [...] se trouvent au seul dépôt pour la France, OFFICE GÉNÉRAL DE PHOTOGRAPHIE, 53, rue des Mathurins, Paris," *Le Journal Illustré* 1886: 287.
11. Exposé des motifs pour la revendication de la propriété exclusive du pseudo-nyme Nadar. M. F. Tournachon-Nadar contre MM. A. Tournachon Jeune et Compagnie. Mémoire addressé au tribunal de commerce de la Seine siégeant le 23 avril 1856, Bibliothèque Nationale de France archives: 4-FM-31413.
12. "Lettre critique de M. Gustave Grignan, 25 novembre 1886," cited in Nadar (1886).

BIBLIOGRAPHY

Barthes, Roland (1968/1984) "L'effet de réel", in *Le Bruissement de la Langue, Essais Critiques IV*, Paris: Seuil.
Benjamin, Walter (1936/1996) *Das Kunstwerk im Zeitalter seiner technischer Reproduzierbarkeit*, Frankfurt am Main: Suhrkamp.
Braun, Martha (1992) *Picturing Time, The Work of Etienne-Jules Marey (1830–1904)*, Chicago: University of Chicago Press.
Chevreul, Michel E. (1866) *Introduction à l'Étude des Connaissances Chimiques*, Paris: L. Guérin.
Demenÿ, Georges (1892) "Les photographies parlantes", *La Nature*, n. 984 (9 April 1892): 51.
Duchenne de Boulogne, Guillaume-Benjamin (1862) *Mécanisme de la Physiono-mie humaine, ou analyse électro-physiologique de l'expression des passions applicable à la pratique des arts plastiques*, Paris: J. Renouard.
Dupouy, Stéphanie (2005) "Künstliche Gesichter. Rodolphe Töpffer und Duchenne de Boulogne", in A. Mayer and A. Métraux (eds), *Kunstmaschinen. Spiel-räume des Sehens zwischen Wissenschaft und Ästhetik*, Frankfurt am Main: Fischer Taschenbuch Verlag.
de Font-Réaulx, Hyacinthe (1888) *Les Hommes utiles*, Lille: J. Lefort.
Geimer, Peter (2004) "Picturing the black box: On blanks in nineteenth century paintings and photographs", *Science in Context*, 17(4): 467–501.
Geison, Gerald L. (1995) *The Private Science of Louis Pasteur*, Princeton, NJ: Princeton University Press.
Gosling, Nigel (1976) *Nadar*, London: Secker and Warbung.
Guiard, Émile (1886) *A Chevreul, stances dites...au théâtre national de l'Odéon, à l'occasion du centenaire de M. Chevreul, le 31 août 1886*, Paris: P. Ollendorff.
Jammes, André (1978) "Duchenne de Boulogne, La Grimace provoquée et Nadar", *Gazette des Beaux-Arts*, 6: 17.
Jenkins, Reese V. (1975) *Images and Enterprise. Technology and the American Photographic Industry 1839–1925*, Baltimore: Johns Hopkins University Press.
Lavater, Johann Caspar (1775–1778/1969) *Physiognomische Fragmente zur Beför-derung der Menschenkenntnis und Menschenliebe*, Leipzig: ed. Leipzig.
Le Courrier Français (31 August 1886), 3.
Le Journal Illustré (5 September 1886), 23.
Nadar, Félix (1899), *Quand j'etais photographe*, Paris: E. Flammarion.
Nelms, Brenda (1987) *The Third Republic and the Centennial of 1789*, New York: Garland.
Nochlin, Linda (1972/1990) *Realism*, London: Penguin.

Reynes, G. (1981) "Chevreul interviewé par Nadar, premier document audiovisuel (1886)", *Gazette des Beaux Arts*, Ser. 6, 98: 155–184.

Sennett, Richard (1976/1986) *Fall of public man*, London: Faber.

Wiener, Joel H. (ed.) (1988) *Papers for the Millions. The New Journalism in Britain, 1850s to 1914*, New York: Greenwood Press.

Zola, Émile (1880/1890) *Le Roman Experimental*, Paris: Charpentier.

8 Visual Arguments

The Role of Images in Sciences and Mathematics[1]

Dieter Mersch

THE STATUS OF IMAGES

Scientific and mathematical knowledge is considered well-grounded knowledge. It complies with the structures of both discursive logic and calculability, but also with the structures linking reason and consequence as well as the laws of induction and causality. It is based on determining facts and making declarative statements, whose general form, the predicate, represents the identification of "A is P." The "truth" of these statements is in turn derived from a series of additional sentences, with the same form, even when there is no last sentence which closes the chain by formulating a final reason. No knowledge is complete; it stumbles not only on the provisional nature of the paradigm, and on the sketchiness of the empiricism, but also because it remains embedded in an indefinite ring of ignorance, which can provide *per definition* neither a theory, nor meaning.

Nevertheless, it should not be forgotten how much knowledge equally depends on its manner of presentation and its specific mediality. Whereas language and in particular the predicate form of the sentence are dominant in the conventional understanding of the scientific episteme, the significance of the iconic as an instrument and method of knowledge appears to be only secondary. However, with the flood of visualization strategies, with which scientists in the last decades have begun to make the invisible visible, their outstanding role and relevance can no longer be denied. The production of scientific knowledge has always appeared to be dependent on a multitude of visual techniques and methods such as graphs, models, diagram, illustration, and figures (Holländer 2000: esp. 11, column 2f.). These have rarely been considered with regards to their own mediality or their particular structure and form of representation though. They have instead been ignored within the philosophy of science.[2] How images are produced is not actually considered epistemically relevant. Rather, it tends to be thought of as a mere supplement, as an adjunct to verified knowledge that has been gained elsewhere (see Mersch 2005a) and therefore only emphasizes it or provides a background for it. Images though, are a genuine method of creating knowledge and therefore should be taken seriously as a part of the epistemic process because visualizations, like "visual arguments" in general, claim their remarkable significance in the discursive and perhaps even gain their fundamental parity in the process of generating knowledge. Ludwig Wittgenstein, who like hardly any other philosopher based his reflections on sketches, models, and figures, suggested in his unpublished notes: "Is there a preferable, a particularly direct way of making a picture? I don't think so! All methods of illustrating are equal."[3] This comment from 1929, which stems from his notes on the criticism of picture theory in *Tractatus*, de-hierarchizes picture and sentence, because the word "picture" in this context characterizes every type of medial representation. It gives the iconic the same rank as language.

Wittgenstein, who initiated Rudolf Carnap's logical positivism, which gave preference to protocol statements alone, dedicated his whole life to the description of language. However, according to him, there is no reason to exclude imaging techniques as a form of visual thinking or to attach a lower authority to them in science, because they play a fundamental role in the creation of knowledge and are at times essential for its generation. Both visuality and discursivity share the scientific field with techniques such as scanning, acoustics, numerics, or statistics. Whereas *discursive procedures* work on the creation and verification of *factual claims*, the production of evidence falls to the visual processes. Both therefore, follow different methods of validating. However, when we relate them to one another, this allocation of task makes clear how tightly the evidence effects of visualizations and the truth effects of discursive practices are intertwined in order to first make knowledge valid as knowledge.

Nevertheless, the visual strategies are themselves very disparate. Even when difficult to differentiate from one another, they can be heuristically divided into two basic classes. Hans-Jörg Rheinberger's (2002: 58) so-called "epistemic objects," slide preparations, which are of a completely different nature, could also be added as a third class: The first class includes those presentation methods that essentially function as *testimony* and that use the visual as proof. The second contains those that *arrange* the knowledge in abstract tables or with regards to an underlying volume of data, transforms it into *calculable figures*. The first proceed referentially; they prove an existence or highlight a "trace," whereas the latter argue with diagrams or graphs whose format, a scripture or "marking," takes on a constructive status. We are therefore dealing with at least two disparate types of visualizing, each of which fulfills different purposes and obeys other "logics." On the one hand, we can speak of methods of representation in the broadest sense, and on the other hand of "constructive" or "model-like" visualizations, which at the same time call upon "writing" and "numbers" as basic cultural techniques (Mersch 2003; Krämer and Bredekamp 2004) and thus, function as a hybrid between notationality and iconicity.

VISUALIZATIONS OF KNOWLEDGE

Nonetheless, up until the nineteenth century, images seemed to be matter of art. This applied not only to paintings, but also to copperplate engravings, or the production of all kinds of plans, maps, and diagrams. Here, the creative process was ruled by imagination. Its freedom was to be ensured by the artist who had the authority and competence for the visible—both in reference to techniques of illusions as well with respect to the scientific representation of knowledge acquired through the likes of telescopy, microscopy, or other tools. Hence, science remained dependent on the dialogue with the arts. However, Lorraine Daston and Peter Galison have shown that this relationship began to erode during the nineteenth century due to shifting standards of objectivity (Daston and Galison 2002). The transformation effected not only science, but also the images which from then on had to meet the criterion of a "mechanical record": "Non-intervention—and not similarity—was the heart of mechanical objectivity" (Daston and Galison 2002: 31, 54f., 94). Accordingly, the apparatus moved to the foreground and dethroned the artists' hegemonic knowledge. The artists in turn automatically ceded the precision of their eye to optical and other automatic instruments for recording in order to henceforth leave the perception of measurement to machines. That means that the image tends towards a graphic inscription, to *writing,* which is meant to record and repeat the real as authentically as possible, even retaining errors as a mark of its genuineness.[4] In comparison to the fallible human subjectivity, which was inclined towards being deceived, seduced, and to seeing what it wants to see, the apparatuses retained an aura of strict neutrality and incorruptibility. They were not subject to sudden mood swings, irrational changes of mind, or unconscious prejudice; they also did not need to take breaks or fall asleep. Instead they provided their faithful reproductions absolutely continuously and without complications. The ideal of the images they produced was therefore de-aestheticizing, and conversely left the creation of evidence to automation alone. A new dependency was thus produced, namely on technicians and engineers, who particularly in the late nineteenth century were granted a rank similar to that which the artists had held more than a century earlier. If through this mechanization, the images took on a "graphematic" structure, which was inscribed into them via technical instruments of every sort including photography, polygraphy, phonography, tachistoscopy, and phenacistiscopy as well as many others whose names and functions are almost forgotten, then their correlate formed the elimination of every subjectivity, which had a tendency to reduce the semantic to the syntactic. Even the engraving of such graphics for the purpose of publication was replaced with corresponding photographs, in order to avoid the engraver's intervention. This determined the entire discursive field in which images appeared as "visual arguments." Accordingly, illustrations were only able to claim to be valid when they fully complied with a graphic norm. Unlike the idealized and standardized art images of the past, all of the visualizations that were produced and used within scientific and mathematical contexts were supposed to take on this status. We are thus dealing with a fundamental transformation, a change of places that cast out aesthetics for the benefit of technology and the visual for the benefit of precise optical and mathematical methods. In the end the methods' task was to document a "trace" in which, as it is said, the brushstroke of nature or the "language of the phenomenon itself is revealed" (Daston and Galison 2002: 29, 85). This task was consistent with the observation statement in an experiment; the "protocol" was a simple datum, which has been endorsed as the basis of empirical science since Rudolf Carnap. The statement's basic form "an x is located at point k at time t" corresponded exactingly to the simple recording of an occurrence in a picture. Briefly said, the visualization scheme was the representational writing based on recording instruments and self-recording automats, which translated for example, signals, movements, and stimuli into graphical representations (see also von Chadarevian 1994).[5]

Even when the apparatuses turned out to be anything but impartial, the idea remained undisputed into the early twentieth century. Accordingly, Rudolf Arnheim admitted in his film theory that: "Photography has raised our demands," images should not only "be faithful to the object" like in art, but should "guarantee their faithfullness by being mechanical manifestations of the reproduced object itself. The objects that are photographed impinge their own images mechanically upon sensitive emulsions" ("The thoughts that made the picture move," Arnheim 1969: 137). Furthermore, in the twentieth century there arose a whole new arsenal of up until then, unknown recording devices that were not only based on the graphical reproduction of electrical signals and fluctuations, but also tried to make light impulses, electromagnetic waves, electron fluctuations, and other theoretical constructs visible. The visualizations are increasingly due to digital modeling, which, so to speak as a counterpart to constructive models, made it possible for the mathematical processing and representation of data to triumph over the visible. This development is due to two specific circumstances: On the one hand, that mathematics itself finds its fulfillment in the machine and thus manifests the immanent *telos* of automatic recording and on the other hand, how mathematics and sciences penetrate the fields, which *per definition* evade perceptibility. The compensation for this withdrawal occurs then through mathematics that create virtual images by graphical modeling. These images however, raise the question of *what* they are pictures *of* and *what* we get to see *in* them (see Heintz and Huber 2002).

VISUAL EPISTEMIC

Nevertheless, the debate about visualizations and iconicity in the sciences has up until now ignored the specifics of the image. This includes, first of all, its particular *aesthetic*, second its *logic* of showing, which is responsible not only for the lack of syntax but especially for the lack of negation and the visual non-hypotheticity (concerning the logic of showing and affirmative status of figurativeness see Mersch 2004a, 2007), and *third* its spatial organization, which marks the shape or topological structure in the image (Mersch 2005a, 2006a). At the same time, what we can actually know through the images and their visualization is also put to the test (see Heßler et al. 2004; Heßler 2006). What is clear is that such knowledge can not be discussed without considering the relationships between different visualization strategies. As Bruno Latour rightfully commented "the image" does not exist in science and mathematics, instead there are only synchronic and diachronic image series (Latour 2002) that replace one another and can only be read in relation to each other. Visualizations therefore, refer to other visualizations, as well as—perpendicular to them—to discursive processes of contextualizing and interpreting in which images, letters, and numbers constantly interact with one another. Hence, we have to deal with different types of images, which on the one hand are generated through the optical apparatuses or other recording instruments and whose semiotic format is the "trace" or "index" and on the other hand, with the "abstract" image forms such as graphs, plans, models, or diagrams that resemble theories more than images. In the latter case, the tools are "imaging processes" in the broadest sense, which are transferred and edited until a visualization emerges that seems to be of epistemic substance. Furthermore, their validity itself already has a history. However, like the technical conditions of their creation, it remains hidden in the image and lacks true visibility, because that which makes visible in turn eludes all visibility. We are therefore dealing with the same "medial negativity" that allows the medium to recede to the degree that it constitutes a reality.[6] It implies that we are systematically deceived about the picture status of images.

Moreover, the transformability of images into other images which has always been temporally, technically, and materially restricted, thanks to digitalization now tends to be unlimited. However, because different data sets are based on the same encoding, and the same records can apparently be addressed in multiple ways, on the level of the algorithm they are universally manipulable. The effects of the visualization therefore emerge as surface effects and the visualization itself turns out to be a product of "design." Disparate appearances are possible, insofar as the same data can also be represented differently and there are no criteria that privilege one method of presentation over another. As a consequence, we are no longer dealing with an unambiguous visualization, or a stability in the image that represents *something specific*. Instead, we are only dealing with *possibilities* that can coexist incompatibly and invoke different epistemic forms and whose common denominator is at most, their calculability. In the beginning there are discrete data that are based on measurements, which are not necessarily optic in nature, but rather often rest on already theoretical abstractions and calculations. Therefore, their iconicity, when it is removed from its aesthetic quality, is transformed into text and thus approximates a discursive matrix. Its material consists of signals, impulses, or frequencies; *how* these were gained is irrelevant. Instead, the focus is at best, on how they have been translated into electromagnetic waves and from there into digital codes. Their notational spelling does not set whether they are subsequently modeled and processed visually, acoustically, or in some other mode of perception.

In contrast to classical methods of visualization, no genuine visibility forms the starting point of becoming an image. Instead, it is "information" in the cybernetic sense that is transformed into a visual parameter only at the end of the chain. Therefore, as discrete decision-making measures, their status is something other than a visual picture, even when they are optically generated. In creating them, digital scanners, and sensors—which in the case of encephalography measure electrical currents or in scanning tunnel microscopy, tunnel currents in similar locations—are employed in order to convert them into graphs. When added together as "families" they imitate a three-dimensional structure which closely resembles Euclidian stereoscopic images. Therefore, we are actually dealing with graphs or diagrammatic structures that have to be viewed differently from images or visible objects. Occasionally, cartographic methods come into play, in order to add color. However, they do not show anything visible, but rather at best indicate directions, distributions, spatial orders, or patterns. Regardless of what they might refer to, or be "traces" of,[7] they expose nothing real, at the most topologies and relations which in turn do not function as samples or proof of "something," but rather have to be interpreted independent of their aesthetic, as abstractions to which characteristics such as symmetry or structural similarities belong. Consequently, they do not assume any representational or denotative status, but instead a *diagrammatic* or *graphematic* one. Much more than being "traces" from which *something* can be read, they are *ordered syntaxes*. Their epistemic function is not based on proving an existence that is always pinned to materiality, but is rather the modeling of a geometric or figural structure, which remains completely immaterial (see also Mersch 2005a: 337ff.).

DIAGRAMMATIC AND SPATIAL ORDER

What is to be understood by diagrammatic or "graphematic" visualizations though, then needs to be outlined. First, this includes all syntactic types of images that are based on discrete arrangements, thus also, computer generated visualizations that serve to identify structures or patterns. Whereas, maps, plans, diagrams or models, and graphs are usually differentiated (see, for example, Goodman 1995: 158ff., 163ff.), the terms nevertheless vary in different contexts and disciplines and often change sides. Hence, graphs are labeled "curve diagrams," and logic or mathematic notations as diagrammatic forms (see Pape 1997: 404ff.), and finally, network plans, for example, in informatics, are in turn called graphs (Diestel 2000). Obviously, no strict dividing lines can be drawn between them, at best functions can be determined according to which they are applied, in order to fulfill the disparate purposes in the visual production of knowledge. The technical creation of the image does not stand in the center, even when it sets the limits for mathematical modeling. Instead, it is the structural basis of the modeling itself that corresponds with the representational form's mediality.

The guideline is an image production based on measurements, whose values are graphically plotted or depicted through spatial distributions. Here, diagrammatic generally implies those visual-graphic forms which permit argumentations within the visual medium. Indeed, a general diagrammatic theory still has to be written (for the time being see Bertin 1983; Bonhoff 1993; Bogen and Thürmann 2003; Schmidt-Burkhardt 2005). However, it is clear that in it, scriptural and iconic elements refer to one another so that logical or rational relationships can be made visible through a conventional system of visual parameters. In this sense, diagrammatics signify the *visibility of a thought* (see also Pape 1997: 378ff.). Maps, graphs, nets, and such are *hybrids*, which stand out for their "notational iconicity" (see Krämer 2004, 2006). The pre-condition is that the discursive can be read as iconic and the iconic can become visible as discursive. Here, the expression "hybridity" refers to the mutual crossing of scripturality and pictorality in which their indecidability is characteristic. The scriptural and the pictoral do not appear as separate entities, but rather as one in the other and vice versa. Distinct symbols or logical relations accordingly appear as figures or spatial distributions. On the other hand, the elements of the image function as differentiable "marks." This can, however, only be differentiated analytically. Like the debate about geometric shapes since the time of Plato testifies, these "marks" represent "idealizations," which are as such "repeatable" and whose loci, angles, and proportions are relevant, not their concrete design.

The transition from image to "notational iconicity" requires a transformation of figurality into operationality, just as the transition from writing to "notational iconicity" abandons the register of classic linguistics in that the scripturality serves less to record a language than to configure its own structural space (Krämer 2005). It is based on a "spatial logic" of a distribution of points and their relation to one another, or arrangements, accumulations, directions, or metric proportions that permit concentrations into patterns, position changes, and other spatial actions and make new orders visible without having to draw upon rhetorical tropes such as metaphors, metonymy, synecdoche, or catachresis. An independent form of performance is thus inherent in the diagrammatic; an operative performance which constitutes the core of a visual argumentation, insofar that it carries out actions in the "writing-image space" which demonstrate the dependencies, extremes, and isomorphisms. Furthermore, if the scripts are based on discrete notations (Goodman 1995: 212), as Nelson Goodman rightfully emphasized, the "notational iconicity" simultaneously shifts the aspects of the signs and "letters" to their spatial localization and distribution. Their criterion is consequently "interspatiality" (Krämer 2005: 28ff.). Diagrammatic structures visualize on the basis of this "intersticity." At the same time, they allow the setting and obliteration of relationships in space and grant their matrix a pictorial meaning that points beyond discreteness of writing (Krämer 2005: 38).

That means, to identify spatiality as a principle of the diagrammatic: It makes possible not only to differentiate "marks," but also to make them visible as topological structures through allocating the different positions or places in space as well as mapping logical and deictic functions. A number of diagrammatic baselines can be derived. First of all, it is necessary to format a space in order to fix locations as well as to implement metrics and scalings, which integrate the inscriptions into a fixed reference system. The formatted space thus functions as a basic condition for every diagrammatic. It is based both on the partitioning of relevant fields, zones, or subspaces, in which the graphic elements find their place, as well as on making a space discrete. Only then can the data be arranged and formed into a figure. Curves create discrete arrangements of points in the n-dimensional spaces, which can be transformed into figures through interpolation, and other smoothing methods (i.e., algorithms). What the underlying calculation could not represent in the "number" mode can be read in the figuration. In this way an epistemic role falls to the figurality and consequently to the spatiality: The distinctions are not modeled as differences between the "marks." We are therefore not dealing with a genuine discursive scheme. Rather, they are modeled as spatial differences. Hence, we are dealing with "spatial differentialities" that work with contrasts, gaps, distances, or unoccupied places. Their texture constitutes the "visual operation spaces" which essentially identifies the structure of the visual argumentation as topological.

Nevertheless, such argumentations cannot do without excluding aesthetic functions (see Stetter 2005: 121ff.). We are confronted with graphic abbreviators or drawings, whose basis represents "notational iconicities," which in turn enable the representation of "syntactic iconic structures." It is worth noting though, that in this way, logical relations can be inscribed into the picture, which the image would otherwise lack. Hence, one can say that the diagrammatic hybrid is a new genre, which, in a strict sense, belongs neither to the pictorial nor to the written, and does not lie "between" them. Rather, logic and iconic or visuality and discursivity are entangled. Scientific visualizations, and in particular how they are created based on graphematic methods (e.g., MRI, roentgen spectrum, scanning probe microscopy, and scanning tunnel microscopy) are of this nature, because they come to *conclusions* that can be either right or wrong, at the same time as they operate deictic. In doing so, they adopt various functions such as classification, sorting, typology, or generating matrices, which in the same way serves to compress and bundle data as well as speed it up in a similar manner. At the same time, they confront "cartographies," which make something visible in a paradoxical manner, without any optical correlation.

MATHEMATICAL GRAPHS AND IMAGELESSNESS

In this respect, spatial arrangements determine the diagrammatic's mediality, and constitute both a visual and discursive field of knowledge. However, the diagrammatics that thus originate prove to be fundamentally ambiguous, because their spatial structures allow different "drawings" of the same data material. Their ambiguity is caused less by their aesthetic, which would imply ambivalence. Instead, their "logics" often deviate considerably from one another, because the same elements visualize other content due to their alternative forms of structuring (e.g., metrics and scalings). As a general rule of diagrammatics, it can be assumed that the unambiguousness of knowledge behaves proportionally opposite to the degree of iconicity: the more iconic elements that are integrated into a representation, the more ambiguous it is, because it allows for different alternatives of interpretation. Inversely, the graphism ignores the image's ambivalence. This applies mainly for graphs as subsets of diagrammatic visualizations. According to graph theory, they can be understood as "mathematic models for reticular structures" (Tittmann 2003: 11), which are essentially determined by two types of objects, namely location ("knots") and connection ("edges") (Tittmann 2003; as well as Diestel 2002: 2). The definition is sufficiently general that all network structures result from the combination of both objects. However, in comparison to their visual representation they demonstrate resistance, as their structurality alone decides, not their iconicity (Stetter 2005: 121ff., 125). Graphs are thus distinguished by their extreme abstraction of pictorial elements, although eradicating the aesthetic means represents a necessary, but not a sufficient condition. This results because they could be structural equivalents, even when their pictorial representations differ in details. Consequently, isomorphy is a characteristic of their syntax, not their figurativeness (Tittmann 2003: 24ff.). Accordingly, in view of the network structures that are represented, their corresponding visuality seems to be irrelevant; they rather preserve the minimum of visual evidence that is essential for the spatial order of the relational schemata.

The graphic is independent from the iconic as a result of the pre-eminence of mathematics whose validity does not require pictures. The argument between geometry and algebra proves that mathematics is not imageless. However, geometric objects function only as abstract figures. Their formal construction of circles and lines are of interest and not their "appearance" (*eikon*). Furthermore, since Descartes' algebraization of mathematics and Boole's logic, geometry's influence has been suppressed and with it that of the visual as well. The graphic is therefore determined, not by the pictorial but rather by the respective rules or regulations. Nevertheless, graphs, which have dominated analytical geometry, topology, and combinatorics since then, cannot be totally traced back to the graphism of writing. Mathematics results "from the structure of writing" (see Mersch 2005b), however figural elements remain irreducible due to the iconicity of writing. The *return of the pictures* as well as the significance of visual thinking for mathematics has been discussed in particular with regards to chaos mathematics (Peitgen 1994). That applies though, only to patterns that display graphic iterations and which in turn prove to be secondary in comparison to the rules of the underlying function. Their structures emerge figurally, so that the supposedly visual knowledge in the end is based on figural knowledge, which functions as *indications*, not as proof. The visibility of logic and the visuality of thinking thus find their limits: They refer to figurality and not in the true sense of the word, to iconicity.

Accordingly, it becomes evident that all diagrammatic visualizations, in particular graphs, require conventionality and consistency in order to be interpreted. No scientific visualization manages without a legend or discursive commentary. Writing and images do not just refer to each other, but the diagrammatic itself requires the text that makes it interpretable. Whatever function the iconic occupies within science—whether it is as a sketch, heuristic, representation of a structure, or organization of knowledge—it always fulfills a "program" (see Wittgenstein 2000: 39, 42–44; see also Mersch 2006b), which still has to be authenticated through discourse. Therefore, models, networks, maps, resemble instruments, which, like mechanical instruments, remain related to the entire scientific "dispositif."[8] Hence, unlike art, whose representations constantly remain embedded in the material, diagrammatic and graphematic visualizations de-materialize the iconic and restrict it to the *form*. Accordingly Wittgenstein asks:

> But what makes a plan a plan? (i.) e. what makes it different from any other scribbling? (...) Plans still obey the rules of translation (...). The plan is thus obviously a useful instrument. And that justifies its examination, its effectiveness/function. But (...) that is not all. Seeing the drawings is not enough for me to understand the plan (...). I also have to know what it means to follow a plan. (Wittgenstein 2000: 43, see also 74)[9]

This, however, signals a fundamental problem, because Wittgenstein makes it clear that this operation in the end remains bottomless, because no plan explains itself. "I would like to say: to understand a plan has to mean to use it"[10]; but the plan is "not complete."[11]

> That means: "I do not need any further representations that show me how the representation works, that is how to use the first template, because then I need another template to show me how to apply/use the second and so on ad infinitum. (...) The plan as a plan can therefore not be described. (Wittgenstein 2000: 45ff.)[12]

One has to add: Graphs and diagrammatic structures are not self-explanatory. They do not even function as images that make something visible and that are limited to showing. This is why no high technology image can be decrypted based on its visibility alone. Its readability requires theory as a frame for explaining. It emerges from this theory and cannot exist without it. Thus, we are confronted with the impossibility of a clear differentiation between argument and instrument.

ONTOLOGIZING DIAGRAMMATIC STRUCTURES

This indeterminacy constitutes the precarious status of scientific visualizations in the age of the "calculated image."[13] Associated with this are two fundamental problems, which make the questions of what they represent or what they refer to, unanswerable. The *first point of instability* arises because what they present is based on a mass of instrumentally created data, which has to be translated into pixels by virtue of a graphical algorithm. The calculation determines *how* the pixel appears, however, the origin of the data remains unimportant to the calculation itself. It cannot be seen in the images that assumptions have already entered into their existence. Their informational modeling, for example, generally only registers information that complies with the digitalization schema and is thus suitable for computers. It also cannot be decided what surfaces in the image as interference and what does not, because errors, such as those that are caused by malfunctioning or defect instruments exhibit in principle the same informational status. Furthermore they appear in an image in the same mode (i.e., as a visual element which cannot be differentiated from other visual elements). The transformation of data sequence does not allow any visual difference; only the instrumental creation and the reflection of the conditions of production allow for conclusions, and these are not always easily drawn, as the history of instruments shows. However, insofar that the visible does not allow for any reference to its source material, because it no longer functions as its "trace" but rather as "information," the visual presence of the calculated remains a surface, which remains covered in comparison to its technical and algorithmic deep structure and its sources. We are thus generally dealing with the paradox that a medium makes something visible, without making the method of visualization (i.e., also the conditions of production and their dispositives, visible). Nevertheless, this paradox with regards to technical images produces an opacity that increases with the degree of their programming.

The *second problem* arises from the method of the computer's design and its actual digital image editing because through the specific range of software, the diagrammatic or graphematic visualizations of referential images, are increasingly made to look similar. The need to do so originates from specific traditions of seeing, which facilitate their interpretation. The modalities of data polling, which occur before the actual editing and play a role in generating the images, belong to this without being a part of it. The choice of the relevant data contributes not only indirectly to *what* appears in the image, but also to which effects are prioritized in the process. Furthermore, the eye's history and the adaptation to its ways of perception, are already a part of the criterion for selection. In addition to this there are numerous interpolation methods that compensate for the gaps in data, as well as data compression methods, noise reduction, smoothing, and filtering which already model the image at the level of the program structure. That is what computer generating is about. Data polling itself has to do with *supplementary* methods, which leave the decision to the machine. The question of "intervention" is thus posed right from the start, but is nevertheless irrelevant as a question, because the methods do not tolerate any contradictory terms. The most important point is inscribing styles of images and habits of perception that have always been implemented in visualizations as imaginary procedures and which now belong to the software packets' tools. Hence, programs can no longer do anything but apply perspective grids from projection geometry, vanishing points, coherent lighting and shading, object outlines, color and fluent image transitions etc. Software evokes virtual 3-D effects, [14] even when they are of no importance for the visualization of biological, chemical, or physical events.

There is thus a tendency to inscribe aesthetic rules, which seek to impose a conventional image realism, already at the level of abstract images' algorithms. This is where the "ideology" of iconicity, its implicit illusionism, starts. It reminds us of spatial objects, whose impression is owed to mimetic procedures of evidence. It no longer understands diagrammatic visualizations as representations of a topological order or a graph, but rather as *images of objects with their own identity and contours.* They are promoted to stabile, and therefore objects that can be investigated as such. We are dealing with the transposal of discrete graphic structures to ontological entities, which are supplied with the nimbus of occupying properties, and of appearing as available, changeable, and definable objects in the realm of the visible.

THE INDECIDABILITY OF THE REFERENCE

Digitally editing an image therefore implies a reification. By virtue of its preliminary aesthetic decisions, they convert diagrammatic scripts into mimetic scripts, the main effect being the blurring of digital classification. They extinguish the graphs' inscriptions in favor of a analogue appearance and transform their graphematic into a referential order. [15] Their epistemic status appears to be precarious, because a clear line of separation no longer exists between the two. One of the most convincing examples of this are nanotechnological "images." They form the product of an instrument that caused a sensation in the early 1980s and which does not technically function optically. It rather scans surfaces point-by-point, measuring constant tunnel currents, which only originate through apparatus and experiment. As a result we are dealing with "epistemic parameters," which exist relative to instruments and their experimental arrangements. They therefore do not refer to something "existing in front of them," but instead mix theory and data with one another (Hennig 2004). The question of *what* is visualized here thus seems obsolete, because it does not result in a representation of *something* specific that can be ascribed a pre-existing reality. Rather its existence is due to the "dispositif" of measurement, which is based on metric guidelines and parameters as well as assumptions about probabilities in quantum physics. Hence, in a strict sense, no measurement is repeatable, because what is measured changes in the course of the procedure—an effect that can also be used to manipulate the surface so that we are actually dealing with "'events" instead of stable values. Nevertheless, the measurements/manipulations can be represented as graphs, which, by bundling them with other graphs, can be synthesized into a consistent "impression of an image." What it shows, however, remains uncertain. If the visualization resembles a spatial figure made from measurements, which evokes a graphematic image type, nothing can be said about it outside its structure, because this structure reacts exclusively to the deviations of the tunnel currents, and not to a present object. The interpretation of it as an illustration of atomic structures is a theoretical hypothesis, supported by the quantum mechanical theory of complimentarity and the probability according to which particle currents, which are considered impenetrable, "tunnel through" atoms. Quantum mechanics concluded that subatomic relations, which can only be mathematically derived, cannot be represented. Nanotechnological visualizations in comparison are mere "illustrations" of mathematical functions. Principles of identity, design, and materiality, which are attached to an "object," become invalid. The edited "image" accordingly leads astray. Designed according to the laws of mimesis, it reveals objects which are ordered perspectively and seem to be lit by two imaginary sources of lights. Thus they create the illusion that we are seeing "something" specific, namely atoms on a surface. Stains, dark spots, or holes which can also be seen in the "image," indicate in no way empty spaces, but instead indicate changes in the tunnel current. This makes it impossible to decide whether they are the result of fluctuations in the measurements, the result of instrument error when measuring, a defect, or something similar (see Hennig 2006).

Such imprecisions were preserved in early publications, so that the mediality of the technical creation was stored in the "image" as a reflexive trace. With increasing technical perfection and the use of commercial software packages, an ideal of smoothing prevails which serves to cover the constructed nature of visualizations.[16] Accordingly, the visualized imitates a coherent representation that imposes a *referentiality* on the visible, so that in the end we are dealing with a fundamental *inability to differentiate between denotation and construction*. What remains unclear then is what visualizations refer to or if they even describe "something" at all. An instability between the diagrammatic and referential methods of producing images thus arises. The clear chasm that separates the two loses its validity. It is this instability, which constitutes the core of the epistemic problem for computer generated scientific images, their immanent "delusion," their denotative appearance. It implies a transition from the topological and syntactic to indexicality with all attributes of the "trace," the record or the proof of existence. Not that such "images" refer to nothing. However, the decisive point is that we are no longer in the position to decide what an index or a denotation is, and what is a construction or texture.

Ultimately, this instability is caused by the power of the programs and their implemented aesthetic. A key role falls to the computer's design. Comparing the eighteenth with the late twentieth century, one could speak of a role change between the artist and the software engineer.[17] Daston and Galison argue that before 1800, the artist was considered the undisputed expert on the visual and stood in direct contact with the relevant scientist, in the production of scientific illustrations (see Daston and Galison 2002: 54ff.; Galison 1998; Kemp 2003). Whereas the artist was increasingly imposed with mistrust about the authenticity of the representation, programmers now occupy this place that had been vacated and in turn often understand themselves as artists. Nevertheless, since the mid-nineteenth century, sciences and mathematics have in no way freed themselves from the bonds of ruling aesthetic knowledge by becoming devoted to the ideal "mechanical objectivity." Instead they have subjected themselves to new dependencies, by leaving the creativity of the gaze to apparatuses, and by replacing the inadequate individual with technology. Through the digitalizing of the technical, this has been passed onto programmers. However, their "work" proves to be just as opaque, because the available software packets have to be purchased as bundles and do not tolerate any control or intervention, just as their production, circulation, and execution obeys the logic of an anonymous economic machinery, which follows other than purely scientific interests.

NOTES

1. The following reflections date back to the research project *Visualisierung in der Wissenskommunikation* [*Visualization in the Communication of Knowledge*, Transl.] conducted by Martina Heßler, Jochen Hennig, and myself http://www.sciencepolicystudies.de/projekt/visualisierung/index.htm. The project was part of the Federal Ministry of Education and Research (BMBF) financed initiative *Wissen für Entscheidungsprozesse* [*Knowledge for Decision Processes*, Trans.]. I would like to thank Martina Heßler, Jochen Hennig, and Christine Hanke for our conversations, as well as for inspiration and critique; translation from German by Rett Rossi.
2. One of the recent exceptions indeed are the works of Peter Galison; for example *Image and Logic* (1997).
3. "(G)ibt es eine bevorzugte, etwa besonders unmittelbare Art der Abbildung? Ich glaube nein! Jede Art der Abbildung ist gleichberechtigt." Ludwig Wittgenstein, *Philosophische Bemerkungen*, Viennese Edition Volume 3, Vienna: Springer-Verlag 2000: 4. [Translator's Note: Here and elsewhere in the text, where the English citations are my own translations; the original German citations and bibliographical information are provided in the notes.]
4. This is more elaborately explained in: Dieter Mersch (2005a).
5. "Wie ein zeitgenössischer Physiologe bemerkte, musste sehr bald jede physiologische Behauptung 'mit einem Graph bewiesen werden'" (de Chadarevian 1994: 142).
6. With regards to "negative media theories" for the time being see Dieter Mersch (2004b, 2005c).
7. With regards to referenciality as criterion and the symbolic status of scientific images see also Martina Heßler (2006: 18ff., 27ff).
8. Translator's note: The French word "dispositif", as used by Michel Foucault, has no direct English equivalent; it includes apparatus as well as the whole socio-technical system and its material and immaterial conditions.
9. "Aber was macht einen Plan zum Plan? D.h., was unterscheidet ihn von einem beliebigen Gekritzel? (...) () (Z)u dem Plan gehört noch die Regel der Übersetzung (...). So ist der Plan offenbar ein nützliches Instrument. Und das rechtfertigt seine Untersuchung seiner Wirksamkeit/Funktion. Aber (...) das ist noch nicht Alles. Es genügt um den Plan zu verstehen nicht, dass ich diese Zeichnung sehe (...). Ich muss auch wissen, was es heißt, einem Plan zu folgen."
10. "Ich möchte sagen: einen Plan verstehen muss schon heißen, ihn anwenden" (Wittgenstein 2000: 45, 46 passim).
11. "nicht abgeschlossen" Wittgenstein (2000: 45, 46 passim).
12. "Ich brauche keine weitere Abbildung, die mir zeigt, wie die Abbildung vor sich zu gehen hat, wie also die erste Vorlage zu benutzen ist, denn sonst brauchte ich auch eine Vorlage, um mir die Verwendung/Anwendung der zweiten zu zeigen usf. ad infinitum. (...) Der Plan ist als Plan (...) nicht zu beschreiben."
13. With regards to the term "calculated image" [*errechneten Bildes*] see also Friedrich Kittler (2004).
14. With regards to the individual methods see: Manovich (2001: 184).
15. With regards to the problems of reference and to the difference between readings and observations of such "images" see, for example, Martina Heßler (2006: 34).

16. I am grateful to Jochen Hennig for this and the following consideration. The results of his empirical research were presented at the workshop "Wissen für Entscheidungsprozesse" from the BMBF, Berlin 1./2.12.05.
17. For more information see Martina Heßler (2004: 54): http://www.science-policystudies.de/dok/explorationsstudie-hessler.pdf.

BIBLIOGRAPHY

Arnheim, Rudolf (ed.) (1969) "The thoughts that made the picture move", in Rudolf Arnheim (ed.), *Film as Art*, London: Faber and Faber: 137.
Bertin, Jacques (1983) *Graphische Semiologie, Diagramme, Netze, Karten*, Berlin: De Gruyter.
Bogen, Steffen and Felix Thürmann (2003) "Jenseits der Opposition von Text und Bild. Überlegungen zu einer Theorie des Diagramms und des Diagrammatischen", in Alexander Patschovsky (ed.), *Die Bildwelt der Diagramme Joachims von Fiore. Zur Medialität religiös-politischer Programme im Mittelalter*, Ostfildern: Thorbecke: 1–22.
Bonhoff, Ulrike Maria (1993) *Das Diagramm. Kunsthistorische Betrachtung über seine vielfältige Verwendung von der Antike bis zur Neuzeit*, Münster.
von Chadarevian, Soraya (1994) "Sehen und Aufzeichnen in der Botanik des 19. Jahrhunderts", in Michael Wetzel und Hertha Wolf (eds), *Der Entzug der Bilder*, München: Wilhelm Fink Verlag: 121–144.
Daston, Lorraine and Peter Galison (2002) "Das Bild der Objektivität", in Peter Geimer (ed.), *Ordnungen der Sichtbarkeit*, Frankfurt am Main: Suhrkamp: 29–99.
Diestel, Reinhard (2000) *Graphentheorie*, Berlin, Heidelberg, New York: Springer-Verlag, 2nd edition.
Galison, Peter (1997) *Image and Logic*, Chicago: University of Chicago Press.
—— (1998) "Judgement against objectivity", in Caroline A. Jones and Peter Galison (eds.), *Picturing Science, Producing Art*, New York, London: 327–359.
Goodman, Nelson (1995) *Sprachen der Kunst*, Frankfurt am Main: Suhrkamp.
Heintz, Bettina and Jörg Huber (eds) (2002), "Der verführerische Blick. Formen und Folgen wissenschaftlicher Visualisierungsstrategien", in Bettina Heintz und Jörg Huber (eds), *Mit dem Auge denken*, New York, Zürich: Edition Voldemeer, Springer-Verlag: 9–40.
Hennig, Jochen (2004) "Vom Experiment zur Utopie: Bilder in der Nanotechnologie", *Bildwelten des Wissens. Kunsthistorisches Jahrbuch für Bildkritik*, 2,2: 9–18.
—— (2006) "Die Versinnlichung des Unzugänglichen—Oberflächendarstellungen in der zeitgenössischen Mikroskopie", in Martina Heßler (ed.), *Konstruierte Sichtbarkeit. Wissenschafts- und Technikbilder seit der frühen Neuzeit*, München: Wilhelm Fink Verlag: 99–116.
Heßler, Martina (in collaboration with Jochen Hennig and Dieter Mersch) (2004) "Explorationsstudie im Rahmen der BMBF-Förderinitiative 'Wissen für Entscheidungsprozesse' zum Thema 'Visualisierung in der Wissenskommunikation', in Martina Heßler (ed.), *Konstruierte Sichtbarkeit. Wissenschafts- und Technikbilder seit der frühen Neuzeit*, München: Wilhelm Fink Verlag: 54ff: http://www.sciencepolicystudies.de/dok/explorationsstudie-hessler.pdf.
—— (2006) "Annäherungen an Wissenschaftsbilder", in Martina Heßler (ed.), *Konstruierte Sichtbarkeit. Wissenschafts- und Technikbilder seit der frühen Neuzeit*, München: Wilhelm Fink Verlag: 11–37.

———— Hennig, Jochen and Dieter Mersch (2004), *Visualisierung in der Wissens-kommunikation*, explorative study for the BMBF, Berlin 2004: http://www.sciencepolicystudies.de/dok/explorationsstudie-hessler.pdf.

Holländer, Hans (2000) "Einführung", in Hans Holländer (ed.), *Erkenntnis, Erfindung Konstruktion*, Berlin: Mann: 9–15, esp. p. 11, column 2f.

Kemp, Martin (2003) *Bilderwissen. Die Anschaulichkeit naturwissenschaftlicher Phänomene*, Kölon: DuMont.

Kittler, Friedrich (2004) "Schrift und Zahl—Die Geschichte des errechneten Bildes", in Christa Maar und Hubert Burda (eds), *Iconic Turn. Die neue Macht der Bilder*, Köln: DuMont: 186–203.

Krämer, Sybille (2004) "'Schriftbildlichkeit' oder: Über eine (fast) vergessene Dimension der Schrift", in Sybille Krämer and Horst Bredekamp (eds.), *Bild, Schrift, Zahl*, München: Wilhelm Fink Verlag: 157–176.

Krämer, Sybille (2005) "'Operationsraum Schrift'. Über einen Perspektivenwechsel in der Betrachtung der Schrift", in Gernot Grube, Werner Kogge, and Sybille Krämer (eds), *Schrift. Kulturtechnik zwischen Auge, Hand und Maschine*, München: Wilhelm Fink Verlag: 33.

Krämer, Sybille (2006) "Zur Sichtbarkeit von Schrift oder: Die Visualisierung des Unsichtbaren in der operativen Schrift. Zehn Thesen", in Susanne Strätling and Georg Witte (eds), *Die Sichtbarkeit der Schrift*, Paderborn: Wilhelm Fink Verlag: 75–84.

Krämer, Sybille and Horst Bredekamp (eds) (2004) *Bild, Schrift, Zahl*, München: Wilhelm Fink Verlag.

Latour, Bruno (2002) *Iconoclash. Gibt es eine Welt jenseits des Bilderkrieges?*, Berlin: Merve.

Manovich, Lev (2001) *Language of New Media*, Cambridge, MA: The MIT Press:184ff.

Mersch, Dieter (2003) "Wort, Bild, Ton, Zahl. Modalitäten medialen Darstellens", in Dieter Mersch (ed.), *Die Medien der Künste*, München: Wilhelm Fink Verlag: 9–49.

———— (2004a) "Bild und Blick. Zur Medialität des Visuellen", in Christian Filk, Michael Lommel and Mike Sandbothe (eds), *Media Synaesthetics*, Köln: Halem: 95–122.

———— (2004b) "Medialität und Undarstellbarkeit. Einleitung in eine 'negative' Medientheorie", in Sybille Krämer (ed.), *Performativität und Medialität*, München: Wilhelm Fink Verlag: 75–96.

———— (2005a) "Das Bild als Argument", in Christoph Wulf and Jörg Zirfas (eds.), *Ikonologien des Performativen*, München: Wilhelm Fink Verlag: 322–344.

———— (2005b) "Die Geburt der Mathematik aus der Struktur der Schrift", in Gernot Grube, Werner Kogge, and Sybille Krämer (eds), *Schrift. Kulturtechnik zwischen Auge, Hand und Maschine*, München: Wilhelm Fink Verlag: 211–233.

———— (2005c) "Negative Medialität. Derridas Différance und Heideggers Weg zur Sprache", *Journal Phänomenologie, Jacques Derrida*, 23: 14–22.

———— (2006a) "Naturwissenschaftliches Wissen und bildliche Logik", in Martina Heßler (ed.), *Konstruierte Sichtbarkeit. Wissenschafts- und Technikbilder seit der frühen Neuzeit*, München: Wilhelm Fink Verlag: 405–420.

———— (2006b) "Wittgensteins Bilddenken", *Deutsche Zeitschrift für Philosophie*, 54, 6: 925–942.

———— (forthcoming) "Blick und Entzug. Zur Logik ikonischer Strukturen", in Gottfried Boehm, Gabriele Brandstetter, and Achim von Müller (eds), *Bild—Figur—Zahl*, München: Wilhelm Fink Verlag.

Pape, Helmut (1997) *Die Unsichtbarkeit der Welt*, Frankfurt am Main: Suhrkamp.

Peitgen, Heinz-Otto (1994) "Mit den Fraktalen kehren die Bilder in die Mathematik zurück", in Florian Rötzer (ed.), *Vom Chaos zur Endophysik*, München: Boer: 98–114.

Rheinberger, Hans-Jörg (2002) "Objekt und Repräsentation", in Bettina Heintz and Jörg Huber (eds), *Mit dem Auge denken*, New York, Zürich: Edition Voldemeer, Springer-Verlag: 55–61.

Schmidt-Burkhardt, Astrit (2005) *Stammbäume der Kunst. Zur Genealogie der Avantgarde*, Berlin: Akademie-Verlag.

Stetter, Christian (2005) "Bild, Diagramm, Schrift", in Gernot Grube, Werner Kogge, and Sybille Krämer (eds), *Schrift. Kulturtechnik zwischen Auge, Hand und Maschine*, München: Wilhelm Fink Verlag: 115–135.

Tittmann, Peter (2003) *Graphentheorie. Eine anwendungsorientierte Einführung*, Leipzig: Fachbuchverlag.

Wittgenstein, Ludwig (2000) *Bemerkungen*, Viennese Edition vol. 3, Vienna: Springer-Verlag.

9 Imagination, Multimodality and Embodied Interaction

A Discussion of Sound and Movement in Two Cases of Laboratory and Clinical Magnetic Resonance Imaging

Lisa Cartwright and Morana Alač

In this project we combine an analysis of multimodal interaction (e.g., Goodwin 2000a; 2000b; Heath and Hindmarsh 2002) in a magnetic resonance (MR) laboratory with a reading of a patient experience in the MR clinical context. We consider the co-construction of meaning and experience in these two contexts, focusing on the intersubjective enactment of imagination (Murphy 2004, 2005; Nishizaka 2000, 2003) between researchers, and between patient and clinical technician (for a treatment of imagination in the workplace see also Suchman 2000). We propose that meaning is spatially and intersubjectively enacted in embodied practice through gestures (which in some cases can be acoustic as well as visually perceived events), vocalizations (acoustic events which may be linguistic or nonlinguistic), relationships of gaze and body position, and negotiation of representations and instrumental modalities in the environments of the magnetic resonance laboratory and clinic (Alač 2005, 2006; Alač and Hutchins 2004). Actors in the fields of enactment we consider include bodies, equipment, representations, as well as the space of the imagination.[1]

Our concern particularly rests with what does not explicitly appear in the visual images observed by laboratory researchers, but what is deduced and articulated—in a word, imagined between them—on the basis of images. We propose that these performances by researchers of activity they together imagine the research subject to have performed, but which were not directly observed, are worthy of consideration for what they can tell us about the function of imagination in the production of meaning and in the constitution of the representation of the research subject in the laboratory or the clinic. Our contribution, then, is to introduce the idea of the imagination of the researcher and the research subject as important entities of laboratory study. Further, we stress that the performance of the researcher or technician of an imagined process that another body undergoes inside the vacuum tube—a performance enacted before the technician, apprentice, or collaborator researcher—is another important representational and imaginative aspect of laboratory and clinical work processes. Embodied performance, vocalizations, and speech are elements we can empirically observe and analyze to get at these aspects of making visible researchers shared imagined aspects of another's unobserved behavior or feeling. Our emphasis in part, then, is the researcher bodies that interact with elements in the visual, spatial, and acoustic field to perform an interior image of another, absent body that he or she hopes to better know or understand. This requires an analysis that accounts for the movements, utterances and relationships of bodies in the visual field as well as for data such as images, text, graphs, and computer screens, as observation of these sites helps to understand what lies outside our reach: namely, the nature of the imagination. Movement is important to our discussion not only because the two cases we discuss involve interpretation of fMR (functional magnetic resonance) images,[2] but also because in anatomical MR imaging movement of the subject in the vacuum tube, imagined or evident in image artifacts, becomes a matter of important concern for interpreters of MR data.

When we hear the term "imagination", we typically understand this to mean the images and ideas, percepts and symbolic representations held by or in the mind of the individual subject as he or she comes to understand objects, others, and experiences in the world through embodied sense perceptions. The locus of imagination is typically understood to reside in the mind of the individual, with communication between individuals functioning as a means of externalizing and communicating precepts or symbolic material intersubjectively. Imagination tends to be discussed in relationship to works of art and creative expression, not scientific practice. Finke, Ward, and Smith (1993) describe imagination in a manner that emphasizes the verbal as well as visual nature of imagined entities, and the fact that imagination results in the production of something new and not merely a recollection of something already known (as in memory processes). For them imagination is

> the process by which people mentally generate novel objects, settings, events, and so on. It is more global than mental imagery in the sense that although these imagined entities might take on the form of mental images, they need not. Imaginative products can also exist in the form of verbal description. Imagination is also more restrictive than mental imagery in the sense that it must involve the generation of something new, whereas certain manifestations of mental imagery can be purely recollective. (Finke et al. 1993: 115)

We embrace the expanded focus from mental imagery to the verbal, and the emphasis on the function of imagination beyond memory. However, we orient our use of the concept of imagination differently in a few ways. First, we emphasize that imagination is more radically multimodal, as well as more dynamically intersubjective and co-constructed, than this model of images and words allows.[3] Second, with Murphy (2004, 2005) and Nishizaka (2000, 2003), we challenge the idea that the mind is the holding ground for imagination. The process of the co-construction of imagination, in the settings we describe below, occurs in a highly fluid and interactive manner between researchers or clinical technicians as they produce or analyze MR data; but it also includes other active agents that move imagination's co-construction beyond the models of conversation and human communication. As has been shown by sociologists and anthropologists of laboratory practice,[4] included as agents or actors in the networked processes of laboratory work with living human bodies are not only patients and research subjects, but also representational data, instruments, technologies, bodies of knowledge in the field tacitly and explicitly recalled, and many other entities and artifacts (Knorr Cetina 1999; Latour 1999; Pickering 1992, 1995).

Second, we emphasize that the metaphor of the mental image (Descartes 1642/1984), which is at the root of the term "imagination" leaves us without adequate means to describe other sensory forms of imagination's enactment including and beyond the verbal. These include nonverbal vocalization and produced sound; the tactile and felt aspects of embodied interaction; and the intersubjective performance of gesture, bodily movement, and gaze that give shape to relationships between bodies in space (Alač 2005, 2006; Alač and Hutchins 2004). These experiences are not fully captured in the idea of "image" that is at the core of the concept of imagination. Mark Johnson, in his book on moral imagination and the implications of cognitive science for ethics, notes: "metaphor is one of the principal mechanisms of imaginative cognition" (Johnson 1993: 33; see also Johnson 1987, 1991; Fauconnier and Turner 2002). We suggest that what the researcher "imagines", whether understood as mental image, imagination (Pylyshyn 1973; Fodor 1975/1981; Kosslyn 1980; Tye 1991), verbal description (Finke, Ward, and Smith), or metaphor (Johnson 1987, 1991), takes shape in the space of interaction through modalities and representations that go well beyond these more familiar forms to include nonlinguistic vocalization, gesture, and embodied interactions. Moreover, these may escape signification in the strict sense. Although these processes unfold through social interaction, our point is not the simple one that they are therefore socially or culturally constituted and not strictly "of the mind." We remain concerned with aspects of interaction that occur in material interaction but nonetheless strictly speaking escape the realm of shared representation and meaning. Our approach is to emphasize that what is experienced as "of the mind" or "in the mind," and what is felt without registering as consciously recognized meaning, remains of critical importance even as we shift the focus to the space of social interaction and knowledge and meaning production.

Third, we are concerned with the concept of imagination's historical link to that which is "not real", that which exists only in a location of mind we identify as "the imagination", or to the Platonic idea of mental precepts as representations of the real.[5] Imagination, multi-modally and intersubjectively produced, we propose, is a strong aspect of moving along the production of scientific knowledge and meaning, even as imagination carries these associations of the imaginary and/as the "not real."[6] Our aim is not to show that imagination helps us to arrive at the real, or at knowledge and meaning, but rather to uncover the ambiguous and complex place of imagination as an essential component of the real and of meaning in diagnostic and research-based interpretive practices.

Finally, we suggest that the intersubjective production of imagination need not entail either the conscious involvement or the physical presence of every agent involved. For example, the researcher need not consciously perform a gesture or vocalization for that movement or utterance to take on meaning and have consequences in the shared space where meaning is produced. And a research subject or patient who is not present at the time that researchers or technicians analyze his or her MR brain data is no less present as an agent in the process of co-constructing the meaning of his or her data set. He or she may be constituted—imagined—as living, present, and willful active agent in the process of analysis that unfolds around his or her MR data.

Two of the concerns that emerge in our discussion below are the questions, when does the research subject or patient become visible or present in the process of interpreting his or her MR data; and how does this "making visible" of the research subject occur? Our project differs from other recent work on MR imaging in that we attempt to bring into view the subjectivity of the research subject or patient. However, we do so not to constitute this subject in a realist or humanist sense, as subjects whose humanity is left out of the laboratory picture, for example, but to emphasize the embodied presence and agency of the human subject as always present in laboratory and clinical processes. Scholarly work on MRI in sociology and anthropology has tended to focus on brain imaging, and has taken up the status of the brain as the conceptual locus of what it means to be human in the current research and popular context (Roepstorff 2001; Beaulieu 2002, 2004; Dumit 2004,). In the processes we discuss, the brain and spinal cord are the objects of concern as well for precisely these reasons. In the interactions we describe, however, what we see produced among practitioners is an imagined sense of the human subject as a whole body and a cognizant subject whose movements signify agency, even as the brain or spine are the isolated objects of scientific or clinical concern. Furthermore, practitioners use their own bodies to make sense of and construct meaning and knowledge about the bodies they study. For example, as we will explain below, in interpreting data, researchers often perform and enact what they imagine to be the imaged body's trajectory as it is performed within the imaging process whose documentation they review after the body is gone from the laboratory or clinic. In addition to imagining, on the basis of documentation, what the body actually did, they also introduce ideas about the kinds of bodily behavior and movement one would expect of a research subject. In the examples we discuss, the presence of the whole body and its actions, understood to be motivated by the human subject, are experienced by researchers as constant problems to be negotiated, managed, and ultimately cleansed away in laboratory and clinical analyses.

We discuss the external expression of imagination in two sets of examples, presented in an intertwined way. In one case, researchers or technicians individually or collaboratively (in teams) imagine and perform the experience that another body, the body of the research subject or patient that they work with, has undergone during a magnetic resonance imaging process, in order to interpret and communicate research or clinical images or data. In another case, we narrate a process wherein the research subject inside the tube imagines himself or herself to be constituted in relationship to another human subject, the technician who conducts the imaging process, and who stages and prompts the patient or research subject's performance inside the machine. In what follows, we interpret these two scenes of performance of imagination as they are enacted both multi-modally and intersubjectively. This action takes place in the laboratory or clinic, and importantly includes the space inside the vacuum tube that houses the imaged body. The interior of the vacuum tube is an under-considered site of physical experience, observation, and of activity that is a scene of at times intense imagination and speculation on the part of technicians or researchers, even as it is a space they cannot, in many respects, see.

Embodied performance, vocalization, and speech between subjects who together interpret MR images are elements we may empirically observe and analyze to discern the trajectory of the researchers' individual and collective mental image of another's unobserved behavior or feeling. In our discussion of researchers below, our emphasis is upon bodies that interact with elements in the visual field to externally demonstrate an interior image of another whose movements he or she hopes to better know, understand, or communicate. We interpret the movements, utterances, and bodily relationships performed in the visual field by researchers as they work in teams to view and interpret the images positioned before them, in the absence of the body of the research subject whose images are being interpreted.

In our first case, a discussion of researcher interaction, our emphasis is in part upon the pedagogical aspects of interactions in which the process of enacting the production of an image and its interpretation requires instruction and guided performance between experts and novices. We follow the interactive process in which researchers sit together before a computer screen, looking and gesturing toward their own bodies and the body of their colleague as they collaboratively not only produce but also inhabit the meaning of the MR body image before them. We do not assume that the expert always remains in the instructional mode nor do we assume that the novice refrains from interpretive or definitive acts. The relationship of expert to novice is fluid and shifting. What concerns us is not only the two bodies of the researchers and their fluid interaction, but the multivalent relationship that also includes the image screen and the mental screens of the researchers, as well as the imagined, absent body whose representations inhabit those screens, and whose performance is the object of the researchers' concern. In multi-modal studies of conversation in workplace and laboratory practice, applied linguist Charles Goodwin has emphasized dialogic, multiparty activity, gesture, and performance as aspects requiring careful consideration. His studies of multimodal semiotic interaction in the workplace and in laboratory practices have included visual documentation as an empirical record of interactions, and in some cases graphs, charts, and pictures have been taken into account as modalities requiring consideration in the analysis of these contexts. We draw on the technique of multi-modal laboratory analysis used by Goodwin (2003a, 2003b, 2003c). Our approach differs from his, however, in that we emphasize the imagination as a crucial element in laboratory and clinical communication. We draw upon our respective areas of expertise as a researcher trained in semiotics and cognitive science who works in the mode of ethnographic documentation of laboratory work (Alač) and a communication and media scholar who works in qualitative methods of visual analysis (Cartwright) to render a discussion across these two domains.

Bodies, in our account, are elements that perform semiotically and affectively in the visual field, interacting with nonhuman elements in that field to bring meaning to the MR image. Our account is based on empirical observation of those aspects of the clinical and laboratory experience that are most intangible. But we wish to approach these through the tangible aspects of the psychic relationship of interlocutor and actors, human and nonhuman, and the tangible enactment of the imagined activity or feeling co-constructed in that setting. To do this requires attention not only to speech and data representations, as would be expected in a semiotic analysis, but also to vocalization, gesture, bodily positioning, and intersubjective movement—elements of the dialogic that go beyond representation and conversation.

Sound and movement present useful entry points for the discussion of visuality in scientific practice. Magnetic resonance has been associated with imaging for most of the technique's history. Sound, however, is a phenomenologically pervasive, one might even say intrusive part of this modality. It is emphatically present in the production of MR data. Studies of MRI have emphasized the interpretation of recorded image and numerical data reports, with little discussion of the acoustic experience of the MR data production process as experienced by the researcher, the technician, the patient, or the research subject. Sound is a significant absence in social studies of MRI. It is an important aspect of experience for researchers and technicians who work with this modality. Sound is an aspect of experience with the technology by patients and research subjects, who may be unaware of the extent of the experience with sound that they will encounter with the process.

Below we introduce a discussion of the acoustic experience of MR data production, and the role of sound in the enactment of coaching the

patient through the imaging process, as well as in the researcher's process of explaining and classifying MR data for the benefit of a co-worker.
Our point about sound is that it is viscerally present and physically experienced, though rarely addressed as a significant factor in social studies of
MR laboratory work. Second, we emphasize that sound serves a role as a
representational and explanatory model in coaching the patient or research
subject, and in teaching or explaining the MR process from researcher to
researcher. Sound is thus an empirically felt disturbance in the production
process of MR "imaging" data for researchers, technicians, patients, and
research subjects. But it is also nonetheless pragmatically enlisted as a semiotic resource in this process of relating to one another about the process
among all of these interlocutors. A "disturbance" thus becomes a semiotic
and mediational agent.

Regarding movement, which we also suggest is characterized as absence:
In discussions of cinematic movement and temporality it is common to
refer to the function of the sequential ordering of static frames and the
incremental differences between those frames as significant factors. Meaning, if we follow Sergei Eisenstein's use of the dialectic for film study, is
produced in the dialectical collision of frames, and not in a single image.
Researchers using structural magnetic resonance images to produce a set
of anatomical images that offer a representation of a given structure, such
as a brain, at times follow a similar logic of reading differences between
images for meaning. Some important information may reside in differences
between frames and not within a given image: The appearance of spatial
disjuncture within a given structure may, for example, be a product of the
subject's unwanted movement in the scanner, leaving the researcher in the
position of having to account for apparent movement in an imaging modality designed to document static anatomical form. Below we use the case
of an interaction among researchers interpreting disjuncture between MR
images in a series as a means of demonstrating documentation of something (namely, movement) that was neither intended to appear in the image,
nor present in any one of these images per se. We emphasize, below, that
movement is a visual absence, a troubling element hinted at in artifacts.
Imagining the research subjects' unauthorized movements inside the tube
thus becomes paramount to the researchers' process of seeing. It becomes
of crucial importance to the researchers to identify in the image, and represent, perform, and experience in their own bodies, the aberrant movements
they imagine the patient to have performed inside the tube—movements
the researchers neither saw nor clearly documented, except as artifact.

Our point, then, in describing this process, is to demonstrate that meth-
odologically when we analyze the visual we need to take into account not
only aspects such as vocalization, gesture, bodily movement, and sound but
also the imagination as it gives life to events unseen, unheard, and undocu-
mented except as absence—in the differential between visual frames, for
example, or as, in the case of sound, by-product of a process that produces

numerical or visual data. Our point is not the familiar one that science strives to make visible and concrete that which is unseen, or that meaning is co-produced (this has by now been widely demonstrated). Rather, we emphasize that imagination, enacted intersubjectively through embodied performance and expression and interpreted through processes of identification and projection is a crucial aspect of the processes of interpretation for the researcher, technician, and patient. While we will not elaborate this point here, this suggests that studies of scientific visuality might further consider issues such as identification, spectatorship, and subjectivity as aspects of laboratory and clinical experience.

SOUND AS PEDAGOGICAL AGENT IN THE INTERACTION AMONG RESEARCHERS IN THE LABORATORY

In a functional MRI scanning process, the attainment of experimental data is not directly dependent upon sound *per se*. The high decibel "acoustic noise" is due to the vibration produced by the interaction between the static magnetic field and the time-dependent currents in gradient wires. Such noises may become a severe impediment for successful research in cognition, especially when acoustic rather than visual stimuli are used (e.g., research on language comprehension). However, the following excerpt from a functional (time-based, not static anatomical) magnetic resonance image (fMRI) apprenticeship (observed by Alač) shows how reenactment of sound can play an important role in the acquisition of expertise and in the construction of a shared *professional vision* (Goodwin 1994).

The interaction described took place between a new member of the laboratory—Nick (N), and the experienced practitioner Eric (E). The new member was being taught the procedure of fMRI data analysis. As a part of the apprenticeship, Nick played the role of an experimental subject: Before the analysis of the experimental data, he was scanned in the fMRI facility. In this process, lab members have an experimental subject, while Nick is more engaged in the viewing and analyzing of data, in order to further his understanding of the procedure.

The excerpt below details a part of the interaction where the two practitioners are seated in front of a laboratory computer. Eric explains to Nick how the data are organized:

Excerpt 9.1

1. E: Ok there are <u>three</u> studies (0.2) tha:t (0.1) were in this
2. directory. //There was,
3. N //Three?
4. E: Yeah. (There) = cause there is localizer. Do you remember at the
5. very //beginning

6. N: //At the beginning of functionals?
7. E: No at the beginning of the //MP-rages
8. N: //Oh ok
9. E: You remember you've heard sort of clicking ch-ch-ch ((*imitates the*
10. *sound*)) and then you got just a quick ah ba:um, ba:um, ba:um,
11. ((*imitates the sound*)) and then it was quiet and G. said this
12. was the warm up and here is the real one? That's the localizer.

In the beginning of the scanning process, the experimental subject is placed in the MRI scanner. After she or he is *aligned* so that the head is at the center of the magnet, a *localizer scan* is collected. The scan gives the exact location of the brain. Nick queries the number of studies in the directory (line 3), demonstrating his uncertainty about the meaning of the localizer scan.

Eric, rather than providing a detailed linguistic and technical explanation of what a localizer scan is, and when it is taken, chooses to imitate the sound produced by the scanner (lines 9 through 10). His voice, by bodily re-enacting the *acoustic noise* that accompanied the scanning process, arouses a recall of the experience for Nick. Eric involves his body to provide an explanation of what is a localizer scan. Nick's comprehension is then achieved through an embodiment of Eric's embodied understanding: Nick re-enacts the experience of being in the scanner, following Eric's vocal prompt.

Although Eric's vocalizing constructs objects of knowledge and rationality, it is neither linguistic nor graphical in form. Rather, it is a non-linguistic acoustic event. In this manner, noise, what was previously regarded as an undesired and undesirable aspect of the scanning session, is reintroduced into the process of data analysis as a semiotic agent. Noise, reproduced through the vocal performance of the researcher, assumes an explanatory role that is strongly affective. It stimulates a kind of bodily recall in Nick, who then is prompted to re-enact his own experience in the scanner. The confounding factor of noise becomes an important element in the embodied procedure of sense-making between researchers.

SOUND AS MEDIUM BETWEEN PATIENT AND TECHNICIAN

Here we shift to a different context—our second case in point, that of a patient undergoing a clinical MRI process. We introduce the term "acoustic envelope" to describe the enclosed space of the scanner. The term "acoustic envelope" was used in the writings of film theorist and historian Mary Ann Doane (1980) to describe the experience of the spectator as listener enveloped by sound in the space of the movie theater. Among sound technicians of the mid-twentieth century, acoustic envelope is a term that refers to the boundary of a listening or recording site and the sound of the ambiance, the air, within it (Burris-Meyer 1941).

The acoustic envelope can also be thought of as a cue, in addition to the spectrum, that allows us to distinguish one sound from another. We use the concept of the acoustic envelope in both senses here, to read the patient or research subject's experience in the MR image-making procedure. Doane (citing psychoanalytic theorist Guy Rosalato) notes that the means of sound deployment in cinema establish "certain conditions for understanding" that obtain in the intersubjective relation between film and spectator (Doane 1980: 380). We borrow this concept to adapt it to the space of the MR imaging process, where the means of *producing* (not deploying) images establish "certain relations of understanding" between subjects (technician, patient) and other elements of the apparatus, establishing for these human subjects, we will suggest, an "acoustic envelope" that offers both the alienating loss of embodied self and the shoring-up of a safe, reflective space of self-recognition through such techniques as piped music and the wiring of the MR space for closely miked conversation between the patient in the vacuum tube and the technician seated outside the room, behind a protective windowed wall. Our discussion is performed through a close reading of the MR clinical examination.

We first describe the nature of sound in this environment. Sound is introduced early on in the imaging process. This is indicated on a symbolic level in the product names of Siemens MR machines: Symphony, Allegra, and Harmony are three familiar models. Sound is also introduced early on in the imaging process as a means of alleviating the stress of the noise produced by these machines. Warning the patient that the imaging process will be cold and noisy, the hospital MR unit's operating system technician offers the patient a blanket and headphones, and assists him or her in the selection of a radio station or musical genre. Listening is one of the components of comforts offered in the MR tube.

The technician instructs the patient in the precise sequence of events he or she will experience. Measures of time for each aspect of the process will be signified by changes in qualities of sound. The technician explains that he will exit the room but remain in close touch through headphones wired to his station outside the room. Inside the coil, for a minute, the patient will hear a slow deep banging, then for three minutes a series of rapidly pulsed high short signals. This will be followed by a pause, and then another three-minute sequence of a different tone and rhythm would ensue.

Although the patient might mistake the sound sequences as evidence of the image "take", it is in fact the quiet time after the pulsing of sound during which the protons "relax" or decay, and it is during this "down" time that significant data is registered for measurement. Imaging (signal recording) takes place, then, in a period of silence after a long noisy period of proton activation. Movement comes into play here in relationship to sound. The patient is told how long after the pulse sequence it is necessary

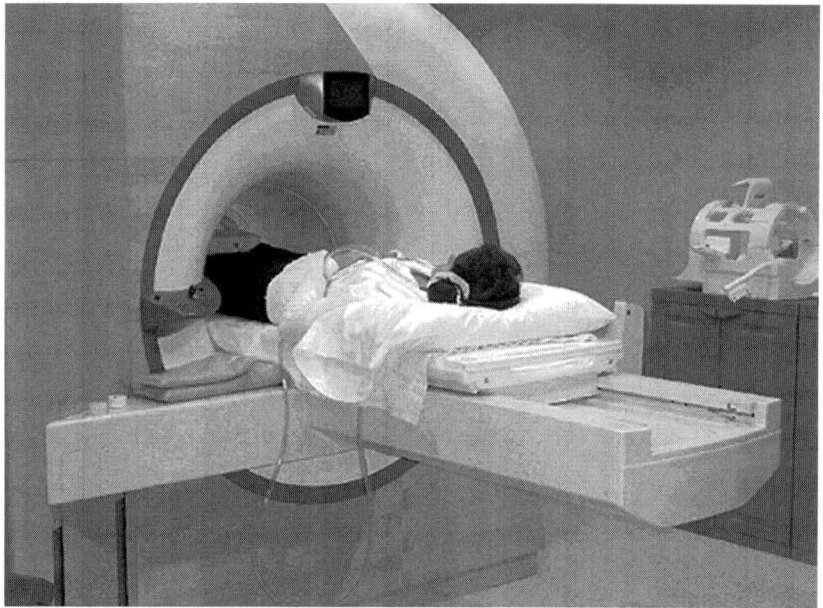

Figure 9.1 The Siemens 1.0 Harmony.

to keep one's body immobile in order for data to be acquired. The technician emphasizes that immobility during these periods of is essential to a successful take.

Some people, the technician explains, find the loud sounds disturbing. The additional sound offered by the headphones and music are meant to alleviate that problem. This technician then places in our subject's hand another sound apparatus designed to manage an overwhelming experience of disturbance: a rubbery orb—a pressure-sensitive signal device wired to the computer-control booth outside the room. This is the patient's signaling device, the hotline to report any experience of distress to the technician. This orb will, in fact, be the patient's sole means of communicating to the technician for the duration of the exam. She is to squeeze it only in the event that she wants to abort the procedure—if she feels pain or overwhelming panic, for example. The technician then points to the large window opposite the machine. He will be positioned there, right outside that window. The patient will be audible to him through the panic button only during imaging passes, and otherwise through an intercom that he switches on in between passes. Near the computer positioned on the other side of this window the patient may be able to see an office chair placed in just the right spot for the technician to keep an eye on the machine while manning the controls of the computer which affords him a view into the tube, and onto the image generated as each phase of the process is complete.

Once slid inside the MR coil, the patient, if her eyes remain open, might notice the smooth, borderless white plastic that curves closely about her body. This surface houses the vacuum layer that shields her from the intense cold of the superconducting magnet coiled around her body. This is a pure modernist space devoid of representation that is not only literally but also figuratively cold. Devoid of edges, of shadow and light, of texture, and of color, the inside of the coil is the logician's dream of empty space. The magnetic coil is bathed in liquid helium kept at 452.4 degrees below zero, a temperature that keeps the wire's resistance low, reducing the electricity needed to make it operate. The patient might close her eyes as her attention is drawn from this blank white surface to the music that passes through the headphones into her ears, a familiar signifier of culture and sociality, of pleasure and warm feeling.

Ensconced face up in the chilly coil, wrapped in a blanket and the scratchy sound of loud radio, the patient might wait a few moments for the other radio process to begin, the one she had been warned about, the loud pulsing of radio waves in conjunction with the scanner. Her thoughts might turn to the technician perched outside the observation window, glancing in at her and the machine periodically as he programs the scanner's functions on the computer. That piece of equipment was turned so that its back faced the window, making its screen a corner insert in the frame that was his view onto the room.

As the Siemens machine whirs into action, our subject might recall that the MR process is loud; that the sound is *noise*. But chances are that nothing she had read or heard reported had prepared her for the experience of MR sound beyond volume, as rhythm, timbre, pitch. Its precision, its intensity, and its organizational logic are remarkable, though unmarked, uncharted in advance for the lay listener.

At the end of the first session, the technician's voice is likely to pass into the coil, much like voice-off narration in a cinema house. From the technician's desk his or her voice passes directly to our subject's ear, through her headphones. The voice in this figuration ensconces the hearer, wrapping the patient in a spatial envelope that may imply safety (the technician is close, is in control, is watching over her). And yet the voice is potentially threatening in that it is so intimate yet the voice of a stranger in a setting where a strange and potentially revealing medical procedure has been undertaken. Sound and voice thus become the devices of spatial configuration and interpenetration and of intersubjective coupling and separation between technician and subject, making for a nervous proximal affinity between strangers. This bond is, antithetically, also characterized by professional distance: The technician observes from outside the protective glass window, the images he observes may already offer him restricted knowledge about something she is still unaware of regarding her own body. Yet it is also characterized by empathetic closeness: the technician not only sees into the patient's interior, he is present there with her in the coil through the sound of his voice, bringing human warmth into a cold and solitary space.

MOVEMENT

We refer to the movement of the research subject in the MR machine, which was not directly observed, but is recognized by the misaligned consecutive images and constructed in the publicly performed imagination of the researcher(s). We also consider the bodily movement of the technicians or researchers, who performed the movement of the subject as they imagine and feel it themselves, to make sense of the data, to practically engage with the problem at hand, and to enact it in a pedagogical manner for the other researcher, who may be a junior colleague or student. Finally, we consider the movement from the screen image to the body of the researchers and vice versa as they collectively perform the movement that wasn't observed when it took place in the MR machine, but can now be seen, experienced, and dealt with through the researchers' enactments of the imagined movement.

In addition to sound, another confounding factor—motion, and its role in understanding and practical problem solving—is described in the text that follows. Similarly to the previous section, this portion of the text reports on an apprenticeship interaction. Once again, the experimental data were previously collected, and the apprenticeship takes place through the process of data analysis. However, the activity described here is situated in a cognitive neuroscience laboratory that largely practices scanning of a brain damaged patient population. Hence, the experimental subject of the scanning session was not the practitioner herself, but a clinical patient. The interaction reports on practitioners' involvement in assessing the existence of artifacts in the experimental data.

Over the course of the experiment a series of brain images are recorded. Each scan or image represents a brain slice. During the preparation of the experimental data for statistical examination, among other types of manipulation, researchers need to assess if the brain slice representations can be aligned. The assessment is achieved by viewing slices (in the axial, sagittal, and coronal view) represented on the computer screen over the time-course of the experiment. The researchers, by using mouse commands, alternate the view of individual images on the screen. The goal is to check if the representations of the brain slices align with each other. When nonalignment is detected, the researchers try to correct it. This is done in order to prevent the appearance of visible defects in the images (i.e., blurry splotches) once the statistical analysis has been performed. When the nonalignment is pervasive throughout the data, the data set has to be rejected.

As the excerpt from interaction will attest (Excerpt 9.2), the nonalignment between the brain images is explained in terms of the subject's movement that caused it. The artifact, caused by the subject's movement, is called the *motion artifact*. In motion artifact the movement of the subject is considered to be an *artifactual cause*, rather than the *visible regularity of independent "natural" phenomena* (Lynch 1989). While the subject's

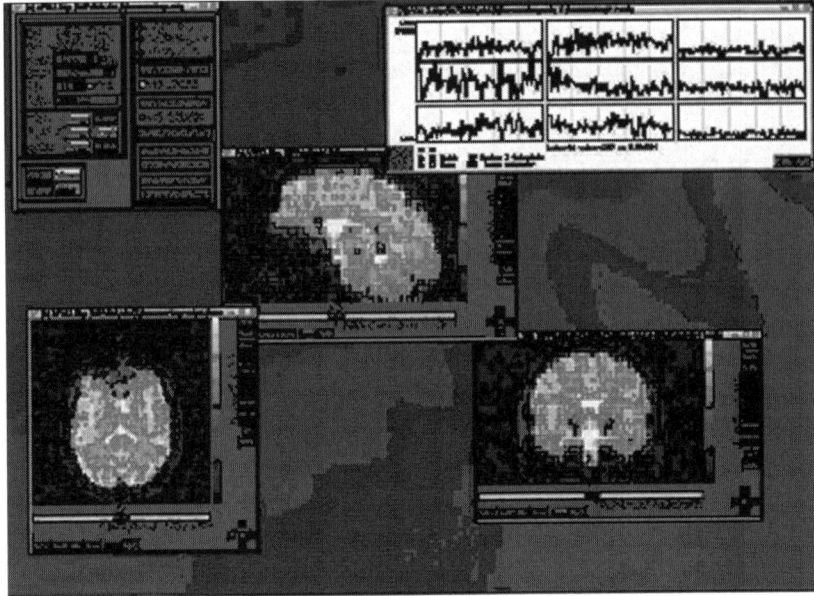

Figure 9.2 Computer screen as it appears during the motion correction activity.

brain and its function (evoked by experimental stimuli) are seen as independent natural phenomena, the potentially unintentional movement of the body is considered to be the cause of an intrusion or distortion in the visibility of fMRI images.

Notice, however, that what is believed to be the cause of the nonalignment (the movement) can only be inferred from its effect (the nonalignment of images). Rather than simply anchoring this nonalignment in the movement that caused it, the subject's movement is inferred from nonalignment. While the images are present on the computer screen and can be directly compared, the movement of the subject could not have been seen or experienced by the researchers while the subject was lying in the scanner. The movements that the subject produces during the scanning section are usually too small to be seen on the computer monitor where MRI technicians and researchers supervise what goes on in the scanner. While in the previous section the apprentice's prior experience was potentially evoked, in this example the experience is entirely created during the data analysis activity. Since the subject of the scanning session was a clinical patient and not the apprentice herself, the practitioners create, rather than re-create the experience of being in the scanner. The excerpt from interaction will illustrate how practitioners to recreate such an experience and to ground the circular explanatory flow between the images and the subject's movement (the images are not comprehended alone; they are understood in respect to the subject's movements; simultaneously, the researchers imagine such movements by grounding them in the brain images) engage their own bodies in the production of meaning.

The excerpt reports on an interaction between two fMRI practitioners. The practitioners are seated in front of a computer screen in the cognitive neuroscience laboratory. The person on the left is an advanced graduate student, Gina (G). Gina teaches Ursula (U), a newcomer to the laboratory, seated on the right, how to analyze fMRI experimental data. The excerpt from the interaction was recorded during an early stage in the practice of data analysis where experimental data need to be prepared for statistical examination. Each image on the computer screen represents a brain slice at a particular moment of the experiment. Ursula's task is to check for alignment between images in the series (representing the time-course of the experiment). This is done in order to keep experimental data clean from motion artifacts.

Excerpt 9.2

1. U: Up?

Figure 9.3a ((Stretches up.))

2. G: Yep again *((Nods))*
3. And then actually it looks like there is more involvement,
4. I am thinking there is more involvement=
5. in the front or in the back can you tell? [see Figure 9.3b, 9.3c]

Figure 9.3b and 9.3c ((Taps the front and the back of her head with a pencil.))

6. U: ((Works with the computer))
7. Back?

Figure 9.3d ((E. keeps touching the back of her head.))

8. G: Yeah right so you can see the voxels=

Figure 9.3e ((Points with the pencil towards the screen and moves the pencil up and down.))

9. in the back moving a lot more that the voxels in the front

Figure 9.3f ((Turns the pencil in the vertical position and moves it now up and down. At the same time she nods. The novice approves by nodding.))

10. so she is (.) this is an interesting movement because most people would=

11. if you think if you think laying in the scanner=

Figure 9.3g ((Pushes against the back of the chair.))

12. most people would nod=

Figure 9.3h ((Nods with her head.))

13. before they

14.

Figure 9.3i ((Movement of her neck against the back of the chair. The novice executes in parallel the same type of movement.))

15. cause your neck doesn't=

Figure 9.3j ((Points to the back of her neck.))

16. really stretch quite that way=

Figure 9.3k ((Moves up and stretches her neck.))

17. right?
18. U: (No)
19. G: So I don't know what she is doing ((Gesture of surprise))
20. but she is doing this. Hhhhh ((Laugh)) for some reason

Practitioners involved in the process of detection of nonaligned brain images need to identify motion artifacts in the images in order to exclude them from the further data analysis. A pervasive component of the process is the embodied performance of the experimental subject's movements. To understand the nonalignment between images the practitioners use their bodies to model the hypothetical movements of the subject. In line 1 Ursula stretches up to perform the subject's movement. Her stretching makes the feature present in the experimental data available in the public space of action. At the same time the performance signals Ursula's capacity to look at the images as an fMRI practitioner. In line 2 Gina approves.

Lines 3–7 show another multiparty interpretation moment. Now Gina takes the lead guiding Ursula's understanding of digital images. She does so by enacting the linkage between the images with the imagined and enacted movements. In line 3, by saying "it looks like", Gina first directs the attention towards the screen. Subsequently, she points to her own head (line 5) as the imagined head of the experimental subject. The indexical gesture links the semiotic field of images and the field of the embodied performance, and guides Ursula in answering Gina's question as to whether the difference between the images corresponds to a movement of the front or the back of the subject's head (line 5). To answer the question, Ursula directs her gaze from Gina's head to the computer screen and back (line 6). In line 7 she expresses her hesitant answer verbally: "Back?" During her search for the answer Gina keeps touching the back of her head thus guiding Ursula's reading of the digital images.

To strengthen Ursula's answer Gina further elaborates the performance of the *subject's movement in the images*. In line 8 she uses the pencil to point toward the screen, and moves it up and down. In such a way she projects the movement onto the images while enacting the selected features of the images in the public space of action. While the indexing portion of the gesture points towards the voxels that exist in the digital realm, the iconic/symbolic character of the sign recreates the voxels' movement in the physical space of action. As if *pulling the voxels* from the digital image onto the enacted movement gesture, the indexical gesture facilitates the *semiotic contagion* between the digital screen and the public space of embodied action. Subsequently (line 9) Gina slightly retracts her gesturing hand toward herself. She turns the pencil in the vertical position and moves it again up and down. This change in the pencil's position strengthens the iconicity of the gesture while it weakens its indexical character. The focus of the action is now on the enacted movement rather than the digital brain images. While Gina performs the voxels' movement through the pencil movement, she nods with her head thereby performing the subject's supposed movement through the movement of her own head. At this point, in contrast to her hesitant answer in line 7, Ursula can confidently confirm her understanding by nodding (line 9).

Gina liberates the movement from the flat world of the computer screen, incorporating it into her own body, and performing it and feeling it as a means of knowing it.

Now that the movement has been liberated from the flat world of the computer screen, Gina can further elaborate and comment on it (lines 10–17). Of particular interest is the performance and elaboration of the movement in a hypothetical space of action where the practitioners' bodies assume multiple discursive roles of individual, general, and imagined bodies. In order to underline the unusual nature of the subject movement, Gina asks Ursula to imagine herself in the scanner. Up to this point the practitioners were basing the supposed subject movement in the brain images on the computer screen. This anchoring of the subject's movement in the images gave the movement a sense of reality. In contrast, by evoking the hypothetical scenario, the practitioners are now deliberately creating an imaginary movement that cannot be seen in the images. Yet, this imaginary movement represents the space of the expected, correct, and hence in some sense real movement.

Such a hypothetical space of the correct movement is largely constructed through Gina's use of her own body. Through the multimodal discursive practice, Gina's body assumes the role of Ursula's body or the *human body in general* imagined to be laying in the scanner: "if you think lying in the scanner"; "most people would." Moreover, in creating her explanation, the expert explores the resources directly present in the material environment surrounding her as potential semiotic objects.

While saying "if you think, if you think lying in the scanner," the expert pushes abruptly against the back of the chair. Through this movement the chair gets *transported* into the hypothetical space where it assumes a crucial role in the discursive action. While the expert's body in the vertical position maps to a human body in the horizontal position, the chair on which the practitioner is sitting stands for the scanner table on which the subject is laying. Because the chair is in some ways like the scanner table (i.e., it is flat, it supports a person's back, etc.), and it happens to sustain Gina's body during the teaching session, Gina is able to enact the movement that *most people would* (line 12) produce in the scanner and the one that is unexpected. While common movements are those constrained by the physical characteristics of the scanner table, an *interesting movement* (line 10) is not.

When uttering: "Cause your neck doesn't= really stretch quite that way" (lines 15–16), Gina points at the back of her neck (line 15). By indexing the specific part of her own body, she inscribes ephemeral marks onto the general, *most people's body* that lies in the scanner. In line 16 she briefly moves up and stretches her neck in order to enact the movement that the subject performed in the scanner. At the same time she provides an embodied demonstration of the way in which "your neck doesn't really stretch": she "moves" in an "interesting" way.

Ursula, who is looking at and manipulating the screen, executes the same type of movement performed by Gina in line 14. She understands Gina's explanation through her own embodied performance of it. Rather than passively observing Gina, she coordinates with her and performs the *impossible* movement herself. This allows her to promptly answer Gina's question (line 18), verbally confirming the impossibility of the movement. By translating the difference between brain images into an effective physical motion, she learns that "your neck doesn't really stretch quite that way" (lines 15–16).

By comparing the general body and its movements constrained by the physical structure of the scanning environment to the subject's body, the expert labels the subject movement as *interesting* (line 10) and incomprehensible: "I don't know what she is doing but she is doing this ((laugh)) for some reason" (lines 18–19). In approximately 15 seconds of action, the expert's body *travels* from being a prototypical body (lines 11–15), to the subject's body (line 16), and finally becoming again a signifier of her own body (lines 18–19). At this point she ceases enacting imaginary bodies, and comments on their behavior. But even so, when expressing surprise (line 18) and laughter (line 19), her body is not just an envelope for internal, abstract thought, but a critical component of meaning production and practical problem solving.

CONCLUSION

Earlier in this paper we posed the questions, when does the research subject or patient become visible or present in the process of interpreting his or her MR data; and how does this "making visible" of the research subject occur? Above we have tried to show, through a microanalysis of a research laboratory and a sketch of a clinical MRI procedure, that "becoming visible" is not only, and is not even primarily, connected to the faculty of sight, or to the aspect of imagination captured in the concept of mental imagery or verbal representation. Rather, making visible entails aspects of embodied feeling and enactment. Furthermore, the experience of "making visible" occurs not most importantly in the mind of the researcher(s), but in the interactions among researchers or technicians, patients or research subjects, and the various other actors or agents that comprise the scene. Our point, again, is not simply that "making visible" or seeing is a cultural, social, or shared process and not strictly cognitive in the sense of being located in the individual mind. Rather, we emphasize that imagination is a process that occurs through intersubjective and multimodal experiences that cannot be reduced to the location of the individual or to a dominant sensory modality or paradigm as the primary one in the production of knowledge and meaning. Where, then, do we locate human subjectivity? And who, then, is the subject of these interactions, if not the individual researchers, technicians, and patients or research subjects who each play a part in the interactions documented? To address these questions, we would need to move further into the concepts of intersubjectivity and the intra- and inter-psychic aspects of imagination. While we cannot move forward with this project at the conclusion of our paper here, we propose that a theory of subjectivity that takes into account the specificity of the psyche and the factor of the unconscious is new ground that the cultural studies of science would do well to consider. For this project, we propose a return to feminist theories of the subject, based in psychoanalytic concepts, combined with work in distributed cognition and laboratory analysis. Although we cannot move this project forward here, we propose this constellation of approaches as a fruitful direction for the cultural study of science and communication.

NOTES

1. The literature on imagination in cognitive semantics includes Faucconier and Turner (2002), and Johnson (1987, 1991, 1993).
2. Introduced to MR imaging in 1992 (Connelly 1993), function study came late to this imaging modality as compared to others (ultrasound, PET, and SPECT). fMRI's current primary use is to "map" the function of the various regions of the human brain. This time-based technique was an outgrowth of echo-planar MR processes introduced by Peter Mansfield (1977). In current echo-planar MR imaging, one nuclear spin excitation is required to produce a single image, compared to the minutes required in traditional MR techniques to produce enough data for a single image. In 1987 echo-planar imaging was used for the first time to perform real-time motion studies of a single cardiac cycle (Chapman et al. 1987). By 1992 fMRI echo-planar imaging had been adapted to the imaging of function in the regions of the brain responsible for thought and motor control.
3. On the co-construction of meaning in workplace interactions, see Goodwin (1994, 1995, 1997, 2000a, 2000b, 2003a, 2003b, 2003c).
4. On practice as performativity in ethnomethodological and sociological science studies, see, for example, Lynch (1993), Lynch and Woolgar (1990), and Pickering (1992). Pickering (1995) refers to "the mangle" of human and material agency. In Pickering's "dance of agency," scientists as "active, intentional beings" tentatively construct new machines, adopting a passive role and monitoring the performance of the machine (Pickering 1995: 21–22). Knorr-Cetina (1999) distinguishes between science as practice and science as cognition in ways that are significant to our discussion but will not be developed here for reasons of space.
5. http://www.etymonline.com/index.php?search=imagination&searchmode=none
6. Although we cannot elaborate on this point here, imagination involves the work of fantasy and the unconscious. We propose that this is an inevitable, everyday aspect of meaning and knowledge production, even if it has been under-considered in the study of laboratory and clinical interactions. This angle is not fully developed here for reasons of space.
7. These concepts, identification, and projection, are drawn from psychoanalytic feminist theory of spectatorship and will not be fully developed here for reasons of space; see Cartwright (2007) for a full account of the meaning of these terms.
8. Transcription conventions adopted with some changes from Sacks, Schegloff, and Jefferson (1979), and Goodwin (1994).
   // The double oblique indicates the point at which a current speaker's talk is overlapped by the talk of another.
   = The equals sign indicates no interval between the end of a prior and start of a next piece of talk.
   (x.x) Numbers in parentheses indicate elapsed time in tenths of seconds.

: The colon indicates that the prior syllable is prolonged.
___ Underscoring indicates stressing.
( ) Parentheses indicate that transcriber is not sure about the words contained therein.
(( )) Double parentheses contain transcriber's comments and extralinguistic information, e.g., about gesture, bodily movements, and actions.
Punctuation markers are not used as grammatical symbols, but for intonation:
. Dot is used for falling intonation;
? Question mark is used for rising intonation;
, Comma is used for rising and falling intonation.

BIBLIOGRAPHY

Alač, Morana (2005) "From trash to treasure: Learning about the brain images through multimodality", *Semiotica,* 156–1/4: 177–202.
—— (2006) How Brain Images Reveal Cognition: An Ethnographic Study of Meaning Making in Brain Mapping Practice, Doctoral thesis, Department of Cognitive Science, University of California at San Diego.
—— and E. Hutchins (2004) "I see what you are saying: Action as cognition in fMRI brain-imaging practice", *Journal of Cognition and Culture,* 4, no. 3: 629–661.
Beaulieu, Anne (2004) "From brainbank to database: The informational turn in the study of the brain", *Studies in History and Philosophy of Science Part C: Studies in History and Philosophy of Biological and Biomedical Sciences,* 35(2): 367–390, Special issue on the "Brain in the Vat".
Burris-Mayer, Harold (1941) "Development and current uses of the acoustic envelope", *Journal of the Society of Motion Picture Engineers,* 37, no. 1: 109–114.
Block, Ned (ed.) (1981) *Imagery,* Cambridge, MA: MIT Press.
Cartwright, Lisa (2007) *Moral Spectatorship,* Durham, NC: Duke University Press.
Chapman, R., R. Turner, R.J. Ordidge, M. Cawley, R. Coxon, and P. Mansfield (1987) "Real-time movie imaging from a single cardiac cycle by NMR", *Magnetic Resonance Medicine,* 5: 246–254.
Descartes, René (1642/1984) "Meditations on first philosophy", in John Cottingham, Robert Stoothoff, Dugald Murdoch, and Anthony Kenny (eds), *The Philosophical Writings of Descartes,* vol. 2, Cambridge, UK: Cambridge University Press.
Doane, Mary Ann (1980) "The voice in the cinema: The articulation of body and space", *Yale French Studies,* 60 (1980): 33–60. Reprinted in Leo Braudy and Marshal Cohen (1999) *Film Theory and Criticism: An Introductory Reader,* 5th edition, New York and London: Oxford University Press:. 363–375.
Dumit, Joseph (2004) *Picturing Personhood: Brain Scans and Biomedical Identity,* Princeton, NJ: Princeton University Press.
Fauconnier, Gilles and Mark Turner (2002) *The Way We Think: Conceptual Blending and the Mind's Hidden Complexities,* New York: Basic Books.
Finke, Ronald, Thomas B. Ward, and Steven M. Smith (1993) *Creative Cognition,* Cambridge, MA: MIT Press.
Fodor, Jerry A. (1975/1981) "Imagistic representation", in N. Block (ed.), *Imagery,* Cambridge, MA: The MIT Press, pp. 63-86.

Goodwin, Charles (1994) "Professional vision", *American Anthropologist*, 96(3): 606–633.

—— (1995) "Seeing in depth", *Social Studies of Science*, 25: 237–274.

—— (1997) "The blackness of black: Color categories as situated practice", in L. Resnick, R. Säljö, C. Pontecorvo, and B. Burge (eds), *Discourse, Tools and Reasoning: Essays on Situated Cognition*, New York: Springer-Verlag: 111–140.

—— (2000a) "Practices of seeing, visual analysis: An ethnomethodological approach", in T. van Leeuwen and C. Jewitt (eds), *Handbook of Visual Analysis*, London: Sage: 157–182.

—— (2000b) "Action and embodiment within situated human interaction", *Journal of Pragmatics*, 32: 1489–1522.

—— (2003a) "Pointing as situated practice", in K. Sotaro (ed.), *Pointing: Where Language, Culture and Cognition Meet*, Mahwah, NJ: Lawrence Erlbaum: 217–241.

—— (2003b) "The semiotic body in its environment", in J. Coupland and R. Gwyn (eds), *Discourses of the Body*, New York: Palgrave Macmillan: 19–42.

—— (2003c) "Conversational frameworks for the accomplishment of meaning in aphasia", in C. Goodwin (ed.), *Conversation and Brain Damage*, Oxford: Oxford University Press: 90–116.

Heath, Christian and Jon Hindmarsh (2002) "Analysing interaction: Video, ethnography and situated conduct", in T. May (ed.), *Qualitative Research in Action*, London: Sage: 99–121.

Johnson, Mark (1987) *The Body in the Mind: the Bodily Basis of Meaning, Reason and Imagination*, Chicago: University of Chicago Press.

—— (1991) "The imaginative basis of meaning and cognition", in S. Kuchler and W. Melion (eds.), *Images of Memory: On Remembering and Representation*, Washington, D.C.: Smithsonian Institution Press: 74–86.

—— (1993) *Moral Imagination: Implications of Cognitive Sciente for Ethics*, Chicago: University of Chicago Press.

Knorr-Cetina, Karin (1999) *Epistemic Cultures: How the Sciences Make Knowledge*, Cambridge, MA: Harvard University Press.

Kosslyn, Stephen M. (1980) *Image and Mind,* Cambridge, MA: Harvard University Press.

Latour, Bruno (1999) *Pandora's Hope: Essays on the Reality of Science Studies*, Cambridge, MA: Harvard University Press.

Lynch, Michael (1993) *Scientific Practice and Ordinary Action: Ethnomethodology and Social Studies of Science*, Cambridge, UK: Cambridge University Press.

Mansfield, Peter (1977) "Multi-planar image-formation using NMR spin echoes", *Journal of Physics C-Solid State Physics*, 10, 3: L55–L58.

Murphy, Keith M. (2004) "Imagination as joint activity: The case of architectural interaction", *Mind, Culture, and Activity*, 11 (4): 267–278.

—— (2005) "Collaborative imagining: The interactive use of gestures, talk, and graphic representation in architectural practice", *Semiotica*, 156–1/4: 113–145.

Nishizaka, Aug (2000) "Seeing what one sees: Perception, emotion, and activity", *Mind, Culture and Activity*, 7(1–2): 105–123.

—— (2003) "Imagination in action", Theory & Psychology, 13(2): 177–207.

Pickering, Andrew (1992) *Science as Practice and Culture*, Chicago: Chicago University Press.

—— (1995) *The Mangle of Practice: Time, Agency and Science*, Chicago: Chicago University Press.

Pylyshyn, Zenon W. (1973) "What the mind's eye tells the mind's brain— A critique of mental imagery", *Psychological Bulletin*, 80: 1–24.

Roepstorff, Andreas (2001) "Brains in scanners: An Umwelt of cognitive neuroscience", *Semiotica*, 134: 747–765.

Suchman, Lucy (2000) "Embodied practices of engineering work", *Mind, Culture and Activity*, 7: 1–2, 4–18.

Tye, Michael (1991) *The Imagery Debate*, Cambridge, MA: MIT Press.

Part IV

Science Images and Contemporary Art

10 Neuroscience and Contemporary Art

An Interview

Gabriele Leidloff and Wolf Singer

Leidloff: Imagine a person who is shown nothing yet believes she has seen everything—what will she see? What do we do when we see? What is needed in order to see an image?

Singer: Seeing is a highly active process in which knowledge based interpretations and inferences prevail. Seeing requires a priori knowledge of the visual world. Without such knowledge it would be impossible to interpret the image on the retina because it consists of nothing but a two-dimensional distribution of luminance values. In order to extract from this distribution the information necessary for distinguishing individual objects and separating these objects from the background, the brain needs a set of rules according to which brightness distributions can be segmented into contours, contours grouped into figures, and figures separated from the embedding environment. Only when these steps of a perceptual operation have converged successfully is it possible to proceed with the identification process proper, which also requires a priori knowledge. This leads one to ask how the implicit knowledge required for perception is acquired and implemented in the brain. It is well established by now that all knowledge and the rules for processing this knowledge are located in the functional architecture of the brain's neuronal networks. This architecture is equivalent to the blueprint for the design of the connections of neurons. Connections can be excitatory or inhibitory and they can be strong or weak. These variables determine the dynamics and the structure of the spatial-temporal activity patterns that develop in neuronal networks and are considered to form the substrate of all cognitive and executive processes. The question of how brains acquire knowledge and the rules according to which this knowledge is used can thus be reduced to the question as to which factors specify the functional architecture of the brain.

The most important factor and hence the most important source of knowledge is evolution. The generation of variability through genetic mutations and simultaneous selection pressures has led to the development of neuronal networks adapted to the specific environments of organisms. Knowledge about successful strategies of interpreting two-dimensional brightness distributions of retinal images have been acquired over time, stored in the genes, and expressed in functional architectures of nerve nets. Evolution can therefore be considered a cognitive process. The knowledge acquired through this process is highly idiosyncratic and confined to those properties of the environment that are relevant for survival. As a result, we perceive only a fraction of the world around us, are sensitive to gradients rather than absolute values of physical and chemical variables, and subdivide physical continua into distinct sensory categories arbitrarily. In most organisms, especially in those with highly developed nervous systems, this evolutionary "learning process" is complemented by experience-dependent modifications of neuronal architectures. During this process that starts after birth and in humans ends at the age of twenty, neuronal connections are highly malleable and susceptible to experience-dependent modifications. There is a steady formation of new connections and simultaneous removal of pathways identified as ineffective. This process of fine-tuning is guided by neuronal activity. Connections between neurons with a high probability of temporally correlated activity tend to stabilize while connections are removed between neurons that are coactivated only infrequently. Neurons wire together if they fire together. Because neuronal activity in the developing organism is massively influenced by sensory signals and self-initiated interactions with the environment, this use-dependant selection of neuronal connections leads to an adaptation of the genetically pre-specified functional architecture to the actual requirements of the environment of evolving organisms. This phase of knowledge acquisition contributes to the culture specific shaping of cognitive schemata. Here criteria are acquired for basic aesthetic judgments and the neural networks undergo culture specific modifications in order to permit rapid, automatic comprehension, and the generation of the symbols used in interindividual communication.

Once this developmental phase is complete, large scale reorganizations of connectivity no longer occur. The outgrowth of new connections has come to an end and connections are removed under pathological conditions only. Learning induced modifications of functional connectivity then depend solely on the greater or lesser efficiency of the previously stabilized connections. These changes also have a structural correlate but at this stage the modifications occur mainly at the molecular level.

To return to the initial question: What we see and how we see depends on the brain's ability for interpreting and constructing. This ability is specified by evolution, developmental imprinting, and conventional learning. Thus, when we perceive an image, we compare the sparse signals made available to us through the eyes with a large body of a priori knowledge about the likely condition of the world. We process the signals according to rules acquired through the described mechanisms. In most cases we are unaware of this process and believe we perceive what is actually in front of our eyes. We cannot imagine perceiving otherwise because our consciousness has no access to the mechanisms that generate percepts from sensory signals.

By experimenting with our senses, artists and scientists have contributed to the insight that perceiving is constructing and selecting the most likely hypothesis.

Leidloff: What is an image of the mind? How are mental images transformed into material images?

Singer: Beyond the level of the retina the neuronal correlates of the visual percept become increasingly abstract. They consist of highly complex spatial-temporal patterns of neuronal activity. At early stages of visual processing, neuronal networks extract simple features from the retinal brightness distribution such as the orientation of contours, their spectral composition, or direction of motion. These elementary features are subsequently recombined in neurons responding to perceptual primitives, which consist of the recombination of distinct elementary features. This condensed information is then routed along different processing streams, one of which deals with the identification of objects and the other with the analysis of their location in space and their movement. The "where" and the "what" of an object are analyzed in different and spatially separated neuronal networks. Patients suffering from lesions in the where-pathway develop a syndrome called "visual appraxia." They are able to identify objects but they have difficulties in processing them because the where-pathway conveys the information required for the programming of visually guided movements. The "where networks" analyze shape and motion of objects only to the extent that appropriate reaching and grasping movements can be programmed. By contrast, lesions in the what-pathway lead to "visual agnosia." Patients are no longer capable of identifying objects visually but have no difficulty in locating and moving them appropriately. They need to infer information about the identity of an object from their sense of touch. In addition to this compartmentalization of the visual system into two main processing streams, there are a large number of further subdivisions. About thirty distinct areas of the neocortex deal with the analysis of visual signals, covering approximately one third of the cortical surface. Each of these areas processes different aspects of visual objects such as texture, color, global outlines, local details, and distance. For the representation of a visual object all of these areas cooperate through a large number of reciprocal connections. Hence, there is no single location for the representation of a particular visual object and no single nerve cell which responds exclusively to a particular object. Rather, in most cases, the neuronal correlate of a perceived object and even more so of the percept of a visual scene is an immensely complex spatial-temporal pattern of activity in which myriads of spatially distanced neurons participate. The complexity of these patterns escapes our imaginative abilities. We are still far from understanding the nature of the distributed codes of our percepts but it can be predicted that they will have to be described in abstract mathematical terms as states of a dynamical and non-linear system designed to generate a virtually infinite number of different states.

Also unresolved is the question of how such dynamical patterns give rise to what we experience as the content of our consciousness. The brain's organization provides no hint whatsoever of the existence of a center in which the partial results obtained in the many spread-out processing areas would be bound together into a coherent percept. The information that defines a particular object is spread across numerous processing networks and is passed on to distinct structures in simultaneous moves. Even parts of the signals that do not reach the level of awareness can trigger emotions and give rise to actions. There is no site in the brain where an observing homunculus could be located and there is no center in the brain that coordinates the myriads of separate processes that occur simultaneously and in different modalities. The brain is an orchestra with no conductor. This raises the fascinating question of how the brain organizes itself in order to create ordered states which then are the correlate of percepts, decisions, action plans, and motor commands.

Following a discovery made in the late eighties of the last century we believe that the brain utilizes precise temporal synchrony among the responses of separate neurons as a relation defining code. It is our working hypothesis that the gathering of disseminated neurons supporting a particular percept is temporarily bound together through the precise synchronization of the activity of the neurons participating in the respective gathering. When the eyes move to a new target the percept changes and new gatherings are formed through the activation of other neurons. This new constellation is also defined by synchronicity. Evidence indicates that responses tend to show an oscillatory modulation and the binding of these oscillations synchronizes them. The respective rhythms occupy a broad frequency range that is centered around 40 to 60 Hertz. Thus, neuronal representations of sensory objects have no resemblance with these objects. Rather, they are exceedingly complex, dynamical, and ever changing patterns of activity resulting from the self-organizing coordination of large numbers of spatially spread neurons. Most of the information encoded in these patterns lies in the spatial and temporal relations among the responses of the neurons that contribute to a particular representation.

Leidloff: What is the information needed to distinguish between an abstract and a figurative image? What do you see when you look at my work?

Singer: In order to distinguish between abstract and figurative paintings we need to call upon our a priori knowledge. If we had no preconception of the identity of objects in a figural painting because the objects are unfamiliar to us, even a figural painting would appear abstract to us. When opting for abstract painting, artists deliberately choose to avoid figural elements that share qualities of real world objects we are familiar with. When confronted with such paintings one often observes that one's visual system engages in a search process applying the full set of inborn and acquired Gestalt rules in order to find ways of arranging colors and contours that correspond to see-

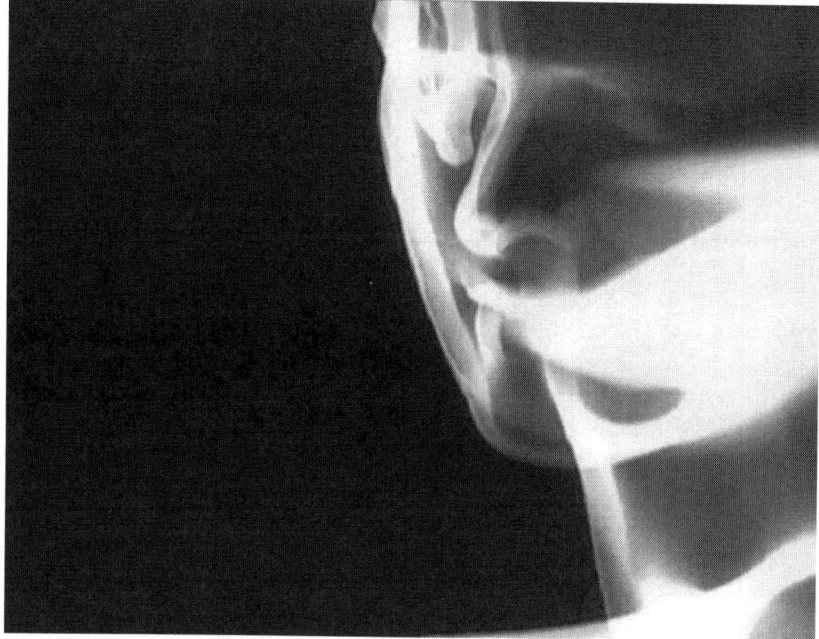

Figure 10.1 Gabriele Leidloff, "Ugly Casting 1.2.," 2000, digital-video installation, video still; radiograph. (© Gabriele Leidloff)

ing familiar objects. Depending on the observer's expectations this search may have frustrating or stimulating effects. In all likelihood, the appropriate reception and contemplation of a non-figural composition is possible only once the observer abandons the fruitless search for familiarity.

Leidloff: What are the origins of illusions and tricks of the senses?

Singer: Illusions of the senses are the result of the inferential nature of sensory processes. Human intuition is unable to question the objectivity of perception. This is what makes sensory illusions exciting. We do not believe in the illusion unless we obtain additional evidence from direct assessment of physical qualities. Even if we fully understand the mechanisms that give rise to the misperception, the illusion will not go away.

The rules that we apply for interpreting the brightness distribution on the retina are adapted to natural conditions and enable us to make educated inferences on what we see. This often requires extrapolations and in particular the computation of relations among features such as wave length, size, distribution of shadows, and many others. The evaluation of relations among attributes rather than the determination of the attributes themselves is of utmost importance for the generation of invariance, because the attributes change with distance, rotation, and illumination while relations remain constant. In most cases we are unaware of the sophisticated computations that underlie the evaluation of relations that play a crucial role in generating the neuronal correlates of percepts. These processes can only be experienced by creating artificial conditions which lead to illusions of the senses. Scientific experiments create such conditions because illusions make it possible to observe the neuronal mechanisms applied in the construction of the visual world. Artists have ingeniously played with these mechanisms. This intentional experimentation with cognitive mechanisms is particularly evident in the masterpieces of trompe l'oeil and in the paintings of impressionists and artists of the op-art movement. However, even seemingly "naturalistic" paintings achieve realism only by making use of the synthetic capacities of the observer's visual system. Scientific analysis makes explicit what artists have known implicitly for a long time, namely that seeing is based on construction, interpretation, and inference. In essence, all perception is illusory but this becomes evident only when additional information is gathered that turns out to be of a conflicting nature.

Leidloff: My film scenes constructed through image generating techniques such as radiography, ultrasound, magnetoencephalography, computed tomography, and eye-tracking reveal a paradox. Through an inversion of representations, expected views are estranged and supposedly familiar images turn into surprises. Images that seem erroneously familiar, a face or a body movement like that of a fashion model are exposed to an estranging gaze that opens surfaces. Viewers are often stunned and experts claim that medical apparatus is unable to produce such effects.

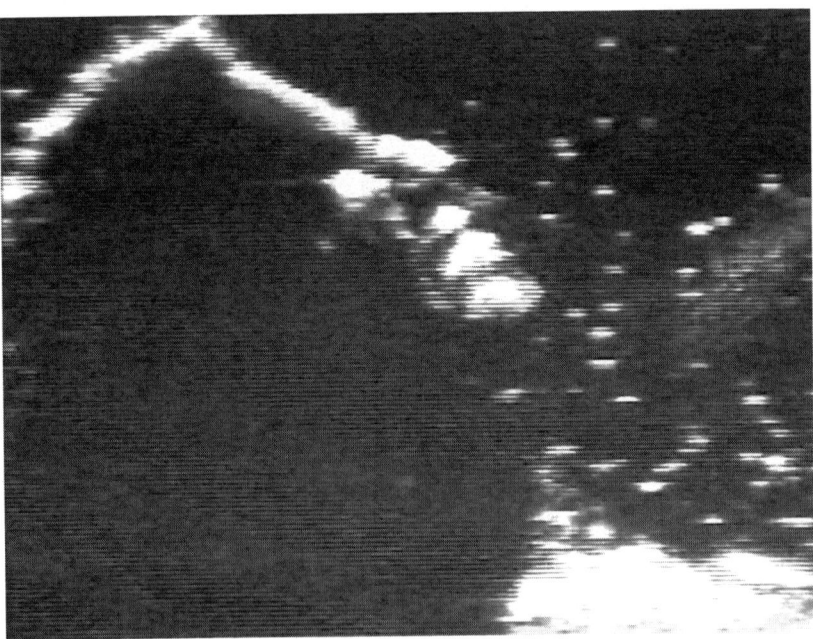

Figure 10.2 Gabriele Leidloff, "www.pussylink.com," 1998, digital-video installation, video still; ultrasound picture. (© Gabriele Leidloff)

Singer: By reversing the contrast of pictorial representations of real world objects, a conflict is generated between expectation and what is actually seen. Consequently, the visual system searches for the most plausible solution and this often implies assumptions like seeing a surface from an unusual angle, seeing a face from behind, assuming the source of illumination to be inside rather than outside an object, and so on. Inverting the contrast often also causes an initial reversal of the association of contours with figure or background. In order to be able to identify a figure it is therefore necessary to reverse the initial association. A resolution of these conflicts between knowledge-driven expectations and actual conditions requires an increased investment in attentional processes and search strategies. Such pictures therefore evoke feelings of insecurity, surprise, strangeness, and even shock.

You seem to be aware of these effects because the application of x-ray techniques in their initial film based version inverts contrasts. The solid appears bright, the empty dark. Clinicians have become adapted to these inverse representations to a degree that even the modern digital imaging techniques, x-ray tomography, functional magnetic resonance imaging, and sonography are adjusted to produce pictures with inverse contrast, even though the generation of images with normal contrast would require no additional computation. This is yet another example that illustrates how much our perception depends on prior experience and learning.

Leidloff: How do you interpret these unintended and new applications of the image generating techniques? Do you think that they challenge the alleged objectivity of the scientific images? How would you describe a scientist's approach to images in contrast to that of an artist?

Singer: Since the development of telescopes and microscopes has extended the reach of our eyes to the very distant and the very small, scientists have been fascinated by the strangeness and beauty of a world that extends beyond the range of our natural senses. This world is by no means stranger than the mesoscopic world to which our senses have adapted through evolution. Before the invention of magnifying lenses, we had no way of familiarizing ourselves with the morphology of galaxies and bacteria. Hence, newly discovered structures, whether aesthetically appealing or not could be interpreted as the product of an artist expressing his or her fancy of an imagined world. From early on, scientists and science illustrators have been tempted to exploit the faculties of their tools to create images that look like a piece of art. At present, it is not uncommon to extend scientific conferences by including special exhibitions of photographs and images of objects not perceivable by the naked eye. There are competitions and prices for the best artistic product of this fusion of arts and science images. Not unexpectedly, artists have also discovered the potential of imaging tools for their purposes and your artwork and your Forum of Art and Neuroscience l o g - i n / l o c k e d o u t—http://www.locked-in.com is a magnificent example of the artistic use of sophisticated scientific instruments.

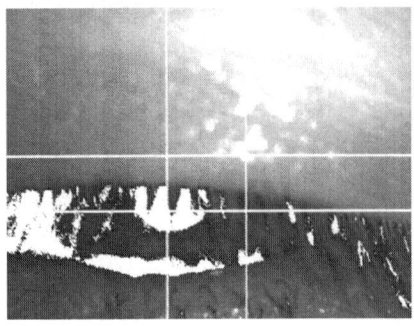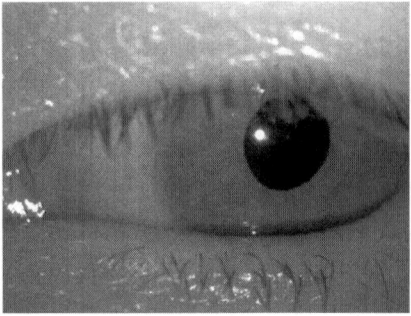

Figures 10.3a and 10.3b Gabriele Leidloff, "In Pursuit," 2004, digital-video installation, video stills; eye-tracking. (© Gabriele Leidloff)

This raises the interesting question as to the dividing line between a scientific and an artistic exploration of reality. The primary goal of the scientific endeavor is to adjust the numerous parameters of measuring devices in such a way that they produce reproducible and undistorted pictures. However, since the invisible remains invisible without the assistance of tools, the adjustment of the respective parameters is, to some extent, left to conventions. There are cases where it is impossible to decide unequivocally which one of many possible representations is close to "reality." Often, it is the scientist's choice which of the representations will eventually be considered canonical. Thus, the way in which the invisible is rendered visible is subject to construction and interpretation, very much in the same way as is our primary perception of the perceivable world. I think, one of the great virtues of your art is to make this process transparent and accessible to observation.

Leidloff: You are a participant in the international Forum of Art and Neuroscience l o g - i n / l o c k e d   o u t— http://www.locked-in.com. In your opinion, what could be achieved for the arts by an engagement of neuroscientists? Do you consider a public dialogue between artists and scientists desirable and productive? What could be a result?

Singer: By applying the imaging tools developed for scientific purposes to objects of the everyday world, you demonstrate the constructive nature of the allegedly objective imaging technologies. Your x-ray works make it intuitively understandable that the world as we perceive it is the result of construction and that what we perceive is nothing but an idiosyncratic interpretation of the world. These pieces of art demonstrate that the world could look different if we were different, if we had different sensory organs and brains. But as we are what we are, it is impossible to imagine how the world would look if we were different.

I would like to leave it to the reader to decide whether there is a difference between a creative artist and a scientist who develops a complicated tool for the exploration of the invisible and deliberately adjusts the param-

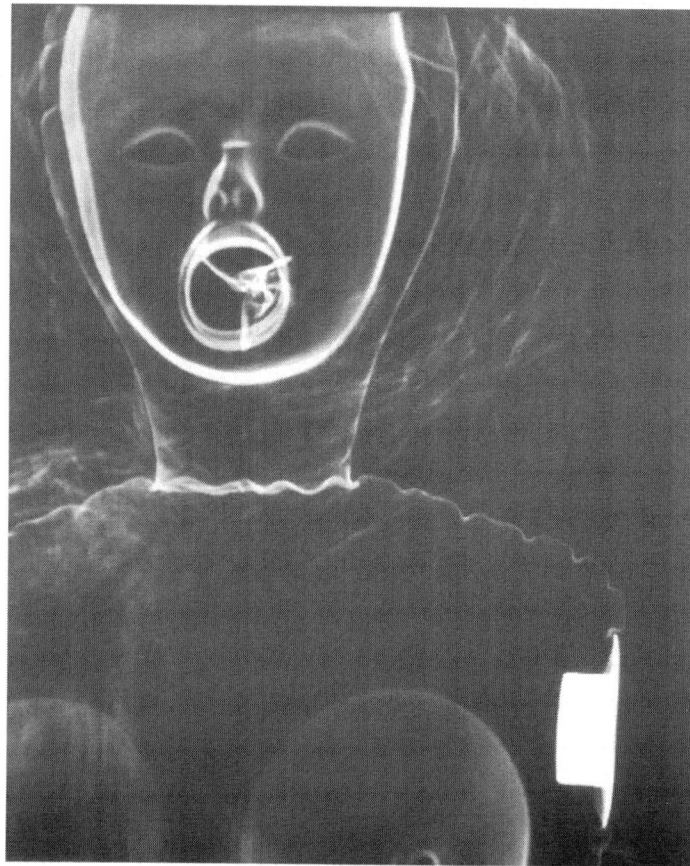

Figure 10.4 Gabriele Leidloff, "Girl 1," 1996, radiograph. (© Gabriele Leidloff)

eters and the selection of the field of vision in order to obtain an image that is provocative or reveals a hidden aspect of reality.

In exploring the world, scientists, like artists, work in a concept driven way and both need to acquire and master specific skills for the exploratory process and its documentation. Both can provide new insights into our condition and neither has access to absolute truth. Finally, in both cases the evaluation of the validity of insights and their quality depends on the view of others who consider themselves competent to judge. In both cases aesthetic criteria enter into these judgments. It is often held that scientific and artistic creations differ because scientific theories can be verified through experiments. However, this applies only to some scientific theories. The validity of others is based on aesthetic criteria such as coherence, consistency. There are, in turn, artistic artifacts with predictive power. We know works of art that envisioned changes and affected our views on the world and our self-perception which, in retrospect, can be verified. Is science a branch of the arts? Are changing views of the world prepared by artists' foresight and do they influence theory construction in the sciences? Do the views of the world imposed upon us by scientific discoveries change the way in which artists explore and represent the world? I think these questions deserve as answers at least a "in all likelihood." It is indeed puzzling that at the turn of the nineteenth to the twentieth century the Cartesian coordinates of space and time were challenged by both scientists and artists. Relativity theory disputes the absolute nature of space and time, quantum theory questions the linear causality of the dynamics that rule the world, the inner monologue in literature disputes the linearity of time, the cubist and eventually abstract paintings break with the seemingly absolute coordinates of space, and composers dare to transcend the harmony relations of tonalities that dominated hearing traditions for centuries.

Leidloff: You are a neuroscientist and director of the Max Planck Institute for Brain Research and I am an artist whose work responds to technical innovations and reflects on theories in the field of the neurosciences. Would you give me a position in your institute? The program of the Frankfurt Institute for Advanced Studies invites innovative thinking without including, however, artists, writers, and filmmakers. Why?

Singer: In principle, I would be able to employ you at the Max Planck Institute or in the newly founded Frankfurt Institute for Advanced Studies (FIAS), at least for a limited period of time. A condition would be, however, that I demonstrate to the selection committees that this cooperation would benefit the sciences and lead to discoveries that are publishable in peer reviewed journals, and the peers are not artists or filmmakers but scientists. Unfortunately, our disciplines address different audiences, are directed at different markets, and apply different evaluation mechanisms. Your own as well as my success depends mainly on the judgment of others. It has effects on our respective work whether others appreciate what we do and whether they consider it valuable. My forum is peer reviewed journals, yours is exhibitions. These worlds are by and large not connected and our respective products are not considered to be of the same currency. Rare are scientific discoveries that make a fortune through patents and rare are artistic products that make the artist famous and rich. So, for the time being we are dependant on financial support and our collaboration will rely on personal relations.

Leidloff: What is, in your view, intuition? And what is intelligence?

Singer: Intuition is implicit knowledge that is not amenable to conscious reflection. It is essentially a knowledge that has been acquired during evolution, during early imprinting, or as a result of learning processes that did not involve conscious reflection. The knowledge base from which we derive our intuitions is apparently broader than the explicit knowledge that we are aware of and can communicate in conscious discourse. Because we have no conscious recollection of the acquisition of our implicit knowledge, the contents of this knowledge often take the form of beliefs, convictions,

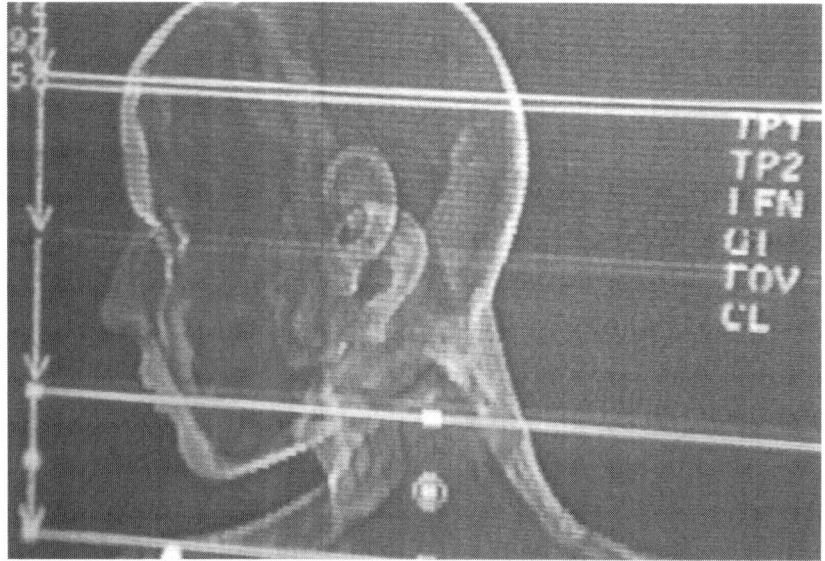

Figure 10.5 Gabriele Leidloff, "www.pussylink.com," 1998, digital-video installation, video still; CAT scan. (© Gabriele Leidloff)

or faith. We are not prepared to discuss their validity because they are an integral part of our idiosyncratic way of perceiving the world and attributing values. Aesthetic judgments are of this nature.

Leidloff: The arts are often seen in a context of madness rather than scientific rationality. What is, in your view, madness?

Singer: Madness is a state of mind that deviates from the average distribution of states of mind and is not well adapted to the constraints of daily life. Nonetheless, it may be a lucid state of mind that provides unconventional insights into what may be aspects of reality not perceived under conditions of normality and by the majority. The distinction between mad and normal is a matter of conventions and the dividing line is subject to cultural modifications. Art history is particularly rich in examples. How often have works of art been characterized and disqualified as the product of a mad or distorted mind and were later valued as seminal for the initiation of a new view of the world. Here, again, artists and scientists share the same fate.

Leidloff: What is, in your view, genius and ingenuity?

Singer: Artists and scientists are geniuses if a majority of competent judges consider them as such and if this judgment survives the test of time. Ingenuity is probably the result of a highly idiosyncratic mixture of gifts that can only be described individually and resists generalization. Necessary ingredients for the expression of ingenuity are certainly independence of immediate recognition, the courage to go long ways alone, and the gift to view things differently, and to bind the previously unbound.

RELEVANT BOOKS BY WOLF SINGER

(2002) *Der Beobachter im Gehirn. Essays zur Hirnforschung*, Frankfurt am Main: Suhrkamp Verlag.

(2003) *Ein neues Menschenbild? Gespräche über Hirnforschung*, Frankfurt am Main: Suhrkamp Verlag.

Part V
Images of Science

11 Women Scientists in Mainstream film

Social Role Models—A Contribution
to the Public Understanding of Science
from the Perspective of Film Sociology

Eva Flicker

More dramas starring women scientists must be shown on television to
encourage female students to follow careers in science and engineering.
(Sarah Cassidy, *The Independent*,11 April 2005: *Home News* 11)

The larger research context of this work can be found in the thematic field
"public understanding of science," exploring the relationship of science and
its public image in society. Based on selected cinema and television films
of the past seven decades, I will analyze which changes and also which
recurring cinematic stagings of scientists are applied and which social functions these might have. The text investigates previously published material
(Flicker 2003, 2004) in order to present the development of various typology models on the basis of more recent films. Following an introduction
to the study's central line of questioning is a short description of the methodological approach. A sketch of social clichés of (male) scientists then
precedes the cinematic presentation of women scientists. Here, previously
published analyses are briefly sketched out in order to more clearly describe
a new type of woman scientist who in recent years has been representative of strong women's roles, which are already evident as new, ritualized
forms.

SOCIOLOGY—FEATURE FILM—GENDER—SCIENCE—
SOCIETY: OUTLINING THE MAIN ISSUES

Sociology does not have exclusive rights to the task of observing and
describing society; the various mass media are also involved in this role.
As functional systems within society, they naturally follow their own laws
with distinct organizations and roles, including links to the social environment (Luhmann 1997). In this, distribution media take more liberties in the
construction of pictures of reality, yet their reference system nonetheless
remains, more or less, social reality. Film, cinema, and television, as popular mediating authorities in documentary forms as well as in the area of
fiction, contribute to the general picture and also the public understanding

of science. The social ambivalence between trust and mistrust in science, between a belief in progress on the one hand, and skepticism with regard to technology or fear of the uncontrollability of technology on the other, are well suited for popular cultural depictions of science in feature films.

"A lack of public understanding about jobs in science, engineering and technology is thought to be a large part of the problem. Television drama is recognized as one of the most powerful ways of changing perceptions" (Cassidy 2005: 11). The reality of everyday life in the scientific world is shaped by the complexity of the specialized subject areas, which are even beyond the grasp of the scientists within a discipline. Mass media play a crucial role in this process in that they contribute to science's construction of meaning. The pictures, stereotypes, and myths of science and scientific work that they transport are all culturally anchored.

Drawing from the broad area of mass media's communication of science, in the present work we will focus on the medium of the fiction film. "Fiction film with a scientific theme" is a label that I have introduced. It is difficult to categorize within standard film genres. What is meant here are films whose plots are set in a scientific milieu. Films in this category can be categorized from within the various common genres, including comedy, crime, action, and science fiction.

The main criteria in putting together the sample was the presence of a woman scientist in a crucial role in the film. There was a clear surplus of science fiction films available because this is the most common genre for scientific themes.

In the following film analyses, gender constructions are examined at several different levels, such as status, roles, body, language, and attributes. The stereotyping of gender through doing and setting gender in fiction films proves to be a recurring medial staging (Keppler 2004), which has been collectively reinforced as a stock of knowledge and values that have become social tradition. In a current study on gender stereotypes in TV series,

> it becomes evident that the leeway of acceptable behavior for women is much narrower than for men. Men are allowed almost everything: one practices a well-meaning leniency with male chauvinists, little creeps, major bad guys, and disoriented softies. On the contrary, women must display the currently appropriate behavior; they must be self-confident and tough, possess a healthy egoism and have interesting, minor weaknesses, but on no account can they forfeit their femininity. They must please their men in gender battles and at the same time, be good, loyal girlfriends and (if relevant) good mothers. (ORF 2005: 65)

In terms of the interconnection of genre and gender, on which there has been a lengthy debate and numerous publications within feminist film studies (Braidt 2004, Gledhill 2004, Kuni 2004), I would like to single out the following two aspects: *first,* it touches upon the issue of cinematic modes of representation and thus the analysis of a connection between what is shown and what can be perceived as real in the environment—concentrating on the relationship of "reality," "realism," and "representation" in connection with genre and gender constructions. Representation in a medium follows a cultural process of negotiation.

> From this perspective, media's fictions—contrary to socialism, feminism, or psychoanalysis—are not conceived as either "reflection" or "distortion," which can be measured on a set package of primary "real" relations, but rather, as a site of cultural circulation and secondary creative processing. (Gledhill 2004: 200)

Representations in film are thus not to be examined in terms of their reference to social reality alone, but should also be interpreted within their own medium of film, its practices and mutual citations. *Cultural circulation* signifies "the interaction between diverse discourses, aesthetic practices, and cultural imagined worlds—processes that transport gendered ideologies and construct narrative worlds, which we call genres" (Gledhill 2004: 200). On the one hand, in their intertextuality, films within a film genre create references to one another through their creative practices (or quotations) and, on the other, have interfaces with social discourses. For the present study, this means that we must pay attention to the artistic dimensions of a feature film's storytelling and also grasp the connections to social discourses about science. Gender constructions are central categories in both dimensions.

Second, it is necessary to work out patterns of recognition and repetition. Mainstream genres work with the audience's experiences and expectations (among other things) and play with a mixture of known and novel elements. This aspect is particularly important when the aim is to ritualize something. Repetitions are an essential structural principle common to all genre films. "'Genre' is therefore to be understood as a process, during which stereotypes are communicated and mediated to the audience" (Hölzer 2004: 211). Genre-typical characteristics are therefore repeated, varied, or also employed ironically. This applies to serial films in particular, where a part two refers to a part one (for example, *Tomb Raider: Lara Croft* from 2001 and *Lara Croft Tomb Raider—The Cradle of Life* from 2003). The audience must be able to recognize and decode this form of *quotation.* In the present text, I will flesh out several forms of repetition of women scientists' traits that have been effectively deployed for many decades in film.

Box office hits or blockbuster films are lined up every year according to their box-office takings. The feature films presented here all belong to this category of mainstream film with widespread international distribution. Their popularity makes them especially relevant for media sociology.

METHODOLOGICAL REMARKS

The relevant academic literature does not adequately develop the film-analytical method of sociological interpretation of films. I took over the first borrowings and definitions from Werner Faulstich (1988). Concretely, I must further develop the methodological procedure. Sociological film interpretation enables a comparative analysis of feature films in a larger sample in relation to their social ties and cross-references. Films, as representational systems and social means of communication, must be analyzed with reference to their social context of production and reception. Film interpretation inquires into a film's message. Films are examined in terms of their position with regard to such issues as relations of dominance, problems, and marginal groups, without relying on simplified concepts reflecting on films as 1:1 depictions of society. Sociological film interpretation occurs primarily for large samples, which are systematically occupied with focused inquiries (e.g., certain epochs, professions, socio-political themes). The principle of intertextuality is central to film interpretation. In the present study, references and quotations in and from films to other films are analyzed both along chronological lines from the 1930s to early 2000, as well as vertically in the context of a single genre, such as science fiction.

> In that film and television texts refer to lines of tradition, which are manifest in other texts, they not only integrate into a culture's universe of texts, but precisely through the knowledge, emotions, and expectations associated with film and television texts, also into the viewers' various horizons of life. (Mikos 2003: 262)

Feature films from various genres will be examined, including romance, drama, action, science-fiction, and horror. The sample of examined feature films will be compiled based on contextual and pragmatic criteria. First, there must be a woman scientist among the protagonists. Not included are academics and other non-scientific professional roles, such as doctors, lawyers, or agents. Second, the feature films must be available in VHS.

The stages of analytical work comprise (1) viewing the film in as normal a viewing situation as possible; (2) description via sequence protocol (minutes, picture, sound, music, dialogue) with special attention given to character development; (3) analysis of the film plot and the film characters based on selected categories; (4) interpretation at the level of film; and (5) interpretation in relationship to the total sample.

Although this study should be built upon qualitative interpretations, it is also meant to mediate a brief impression of the quantitative distribution of women scientists in feature films.

This study is based on a large sample for a qualitative study: over seventy feature films from the very beginning of talking movies in the 1930s to the present. The random sample can never be entirely representative as the base total of all feature films with scientific themes is unknown. The sample is, however, compiled and carried out based on all possible means of compiling a controlled random sample (Flicker 1991).

The science fiction film, which comprises approximately 50 percent of the sample, presents the largest genre group of the examined films featuring women scientists.[1] According to other studies, this is also the genre in which science is most commonly thematized in fiction films. The proportion of women scientists' disciplines is distributed two-thirds in the natural and technical sciences and a maximum of one-third in the social sciences and humanities. Aerospace scientists/cosmonauts are remarkably well-represented. (Job titles are assigned approximate scientific categories.) This depiction of the share of women in technical disciplines does not at all correspond with the actual situation. Women continue to be present in much greater numbers in the social sciences and humanities, and there are particularly few women in aerospace research, in particular.

The present work offers the possibility to review film interpretations along a certain line of inquiry and in terms of certain aspects of cinematic structures with the help of detailed and reflected processes. Films are constructed as polysemic and are correspondingly decoded. In the end, every recipient is responsible for a film's interpretation as relevant within the context of his or her own life.

THE MALE SCIENTIST IS THE MODEL IN OUR MINDS

The depiction of women scientists is not the same as the widespread cliché of the (male) scientist, which can be sketched out roughly as follows (Haynes 1994): he is a diligent and hard worker, surrounded by a veil of distraction, confusion, or even madness. In social contacts he is more of an outsider. With regard to other people, he is inconspicuous. Social trends and fads do not interest him. He seems socially displaced. His appearance, with glasses, lab coat, ruffled hair, and the like, is not that of an attractive hero. His diligence is obsessive. His attitude to work can be entirely apolitical to the extent of being unscrupulous. In some cases he even subjects humanity to great risks in the enthusiasm of his scientific curiosity.

Typologies and labels for the spectrum of characters of the (male) scientist in film have been drawn up with various degrees of differentiation (LaFollette 1990; Frizzoni 2004; Haynes 1994; King and Krzywinska 2000; Junge and Ohlhoff 2004): magician, expert, creator/destroyer, hero, inventor, computer freak, heroic adventurer, idealist, unfeeling rationalist, problem maker/solver, harbinger of fear/warning, and many more.

Also more recent block busters such as *The Day After Tomorrow* (USA 2004) are constructed on traditional images of scientists and show humanity's greatest fears and an archaic-patriarchal image of the scientist situated between natural catastrophe and father-son relationship. In this narrative, women are present in secondary roles only. This corresponds to typical and decade-long media practices of gender relations in films with scientific themes.

Feature films, for the most part, show us men in science (82 percent) (see Weingart et al. 2003). Women scientists appear less often and when they do, they appear in different roles from their male colleagues. The audience is confronted with a perspective of science that crosses straight through the natural and social science disciplines and the humanities, according to which the women scientists must adapt to male role models. In the following, I will work out in detail *how* this gender ritualization of men and women as scientists takes place in feature films.

WOMEN SCIENTISTS IN FEATURE FILMS: THEY'RE DIFFERENT FROM (MALE) SCIENTISTS[2]

Constructions of film characters in mainstream cinema are generally subjected to intense simplifications along three basic dimensions: gender, profession, and private life (see Field et al. 1988: 23ff.). The analysis of the female figure of the woman scientist occurs in the present study based on the three dimensions of social status and social relations (professional and private) and attributive gender qualities. The analysis of these dimensions comprises numerous categories, such as:

- Social status (e.g., profession, qualification, hierarchy, team, decisions).
- Social relations (e.g., professional/private; functional/emotional social role models).
- Attributive gender qualities (e.g., bodily traits, qualities, character, clothing, language).

In order to organize the results of recurring forms of depiction of women scientists I formed a typology, whose terms are oriented along the lines of socially established gender stereotypes: "Gender stereotypes are cognitive structures that contain socially shared knowledge about the characteristic traits of women and men" (Eckes 1997: 17). According to Eckes's further descriptions, the gender stereotypes belong, "on the one hand, to *individual* knowledge, and on the other, they form the core of a *consensual,* culturally shared understanding of the respective, typical gender traits" (Eckes 2004: 165). In its social practice, film addresses both levels— on the one hand, it is necessary to confront every viewer with the ability to connect to their own environment, and also, primarily in mainstream cinema, to achieve a large, widespread effectiveness, which especially in a globalized film market provokes a transnational appropriation of social communication (Mikos 2003).

I will briefly recall this typology of six prominent stereotypes of women scientists in feature films (Flicker 2003, 2004), and then expand them with a seventh type based on the most recent analyses of *Tomb Raider: Lara Croft*. The selected film examples and scenes should be seen as representatives of a type, of which a great number of other films are representative.

1. The old maid
2. The gruff women's libber
3. The naïve expert
4. The evil vamp
5. The daughter or assistant
6. The lonely heroine
7. The clever, digital beauty

Already at a visual level, an initial superficial observation of the following characters reveals modifications of the female character over the decades. Beyond the pictorial aesthetics, which would demand a separate analysis, it is possible to recognize social transformations in bodily representations, fashion, and sexualization.

The old maid

In this ritualized depiction of women scientists, mainly from the 1930s to 1950s, their professional status and intimate relations stand at odds with one another; they're either scientifically qualified and a wallflower or erotically attractive and daft. According to this pattern, femininity or sex appeal and intelligence are mutually exclusive traits of a woman scientist. The woman must choose between science and a private life.

A cherished effect, until today, is that of surprising the audience with an (attractive) woman who is also scientifically competent. In an example of the dialogue from a flirt scene in the film *Spellbound* (USA 1945), the character Dr Constanze Peterson receives the compliment: "Professor, I never quite realized how lovely you are!"

The gruff women's libber

The womanliness of this film figure (mainly in films of the 1970s) is not depicted with stereotypical femininity or erotic appeal, but rather, with a dash of austere toughness and "female" intuition in her scientific work. This type of woman scientist, with her apparently asexual character, clarifies typical dualistic ideas of gender. What is shown here is not the conquering of cultural constructions of bisexuality, but more so, the variance of stereotypical female roles. The stereotypical figure of the "male woman" mainly fulfills the dramaturgical function of emotionalizing true rational scientific behavior. The woman scientist personalizes the significance of

intuition in scientific research and also resistance to the malestream establishment, the male-dominated world of science.

In terms of gender relations, she is also a reference to the woman's movement: character of Dr Ruth Leavitt in *The Andromeda Strain* (USA 1971).

The naïve expert

This type of woman scientist is embodied by a very attractive woman who
is unrealistically young measured against her professional status. Indeed,
she has quite a high professional reputation, but her naïve innocence leads
her into a number of difficult situations, from which she can only free herself with the help of a man. This woman scientist embodies serious science,
which she engages in with moral integrity for the welfare of society. Her
scientific activity serves social progress. As a sexually attractive woman,
she is often involved in a love affair with a man who is a scientist and the
story's hero. This is not the case, however, in *I, Robot*, in which the hero is a
policeman. Example: *I, Robot* (USA 2004). The female character, Dr Susan
Calvin, a psychologist, is meant to help robots attain more humanity.

The evil vamp

This type of woman scientist is likewise remarkably good looking and
young, but she acts unscrupulously, egoistically, and is thoroughly willing
to cooperate with the "bad guys." She is corrupt and uses her sex appeal to
trick her male counterparts with the "weapons of a woman," similar to a
vamp. She uses scientific competence either to maximize her own wealth or
out of inquisitive greed. Even clever (male) scientists are trapped by her sex
appeal. Example: character of Dr Elsa Schneider in *Indiana Jones and the
Last Crusade* (USA 1989).

Both of these types of woman scientist, "the naïve expert" and the "evil
plotter" represent the ambivalent relationship of society and science. They
are polarized along the social belief in the use of science and a mistrust
of scientific research. Both dimensions, belief and mistrust, as emotional
categories, are embodied by female roles according to traditional gender
stereotypes, and thereby take up the ambivalent image of science.

The daughter or assistant

The most marked characteristic of this type of woman scientist is that the
female character is dramaturgically set in a structured relationship to a
male scientist. Whether or not she has formal qualifications, her professional activity is one of assisting, either as the daughter of a highly recognized and successful scientist or as his assistant. He hereby represents the
classical cliché of a scientist: brilliant, confused, nervous, and hopelessly
inept in practical matters. Her strength is mainly social competence. Her

task as "interpreter" is to enable a transfer of scientific knowhow for its application in society. Her character remains tied to a male role. This character of the woman scientist takes on a type of bridging function between rationality and emotion. Example: character of Dr Patricia Medford in *Them!* (USA 1954).

However, sometimes they also assist sexually, as a lover or fiancé. Corresponding to the time of sexist female roles in Hollywood films, which in the post-war era presented a clear backlash in comparison to the already relatively emancipated female characters in films of the 1930s and 1940s, also in the sample for this study, the films of the 1950s and 1960s show a particularly questionable picture of the woman scientist: the main task of the assistant is limited exclusively to sexually "assisting" the professor; her workplace is reduced, for the most part, to the bed. Example: character of Sarah Sherman in *The Torn Curtain* (USA 1966). Both character types refer to the male hierarchy in the science industry and do not question it.

The lonely heroine

As a protagonist in feature films of the 1990s, the "lonely heroine" likewise represents the further development of previous stereotypes. What becomes obvious in her is a transformation in the relationship of science to society and in the issues of emancipation in gender relations. This woman scientist declasses male colleagues with her competence. She is a modern, self-determined woman, who moves with skill in a male environment, and has also taken on male behavioral patterns, such as tomboy manners. As a film character, she unites the characteristics of previously described types: an insatiable professional curiosity, moral integrity, modesty, strong visions, intuition and rational judgment, young age, and sexually attractive. As a seemingly strong female character, she needs numerous male counterparts: father, colleague, teacher, mentor, contractor, and lover,[3] in order to ultimately fail. The interplay of male dominated power structures dismantles the innovative woman scientist's success. Only the audience, as an accessory to the plot, knows that her scientific hypotheses are correct. With polysemic constructions of these types of female characters, it is possible to maintain an open range for the viewers' diverse interpretations. This character construction bundles the representation of inner-scientific conflicts of gender, power, and methodology as well as social discourse on the political influence of scientific processes, in which the female character has to keep a low profile. Example: character of the astrophysicist Dr Elleonore "Elly" Arroway in *Contact* (USA 1997).

The clever, digital beauty

The character of "Lara Croft" is important here for two reasons: on the one hand, as an archaeologist she is one of the few scientists in the humanities in feature films, on the other, she represents the "digital beauties" (Pritsch 2004) in films of the late 1990s and early 2000s. Lara Croft can be identified as *the* digital beauty. She is exemplary for the new, seventh type of representation of the woman scientist in feature films and will be described in somewhat more detail than the previously described stereotype with which this type is connected.

The Lara Croft cult began with the computer action game *Tomb Raider* in 1996, long before the feature film. The virtual PC action heroine sold rapidly, attaining vast media presence (advertisements, articles, and cover stories in print media, fashion, music groups, music videos, cinema, and much more) and became something of a "cultural icon." At the end of the 1990s, the

> British minister of science Lord Sainsbury of Turville proposed that Lara Croft be employed as ambassador to promote the extreme importance of British science. Sainsbury presented his idea of exploiting Lara Croft on the occasion of a talk on "science and the knowledge economy" at the Social Market Foundation in December 1998. (Deubler-Mankowsky 2001: 12)

With the invention of the computer figure Lara Croft by a team of male computer scientists, a brilliant coup was accomplished. After the many successful male action heroes, finally a female character was at the center of action, who at first glance seemed to overcome all gender constraints. The phenomenon of immersion; "a form of experiencing, which corresponds with submerging in a virtual reality" (Mikos 2003: 174), is not analyzed here. Immersion is enabled on the one hand by the character, on the other, however, also by narration and dramaturgy as well as figures and actors in the game. Apparently, immersion requires an identification as well as sense of empathy and sympathy with the game characters (Mikos 2003: 174).

The computer Lara, however, is different from the film Lara, and this text refers only to the latter.[4] However, the film character encounters similar expectations: "Lara Croft" is a young, attractive woman of the British nobility and seems to pass her time by training to fight. Those close to her are a young, pale computer freak and a fatherly butler. In addition, she thanks her deceased father, a renowned archaeologist, for her amazing knowledge. She mourns his death as though she were a young girl. Similar to the lonely heroine, "Lara Croft," too, has exclusively male counterparts; her social relations are primarily functional. Her sex life is comprised at best of intimations of an affair with an admittedly good-looking but clearly less clever archaeologist. "Lara Croft" is somewhat like a pubescent wild thing. She drives and parks her motorcycle irreverently; she mingles in society's upper circles clad in a leather suit. Here, as an anti-establishment figure, she offers identification for young people of both genders.

"Lara Croft" appears refreshingly independent and free. Emotionally, however, she is entirely fixated on her deceased father. He is also her driving force. She struggles to complete his work. However, as an action-heroine she seems to be missing her own, personal, inner drive. "Lara Croft" is scientifically highly competent, personable, and financially independent, tried and tested in all styles of battle, self-confident, and extremely attractive. Her feminine figure has almost unnaturally sexualized dimensions. Their appearance turns "Lara Croft" into more than an erotic dream woman—she is a type of "wonder woman" as in Elisa Giomi's analysis (Giomi 2005). "Lara Croft" unites the characteristics of a remote-controlled male action figure and a female sex bomb. Identification and desire are tied together in an overly natural, female character who appeals to both female and male recipients. "Lara is a recipe which works across genders because women want to be her and men want to be with her," says Jeremy Smith one of the directors of Core Design, the site of Lara Croft's discovery (Giomi 2005: 20).

In this construction of desire and identification, "Lara Croft" represents a female character whose scientific competence is a mere secondary attribute. She is stylized as a fighting machine, a sexual event, and socially utterly immature and impoverished. With "Lara Croft," science disappears in a virtual game of male sexual fantasies with a woman as an emotional, remote-controlled action game figure.

CONCLUSIONS

The previous descriptions give insight into the complex results of the film interpretations. Shown is how the pictures of women scientists, at the visual level and in terms of content, have been a mass medial and broadly effective component of social culture for decades. Here, I would like to once again draw attention to the power of metaphors:

> Such images function deeply within us as memories, and also as metaphors. It may be that the science film itself, if it is coming out of Hollywood as an imperfect or sentimental drama, will contain these important metaphors telling us about the world of science. It may be in that situation that such images will in fact become the world of science on film. (Rosenstone 2003: 336)

The analysis of feature films with scientific themes shows that the representations of women scientists are not about chance gender ritualizations. Instead, reference is made to recurring practices of medial gender constructions and to social myths of science. Also evident is how the social discourse of the political women's movement and feminist media and science critique have left their traces in mainstream cinema over the course of many decades.

The cliché description of the "mad scientist" does not apply to the women scientists in feature films. They don't work in secret laboratories or on dubious projects, but instead, remain soberly "with their feet planted on the ground." These female characters hardly contribute to the negative myths of science. The career role of "scientist" is mainly reserved for men; not even a fifth of all scientists are women.

The audience is confronted with the surprise that "the professor" is a woman. The dramaturgic effect of amazement when brilliance is embodied by a woman has endured over the decades. The woman scientist tends to be strikingly beautiful and, measured by her qualifications, unrealistically young. She has the body of a model — slim, athletic, perfect, and is dressed provocatively.

- "Dr. NO," Honey Rider, the daughter of a widowed marine biologist, USA 1962

Figure 11.1 Honey Rider—Bond Girl 1962.

Figure 11.2 Lara Croft 2003.

- "Lara Croft Tomb Raider 2: The Cradle of Life," Lara Croft is archeologist and daughter of the deceased archaeologist Croft (2003).

Whether a marine biologist, security expert, or archaeologist: the outfit makes the scientist successful. When they wear more than a bikini, a white lab coat and glasses are also accessories of the job for women scientists—the woman is consequently de-sexualized or written off as an old maid.

Throughout the examined timeframe of the sample, approximately seventy years, it is possible to interpret clear changes in accordance with a social transformation. If at first women scientists had to choose between a career *or* private life, now they can meanwhile link the two. Sexuality remains fixed in hetero-normality. Although at first, women scientists were dependent on male mentors (father, husband), now they are "only" slowed down by a male-dominated system. "Femininity" *and* intelligence can be united in a female character. However, ultimately, "femininity" and success largely exclude one another in the feature film depiction. Women scientists are also seldom depicted as mothers. Here, it is possible to discern parallels to the depiction of the male scientist.

Woman scientist roles present more of a stereotypical woman's role rather than the profession of scientist. In the portrayal of the role, job stereotypes are combined with gender stereotypes. At the level of scientific activity, women scientists add intuition and emotionality. With these emotional components, they also represent social fears, which thematize different skepticisms according to the decade (e.g., in the face of psychoanalysis and nuclear science before the cold war, and later extraterrestrial life, biotechnology, natural catastrophes, etc.).

At a first, superficial glance, the role differentiation between the male and female scientist appears to have disappeared in feature films since the 1990s. On screen are strong, competent, utterly qualified, and attractive women scientists. The fictional female characters take on parts of male stereotypes by uniting intellectuality and sexuality. Nonetheless, they are still denied complete success—except in virtual games.

In spite of a strong transformation of the image of women scientists in film, the analysis shows that, analogous to other women's roles, they are subjected to a medial sexualization. This ritualization of gender can be recognized by a noticeable "use" of the female body in aspects of aestheticization and in the superimposition of sexual attributes on scientific competence. Feature films with women scientists reveal ritualized gender practices as a praxis of social communication and in their intertextuality. The recurring patterns show professional qualification restricted by a male-dominated scientific system, sexualized physicality, as well as childishly emotional social relations. They are stylized as lonely heroes, and that sells well.

NOTES

1. The research of Petra Pansegrau and Peter Weingart confirms the rough categorization of films with scientific themes, which I have used since 1991, and the relationship between male and female scientists: research seminar *The Perception and Representation of Science by Hollywood*, University of Bielefeld 2000–2002.
2. The results of this study appeared in numerous publications (Flicker 1991, 2003, 2004). Further interpretational perspectives were gained in the scientific discourse coupled with this. Moreover, here we can state that the basic patterns have not changed, but have just become more variegated.
3. A further interpretation arose in the discussion following the lecture. In *Contact*, the astrophysicist comes under attack by a renowned theologian, who was once her lover. Her proof of transcendental experiences is not recognized. In a further sense, it is possible to interpret from this that in the conflict over a transcendental realm, God is another male counterpart for the woman scientist.

4. The differences between the computer game and the film are presented by Deubler-Mankowsky (2001).

BIBLIOGRAPHY

Braidt, Andrea B. (2004) *Gender und Genre in den Filmwissenschaften*, in Monika Bernold, Andrea B. Braidt, and Claudia Preschl (eds), *Screenwise. Film. Fernsehen. Feminismus*, Marburg: Schüren: 196–199.

Cassidy, Sarah (2005) "Female scientists need screen role models, TV producers told", *The Independent*, 11 April 2005, *Home News* 11.

Deubler-Mankowsky, Astrid (2001) *Lara Croft. Modell, Medium, Cyberheldin*, Frankfurt am Main: Edition Suhrkamp, Gender Studies.

Eckes, Thomas (1997) *Geschlechterstereotype. Frau und Mann in sozialpsychologischer Sicht*, Schriftenreihe des Zentrums für interdisziplinäre Frauenforschung der Christian-Albrechts-Universität zu Kiel, Vol. 5, Pfaffenweiler: Centaurus.

——— (2004) *Geschlechterstereotype: Von Rollen, Identitäten und Vorurteilen*, in Ruth Becker and Beate Kortendiek (eds), *Handbuch Frauen- und Geschlechterforschung. Theorie, Methoden, Empirie*, Wiesbaden: Verlag für Sozialwissenschaften: 165–176.

Faulstich, Werner (1988) *Die Filminterpretation*, Göttingen: Vandenhoeck & Ruprecht.

Field, Syd, Peter Märthesheimer, Wolfgang Längsfeld et al. (1988) *Drehbuchschreiben für Film und Fernsehen*, 2nd edition, München.

Flicker, Eva (1991) *"Professor, mir ist nie aufgefallen, wie reizend Du bist!" Eine film- und wissenschaftssoziologische Untersuchung zur Darstellung der Wissenschaftlerin im Spielfilm*, Diploma thesis, Wien: Universität Wien.

——— (2003) "Between brains and breasts—women scientists in fiction film: On the marginalization and sexualization of scientific competence", *Public Understanding of Science*, 12: 307–318.

——— (2004) "Wissenschaftlerinnen im Spielfilm. Zur Marginalisierung und Sexualisierung wissenschaftlicher Kompetenz", in Torsten Junge and Dörthe Ohlhoff (eds), *Wahnsinnig genial. Der Mad Scientist Reader*, Aschaffenburg: Alibri Verlag: 63–76.

Frizzoni, Brigitte (2004) "Der *Mad Scientist* im amerikanischen Science-Fiction-Film", in Torsten Junge and Dörthe Ohlhoff (eds), *Wahnsinnig genial. Der Mad Scientist Reader*, Aschaffenburg: Alibri Verlag: 23–37.

Giomi, Elisa (2005) "The Craft of Lara Croft. A New Role Model for Action Heroines and Women?" paper presented at Conference of the *Sektion Jugendsoziologie und Medien-und Kommunikationssoziologie der Deutschen Gesellschaft für Soziologie*, Paderborn, February 2005. (Planned conference volume on: *"Media—Identity—Identification"*.

Gledhill, Christine (2004) "Überlegungen zum Verhältnis von Gender und Genre im postmodernen Zeitalter", in Monika Bernold, Andrea B. Braidt, and Claudia Preschl (eds.), *Screenwise. Film. Fernsehen. Feminismus*, Marburg: Schüren Verlag: 200–209.

Haynes, Roslynn D. (1994) *From Faust to Strangelove—Representation of the Scientist in Western Literature*, Baltimore and London: John Hopkins University Press.

Hölzer, Henrike (2004) *Der Schatten des Spiegelbilds. Action-Film Sequels als unheimliche Doppelgänger*, in Monika Bernold, Andrea B. Braidt und Claudia Preschl (eds), *Screenwise. Film. Fernsehen. Feminismus*, Marburg: Schüren Verlag: 210–220.

Junge, Torsten and Dörthe Ohlhoff (eds) (2004) *Wahnsinnig genial. Der Mad Scientist Reader*, Aschaffenburg: Alibri Verlag.

Keppler, Angela (2004) *Zum Ritual politischer Talk-Shows*, in C. Wulf and J. Zirfas (eds), *Die Kultur des Rituals*, München: Wilhelm Fink Verlag: 293–302.

King, Geoff and Tanya Krzywinska (2000) *Science Fiction Cinema: From Outerspace to Cyberspace*, London: Wallflower Press.

Kuni, Verena (2004) "Nach allen Regeln der Kunst: Gender is a Genre is a Genre? Cut up! Versuch über Verfahren, einen gordischen Knoten zu durchschneiden", in Monika Bernold, Andrea B. Braidt, and Claudia Preschl (eds), *Screenwise. Film. Fernsehen. Feminismus*, Marburg: Schüren Verlag: 221–232.

LaFollette, Marcel (1990) *Making Science Our Own. Public Images of Science 1910–1955*, Chicago: University of Chicago Press.

Luhmann, Niklas (1997) *Die Gesellschaft der Gesellschaft*, Frankfurt am Main: Suhrkamp.

Mikos, Lothar (2003) *Film- und Fernsehanalyse*, Konstanz: UVK Verlagsgesellschaft.

ORF (2005) *Zur Wahrnehmung von männlichen und weiblichen Rollenbildern in den ORF-Unterhaltungsserien und deren Identifikationspotentiale aus der Sicht der ZuschauerInnen, Studie für den Publikumsrat 2004, ORF Markt- und Medienforschung.* Wien: ORF.

Pritsch, Sylvia (2004) "Virtuelle Gefährtinnen in der Hyperwelt. 'Digital Beauties' als Allegorien der Posthumanismus", in Brigitte Hipfl, Elisabeth Klaus, and Uta Scheer (eds), *Identitätsräume. Nation, Körper und Geschlecht in den Medien. Eine Topografie, Cultural Studies 6*, Bielefeld: transcript Verlag: 222–241.

Rosenstone, Robert A. (2003) "Comments on science in the visual media", *Public Understanding of Science*, 12: 335–339.

Weingart, Peter, Claudia Muhl, and Petra Pansegrau (2003) "Of power maniacs and unethical geniuses: Science and scientists in fiction film", *Public Understanding of Sciencek*, 12: 279–287.

12 Stereotypes and Images of Scientists in Fiction Films[1]

Petra Pansegrau

THE PUBLIC IMAGE OF SCIENTISTS

The growing social impact of the mass media lends increasing importance to the questions "How are science and scientists represented in the mass media and how are they perceived by the audience?" In accordance with this, the focus of this paper is how science is portrayed to the public (and therefore not a purely media or film studies interest) and which repercussions this portrayal has for science itself. In the broadest sense, the research interest falls under the heading of "Public Understanding of Science" in a critical sense. In the context of the increasing interest in the public understanding of science, it has become largely accepted that the media cultivate and communicate their own special representations and perceptions of the world and reality. The media does not simply translate scientific information but are participants as well as producers of a dialogue about knowledge and have an important function within the public discourse. Not only does this have repercussions for the relationship between science and the media but also for science itself. The media are equal partners in the dialogue with the public and have a significant impact on how the public perceives numerous scientific issues. Most scientific studies on the relationship between science and the media have focused on the "traditional" mass media, that is, television and print media. Here, the research focuses on the portrayal of scientific controversies, the relationship between the media and issue attention, the communication of technological risks and the importance of science journalism in the knowledge society. For some time, however, some sub-disciplines of media studies and sociology of media and of science have increasingly begun asking how scientists are perceived and portrayed in the media.

Until now there have been few studies on the development of the public image of science and scientists. There are some older studies on the cultural image of scientists. In 1990 Marcel LaFollette presented her impressive study on the public image of scientists, *Making Science our Own*, in which she claimed that four types or stereotypes of scientists can be found in magazines such as *Harper's*, *American Mercury*, *Century*, or *Scribner's* from

1910 to 1950: (1) the magician or the wizard, (2) the expert, (3) the creator or destroyer and (4) the hero. These stereotypes represent the distinct social roles that have been assigned to scientists by the media.

> When the scientists were described as wizards, they seemed mysteriously clever, possessing secret knowledge and holding considerable power over nature. As experts, they knew all and could be asked to share their knowledge with society. As creators and destroyers, they bore responsibility, both positive and negative, for the end results of that knowledge. As heroes, they combined an optimistic belief in a better future with insatiable curiosity, restlessness, a drive to explore, and the ability to explore new paths. (LaFollette 1990: 108)

We will see that the stereotypes found in films only partially differ from those constructed by journalism. When one considers the previously mentioned perspective that the function of science is increasingly that of social legitimization and, as such, enjoys a high priority in the public discourse and democratic decision making process, it is surprising that the portrayal of scientists in fictional films receives so little attention. Although fictional films are a very popular medium, the myths, schemas, and clichés of the representation of scientists and their relationship to the public's present image of science and scientist have remained largely unexplored. Opinion polls consistently indicate that the institution of science has a good reputation and enjoys a high degree of public trust (e.g., the surveys regularly done by Science and Engineering Indicators; *Science and Engineering Indicators* 1998–2006) but how are the individual scientists perceived? An opinion poll taken by students in Bielefeld, Germany, although not representative, gives an idea of what the public image of scientists may look like. According to the survey, scientists are seen as socially awkward, out of touch with reality, and sometimes bordering on the insane; they are not family oriented, they're focused on their work, and interested in nothing else but finding the solution to their scientific problems.

Considering the results of the study by Roslynn Haynes (1994) it becomes apparent that while there are only a few real scientists who are well known by the general public (i.e., Stephen Hawking, Albert Einstein, Isaac Newton), fictitious portrayals of scientists such as Dr. Faust, Dr. Frankenstein, Dr. Strangelove, or Dr. Caligari are very well known. Obviously the scientific community's perception and image of itself differs greatly from those of the public. This leads to the project's hypotheses regarding stereotypes about scientists.

1. Scientists have the image of being strange, unworldly and detached from everyday life.
2. Fictitious scientists are often better known than real scientists.

3. It can therefore be tentatively assumed that the dominant public images or perceptions of science and scientists are more strongly determined by fictitious characters than by real scientists.

This leads to two questions: (1) Which schemas and clichés are used in films to portray scientists? (2) What function do these portrayals of science in fiction films have?

TYPES AND STEREOTYPES OF SCIENTISTS IN FILM

In this project we examined and statistically analyzed over 220 fiction films from the twentieth century on the basis of 120 questions. This paper only presents a small portion of the results of the study and will focus on the stereotypes of scientists constructed in fiction films (for additional results, see Weingart, chapter 13 this volume). These stereotypes often go hand in hand with the cultivation of myths. For example the best known type of scientist is the "mad scientist," the crazy, obsessed, wild, and uncontrollable scientist who is determined to rule the world, to create an artificial human being or some artificial life form. In the following examples we find variations of that stereotype. In literature as well as in film there are more evil than benign scientists but, most interestingly, only a limited number of stereotypes. These stereotypes clearly serve as a schema in which scientists and their projects can be inserted.

The eccentric

The first stereotype is the foolish, nerdy, eccentric: This type is similar to the popular image that was reflected by the opinion poll. The eccentric is always somewhat unworldly and confused, wears socks that do not match, always forgets some vitally important thing, has messy hair, and ignores dangers but is generally good-natured. The science portrayed in these films usually takes place at home and involves the realization of ideas that are far from scientific reality. In the end it sometimes turns out, however, that the scientist was actually working on an important scientific project. This form of satirical portrayal of science can also be found in fictional literature as early as the seventeenth century when the founding members of the Royal Society were merciless caricatured and popular figures who more likely sought wonders than the truth. The media love eccentric scientists and give them a lot of attention. A real person who fits this image was Albert Einstein. He played the role of the absent minded, good-natured genius so well (most people are familiar with the photo showing him with messy hair and sticking out his tongue) that one can easily forget the important role he played in the development of the atomic bomb (Haynes 2003). A popular example of a fictitious eccentric is the character Doc Brown in the film *Back to the Future*. The character of Doc Brown is so exaggeratedly unconventional, curious, and out of touch with reality that he playfully fulfills the criteria of the eccentric. His nerdy appearance reminds one of Albert Einstein as well as of Gyro Gearloose; and, in spite of his unworldliness and eccentricity, he is completely harmless.

The scientist as hero or adventurer

The second stereotype is that of the hero or the adventurer. The literary origin of this stereotype can be found in novels by Jules Verne (*Voyage to the Center of the Earth* or *Around the World in Eighty Days*). From the beginning this stereotype was the embodiment of the conqueror myth and represented the free spirited heroic scientist who mastered the wonders and mysteries of nature through bravery, conquest, and the use of science. Marcel LaFollette also identified this type as a central category used in the portrayal of scientists in journalism (see above). Newer cinema heroes of this kind are, for example, the scientists in *The Thing from Another World*, in the *Indiana Jones* films, in *Jurassic Park,* or in *Medicine Man*. In films which portray scientists as heroes or adventurers, science is usually performed in the field by athletic, attractive, pragmatic personalities that are willing to risk their own lives or save others in the service of science. Scientists in this category have warmhearted and compassionate emotional dispositions. In addition to the films already mentioned, this type of character can also be found in many science fiction films and in the *Star Trek* series. It appears as if the number of films portraying scientists as heroes and adventurers increased in the 1990s. This possible trend toward athletic characters and morally uncomplicated action scenes may mean that there is a tendency in times of crisis to reinforce the idea that most problems can be solved through bravery, optimism and stamina. The popularity of *Star Trek* and similar films indicates that there is a continuing demand for morally uncomplicated scientist heroes. One genre that has not proven particularly relevant for the database underlying this analysis is that of biographical pictures (BioPics). Some films about real people and real life stories have, however, been included in the corpus of our study, because they include fictional elements; for example those about Thomas Edison, Marie Curie, or other Nobel Prize winners. In nearly every film in which the scientist is based on a real person, one finds the hero stereotype. It should also be noted that these (so called) authentic portrayals of scientists primarily represent scientists positively and as morally sound heroes.

The professional scientist

The third stereotype is the professional scientist. Although this type does not appear very frequently in films, it is related to the hero. The actions of this type of scientist are portrayed as morally flawless. Professional scientists are depicted as genteel, ambitious, unshakable, peace loving, and as experts in their fields. They tend, however, to be indifferent towards their fellow man. Portrayals of this type are close to the self-image that real scientists have and farthest away from the cultural image of scientists described earlier. Clearly this stereotype plays an important role in the production of authenticity. Nearly every film that represents this stereotype is based on a real life person or belongs to the BioPic genre. The film characters of this type can be found, for example, in the films *Marie Curie, The Story of Louis Pasteur,* or *The Secret Passion of Freud.* Because the analysis of the films, for which the program *Statistical Package for the Social Sciences* (SPSS) was used, involved three consecutive steps of data analysis, some films fall under more than one category. In one step, categories were formed on the basis of previously defined variables, in a later step categories were formed on the basis of a cluster analysis independently performed by SPSS. The construction of authenticity is a central motive for such films. For exactly this reason, however, very few films in this category were included in our study because we were looking primarily for the examples of fictitious elements, which are not common in BioPics.

The mad scientist

We now come to the most well represented type of scientist portrayed in films. In Western cultures the dominant image of scientists is that of the evil, dangerous, and insane man. Because the mad scientist appears in numerous forms, we differentiate between the ones most frequently found in the study.

The obsessed mad scientist

The first type of mad scientist is the obsessed, mad character. In films which make use of this stereotype (one of the most frequently found characters altogether) the scientist is obsessed by the desire for power, fame, fortune, or the realization of his ideas. This stereotype first appeared in the literature as Dr. Faust, then in various film adaptations of *Dr. Frankenstein, Dr. Jekyll and Mr. Hyde,* and it continues to be found in films today. The protagonists almost always have very similar moral downfalls. They are unscrupulous, often have delusions of grandeur, are criminal, and are guided solely by their own interests. Many of these films are about the creation of life, from the creation of a single monster, as seen in *Frankenstein,* to the mass transformation of human production, as seen in *Dr. Moreau.* The obsessed scientists, who usually practice their arts secretly in hidden places, see themselves as misunderstood by society. They deliberately disregard ethical and legal norms, thereby causing suffering. The central message of these films is the danger of science and the impossibility of controlling it. There are so many examples that it would be impossible to name all of the variations of the stereotype of the obsessed man who either creates or transforms human life, thereby gaining uncontrollable power over others. A newer protagonist of the stereotype is, for example, *Dr. Moreau* played by the 550-pound Marlon Brando (second remake) on his island with white make-up, dark sunglasses, black doctoral cap, and flowing white cape appearing before the creatures he has created and allowing them to address him as "father." In this allegory of the godlike, he represents how far scientists have distanced themselves from society. The character of Dr. Moreau, who wins the Nobel Prize for genetics in the film, and the depiction of developments made possible through gene technology address the ethical implications and conflicts of gene technology.

The accidental mad scientist

Another type of mad scientist is the accidental mad scientist. Scientists of this stereotype start off with high values and at first keep within ethical and legal limits. They are often likable, doing altruistic research, searching for a cure to a serious illness for example. Usually a hectic or reckless self-experiment goes awry and the scientist becomes the victim of his own research. We can find many well known examples of this stereotype as in *Hollow Man*, *The Invisible Man*, *The Man Who Changed His Mind* or *X—The Man with the X-ray Eyes* who after a self-experiment (eye drops for x-ray vision) becomes addicted to his own discovery and is from then on doomed to lead a horrible and painful life, full of unbearable glimpses into humanity. There are two different kinds of accidents that happen during experiments in films. They either result in a physical transformation or mutation or they effect an irreversible transformation of consciousness or mind (i.e., telepathic control over others). Films which depict this stereotype focus on the unforeseeable consequences of science and play on the fear that scientists can easily lose control of their experiments. In the film, the accidental mad scientist usually develops from the good hero to the lonely, insane victim who is hardly able to survive and is solely responsible for the consequences of his research. One of the most famous examples of this stereotype is *The Fly*. After his unsuccessful attempt to dematerialize and rematerialize himself, the scientist suddenly has the head and arm of a fly. His mental ability becomes increasingly impaired and eventually he commits suicide.

The utopian ruler

Another type of mad scientist is the utopian or world ruler. The utopian ruler wants power, most often he wants to rule the world but second most often he is greedy for wealth. Either way, he has absolutely no moral principles. He is willing to do anything in order to reach his goal. He is irresponsible, ruthless, and evil and he sees people as a means to an end. He is capable of being polite and civilized in order to conceal himself when it serves his purpose. He does not simply want to understand the world. He wants to rule it. These films usually have a happy ending, however. In the end the good guy (usually not a scientist) wins and the evil utopian ruler must die. Examples of this type can be found in numerous horror films and in various James Bond movies. One typical example of this type of scientist is the character Dr. No. Dr. No is a brilliant scientist who lives on his own radioactive-safe island. He is the brain of a terror organization that wants to rule the world. Naturally James Bond wins in the end. The utopian ruler type is very close to another stereotype in our analysis, the Faustian mad scientist. I will, however, only briefly mention the Faustian mad scientist because he plays a significantly less important role in film than in fictional literature (Haynes 1994) and cannot be completely differentiated from the utopian ruler found in our study. Films that characterize scientists as such represent the original mad scientist type, Goethe's *Faust*, who pursues his scientific agenda through a life of conscious agnosticism regardless of ethical principles and moral values. Films such as *Contact*, *Altered States*, *Flatliners*, or *From Beyond* represent various sub-categories of this stereotype.

This cursory presentation shows that some types of scientists that we have found in our study correspond to the popular image of scientists (recall the responses to the opinion poll): the eccentric's bad haircut, the mad scientist's obsession with his work, and the hero/adventurer's lack of interest in family and private life. Science films are nearly as old as film itself—recall *A Trip to the Moon* (1903) or, more popular, *Metropolis* (1926). What all stereotypes have in common is the myth of uniqueness that is represented through the physical and psychological characteristics of the scientists in films. The only exception to this is the character of the professional scientist, who is only found in one (usually non-fiction) genre.

It seems as if academic degrees and prizes such as the Nobel Prize have little or no influence on the prestige of scientists in fiction films. The fact that they represent experts in their fields does not appear to depend on academic degrees and prizes. The prestige and respect of both the scientist as hero/adventurer and the mad scientist are based on scientific expertise rather than on having a certain title. Both types are relatively often portrayed as experts, considering that there are few protagonists with high academic degrees, in general.

"SCIENCE IN FILM" AS A METAPHOR FOR CONTROVERSIAL SOCIAL DISCOURSE

Returning to the study on stereotypes of scientists in magazines (M. LaFollette), it can be seen that the hero stereotype (or the hero/adventurer in our study) can also be found in films. The expert stereotype, as defined by LaFollette, does not exactly match the stereotypes that we have found, however, because over 60 per cent of the scientists were perceived as experts in their fields and therefore could not be considered a separate category. In addition to this, LaFollette was able to identify the stereotypes magician/ wizard and creator/destroyer. To some extent the qualities of these characters can be found in the different variations of the mad scientist, which indicates that the construction and communication of the public image of scientists may follow a specific pattern regardless of the medium. Furthermore, Roslynn Haynes was able to demonstrate in her extensive study on the portrayal of scientists in Western literature over the last 400 years that certain patterns are followed which, to a great extent, are in keeping with film portrayals. However, because much of the literature is significantly older than the films, there are some historical stereotypes which play no role in modern films.

In many cases these newly developed stereotypes appeal to a vague fear of science. Fear of science is fear of power, changes and controls that disempower normal people. Roslynn Haynes writes, "Rulers and military governments can be overthrown, knowledge cannot" (Haynes 1994: 193). Because knowledge cannot be cancelled, public fear of it being in the wrong hands can develop. This is often manifested in controversial social discourses. When one takes a look at which films predominate, then one undoubtedly sees a connection between the fears represented in films and the fears in the public discourse of that time. After World War II and the accompanying technological and scientific advancements, the mad scientist became the demon of urbanization and modernization. Films about scientists not being able to control the atomic bomb followed. Georg Seeßlen writes, "If they don't start a new and final world war, then scientists will at least create gigantic mutant insects, awaken prehistoric dragons like 'Godzilla' or contaminate huge areas of land" (Seeßlen 1999: 47). Every new scientific development involves the formulation or communication of new fears in popular culture. Since the computer became an indispensable scientific instrument, computer criminals and artificial intelligence have rapidly replaced the mad scientist. However, with the development of gene technology, the old image of the mad scientist and his interference with creation has been revived. This motive can be seen in the many new films which directly address societal fears about cloning and gene manipulation.

The film examples presented here and the specific qualities of the scientific characters or stereotypes show that the portrayal of science in film remains a metaphor for dominant fears and problems. At first this metaphor appears quite simple but it actually represents an extremely complex discourse and complicated ideas. The public is no longer afraid that monsters may be created, rather, the monster is a metaphor for the fear of scientific intervention in human reproduction. Nevertheless, although the film world is afraid of the mad scientists, it does not want to forego the promises of science. For this reason, scientists in films sway between the extremes of the demonic and insane scientists, that is, the threatening, fascinating, and tragic aspects of science and good science, performed by the hero and adventurer, the protector of moral and mythical values. The aesthetic portrayal of scientists in popular culture is obviously characterized by this dualism and falls back on the myths and fears that have long been predominant in many societies but also on the hopes and visions that can be transported through the presentation of science.

DISCUSSION

Why is the popular image of scientists important? Does it matter if the public image of science is positive or negative and can one attribute a function to it? The image of an institution or profession is often an important factor in regard to public awareness. Those who are occupied with researching and promoting public understanding of science concern themselves with the distance between science and the public. Science is not only cheered but seen critically and rejected. The popular image of science is, in some respects, considerably different from the self-image of the scientific community. In *Science & Engineering Indicators* it is written about the public image of some individual scientists that, "The charming and charismatic scientist is not an image that populates popular culture" (*Science & Engineering Indicators* 2002: 7–25). The many recent initiatives in Germany and other European countries to improve the image of scientists are possibly insufficient because the myths are old and deeply rooted and the patterns for the construction of such stereotypes in many media have proven resistant to change. These myths and the fear of dangerous knowledge and the loss of control have not yet been taken into account in these campaigns. If the public image of science is influenced by the depiction of science in popular films, it should at least give scientists and researchers something to think about.

NOTE

1. This article is based on the report *The Perception and Representation of Science by Hollywood* prepared in the context of a research seminar at Bielefeld University 2000–2002 by Peter Weingart, Petra Pansegrau, Claudia Muhl, Klaus Brandhorst, Volker Davids, Andreas Lingnau, Christoph Loschen, and Jochen Walter. Thus it is also based on the same data as the article by Peter Weingart (chapter 13 this volume).

BIBLIOGRAPHY

Haynes, Roslynn D. (1994) *From Faust to Strangelove Representations of the Scientist in Western Literature*, Baltimore, MD: Johns Hopkins University Press.

—— (2003) "Von der Alchemie zur künstlichen Intelligenz: Wissenschaftler-klischees in der westlichen Literatur", in Stefan Iglhaut and Thomas Spring (eds.), *Science and Fiction: Zwischen Nanowelt und globaler Kultur*, Berlin: Jovis.

LaFollette, Marcel (1990) *Making Science Our Own: Public Images of Science, 1910–1955*, Chicago: The University of Chicago Press.

National Science Foundation (1998, 2000, 2002, 2004, 2006) *Science & Engineering Indicators* http://www.nsf.gov/statistics/pubseri.cfm?seri_id=2

—— (2002) *Science & Engineering Indicators, Chapter 7, Science and Technology Public Attitudes and Public Understanding*, 25, http://www.nsf.gov/statistics/seind02/pdfstart.htm

Seeßlen, Georg (1999) "Mad Scientist. Repräsentation des Wissenschaftlers im Film", *Gegenworte*, 3: 44–48.

13 The Ambivalence Towards New Knowledge

Science in Fiction Film[1]

Peter Weingart

INTRODUCTION

Michael Crichton, well-known author of bestsellers dealing with science, is a wanderer between the worlds of fact and fiction. Crossing the boundary into "science fiction," combining thrilling action with plausible accounts of scientific advances, several of his books have been made into movies (the most famous being *Jurassic Park*). In a talk before the American Association for the Advancement of Science he presented himself first as an educated scientist boasting degrees in anthropology and medicine as well as publications in the renowned *New England Journal of Medicine*. Then he assumed the position of movie producer and explained to his academic audience that they should not be worried about the negative representation of science in movies, reasoning that since "all professions are depicted negatively why should one expect scientists to be treated differently?" Since there is no match between social reality and the reality of movies there is no reason to be concerned about the depiction of science in movies (Crichton 1999: 1461).

Although we have little doubt about the fact that movies and TV are exceptionally powerful media, we know next to nothing about their actual impact on people's opinions and attitudes toward science. It is an open question if the form of popular critique of science to be found in films is really "extremely effective," as Toumey suggests. Crichton, on the other hand, argues reassuringly. The mass media, he claims, have lost their influence. The film reaches only a fraction of the entire population (*Jurassic Park* was seen by only 8 to 15 per cent of the American people), and the Hollywood version of science is not to be taken more seriously by the public than other media contents (Crichton 1999). Apart from the fact that we know very little about the reception of movies in general and of horror films in particular, Crichton is certainly right in warning against the assumption of a linear causality between watching a movie and believing its contents. However, he would have to explain why the many versions of Frankenstein and Jekyll-and-Hyde type stories appeal to audiences again and again, to have film producers resort to the same kind of (film) stories. After all, they respond to popular demand in order to make a profit. Evidently, this must mean that the images about science depicted in fiction films insofar as they follow the same scripts and exhibit the same patterns, represent relatively stable stereotypes and are icons of popular culture. The issue is not whether the movies are showing a realistic picture of science and whether their audiences believe what they see, but rather why the crowds appear to flock to the same stories again and again.

This question does not seem to have attracted much attention although the first publication of the results of our study has triggered some interest, particularly in Germany. Still, 'science in fiction film' is not much of a topic. Less than a handful of books (and those mostly devoted to specific genres like science fiction and horror movies) deal with the role of science as a subject for movies (Tudor 1989; Skal 1998; Sobchack 1999). The number of articles is equally small, and none of these works looks at the whole range of film genres and their different ways of representing science (Osterland 1968; Gerbner 1987; Back 1995). Where the topic has received attention, as it did at a conference in Bern, Switzerland, February 2006,[2] the primary interest was to what extent the picture of science depicted in the movies was a realistic one. This is typically the interest of academics of film schools on the one hand and scientists on the other. The former are interested in teaching their students to be more authentic as a matter of professional ethos.[3] The latter are interested in having science "shown as it really is," hoping that this may lead to a a more positive portrayal of their profession. As legitimate as these concerns may be they are not what we are interested in.

Our interest in the image of science and scientists portrayed in the many "running pictures" that are shown on the screen extends beyond the immediate PR effects that may contribute to the welfare of the institution of science, nor is it concerned with the films' authenticity to the benefit of film makers' reputations. Rather, the persistence and continuity of the image of science as it is represented in movies is the phenomenon to be explained. The depiction of science, that is, scientific knowledge, seems to suggest that it is an especially problematic element of popular culture. That the gaining of new knowledge may be seen as man's hubris vis-à-vis the Gods is a view that goes back to antiquity and is expressed in the saga of Prometheus. More concretely, modern empirical science is a kind of human knowledge whose legitimacy, although established and substantially expanded since the Renaissance, has continually been contested. Examples of challenges to science's legitimacy are legend, reaching from the trial of Galileo all the way to creationists fighting the teaching of evolutionary theory at the end of the twentieth century and the beginning of the twenty-first. The appeal to "healers" in dealing with AIDS in Southern Africa or the outright ban on Western science by radical fundamentalists of various brands quickly demonstrate that there are still forces that question or even fight the superiority of scientific knowledge. The conflict over the boundaries of, over what are considered legitimate ways to generate and use scientific knowledge is an inherent element of Western culture. Recent debates over the ethical limits of molecular medicine are an illustration of this as well.

Thus, while it could seem that science is far too esoteric a topic for a popular movie one should not be surprised that the dramatization of the processes of creating new knowledge and embedding it in society is exactly what the mass media are interested in. Even a superficial search for movies in which science or scientists play some role produces hundreds of cases, more than 400 in the context of this project. A study of the representation of science on TV has shown that the television audience is heavily exposed to science, technology, and medicine, not through news magazines or documentaries like NOVA in the United States but through prime time dramatic programs (Gerbner 1987: 110).[4] Much of what people see in terms of dramatic entertainment, are movies, be they science fiction or hospital series or regular films. Evidently science and its protagonists are suitable subjects for the dream factory. Although no systematic data are available it also appears that the image portrayed of science and scientists by the mass media does not cultivate a very favorable orientation toward it (Gerbner 1987: 112). What has been stated by Gerbner with reference to TV dramas can most likely by reiterated with respect to fiction film. Just roughly half of our selection of some 200 plus movies (see below) showed scientists in a positive way. To the remaining half they appeared as problematic characters. In order to understand this ambivalence about science and scientists it is necessary to take a brief look at popular myths of scientific knowledge.

POPULAR MYTHS OF SCIENTIFIC KNOWLEDGE

Since antiquity scientific knowledge and its technical applications have been associated both with liberation as well as with enslavement, with the power to exert control as much as with the threat to be controlled, with welfare for the people but also with destruction. Gerbner notes that the "popular market for science is a mixture of great expectations, fears, utilitarian interests, curiosities, ancient prejudices, and superstitions," and that the "mass media appeal to all of these" (Gerbner 1987: 110). This fundamental ambivalence associated with science that communicators have to deal with crystallizes around specific issues that seem to recur again and again and are cast into popular myths whose detailed representation changes with new knowledge. One of these myths, probably the most powerful of all, is the creation of artificial human life or its alteration by intervening in hereditary material (i.e., the creation of hybrids, monsters, and the like). The prototypical figure of this myth is the alchemist Dr. Faustus whom Goethe has creating a homunculus. His most famous literary successor is Dr. Frankenstein, and he has inspired a chain of further stereotypifications:

Dr. Jekyll, Dr. Moreau, Dr. Caligari, Dr. Strangelove, and others (Haynes 2003). Back observes:

> The achievement of the mechanical creation of human life—or even of life at all—looks like a culmination of the acquisition of knowledge and the power that this knowledge brings. Most societies have set definite limits to this extension of human knowledge; modern Western society has been distinguished in trying to obliterate this limit. But the old limits still exert their power and arouse a certain dread of what will be found beyond these limits. (Back 1995: 328)

The ongoing debate over the moratorium on cloning humans illustrates the limits to that hubris, however fragile they may have become under the assault of progress.

If one wants to gauge how deep the roots of the myths critical of science actually reach, one needs to go back to their origins and trace their changes through time. Then it becomes understandable how the representation of science in film follows certain patterns.[5] The persistence of the figure of the alchemist as the embodiment of the scientist is best explained with the deep conflict between modern science and religion. Alchemy is foremost a metaphor for the pursuit of material goods and immortality. Authors of the late Middle Ages and early modernity contrast the "crazy alchemist" with admonitions for a frugal life guided by moral and religious values. In the Christian romantic literature of the eighteenth–century, criticism was directed against the amoral pursuit of mere knowledge about nature. The true alchemy of the search for God is contrasted with the false alchemy of modern science. Goethe's *Faust* represents the limitations of the new experimental science whose far-reaching abilities empower it to manipulate nature but which then loses control over its own products because it lacks the deeper understanding of a holistic natural philosophy.

The division of science into "two cultures" has its origin in this romantic contrast, the core of which is the religiously motivated critique of materialism, nihilism, and hubris. The critique of the materialism of modern science is directed against the fact that it no longer needs a God as creator. Materialist science is atheist. To commit the sin of hubris means to give in to the ambitions of modern science, wanting to unravel the secrets of divine creation. Mary Shelley's *Frankenstein* marks the birth of the mad scientist, whose hubris not only leads him into ruin as was the case with his precursors, but now, above all, also the people in his environment. In the course of the nineteenth century, the critique of modern science's hubris coincides with the moral critique of the obsessed scientist who unscrupulously pursues his goals and knowingly risks endangering other people.

This very brief sketch of the genealogy of the mad scientist from contemporary literary works focuses primarily on the religious roots of the critique of science. Toumey, on the other hand, explains why in his view the character of the *mad scientist* grows increasingly amoral as time passes. He sees the causes to lie in the artistic process of transferring texts into films and the commercial exploitation of characters through the production of sequels. This development, which he illustrates with film examples, takes place outside and independent of real science. It is foremost due to the unavoidable simplification that characterizes the film vis-à-vis the written text (Toumey 1992: 423). Even if one sees the direct influence of the degenerated picture of the scientist conditioned by the medium a bit more sceptically than he, the dynamics inherent in the movie business nevertheless is an important explanation for the independence of the film as medium. It adds to the stabilization and continuity of the myths that determine the social embedding of science.

Thus, it can be expected that these myths play an important role in popular culture in general, and in films in particular. In the great majority of films the depiction of science reveals a deep uneasiness, distrust, and even mystification of science on the part of the filmmakers. In the last analysis this mirrors the sentiments of the audience that watches their films. The images, clichés and metaphors employed by the filmmakers are the popular culture mirror image of science. At the same time the movies enforce these images and provide them with imaginative detail and decorum. Film, as one of the most influential media, interacts in complex ways with its audiences, reflecting, shaping, and reinforcing images and identities. It can safely be assumed that science as one of its subjects is not an exception to this (Turner 1999: 100, 144).

Whether or not the position of science is more precarious now than in the past is a matter of judgement that is regularly skewed by the short memory of the media and their focus on the present. The many attempts on the part of science administrators and policymakers to obtain the public's interest in, understanding of, and even "engagement" with science seem to suggest that science is experiencing a crisis of acceptance. However, the suspicion is that criticism of particular lines of research (e.g., stem cell research, cloning of human embryos) or of the implementation of knowledge in certain technologies (e.g., the genetic manipulation of food) are time-bound expressions of media attention, and that they reflect a much more profound ambivalence toward new knowledge. Thus, it is worth exploring the more stable patterns and stereotypes that are reproduced by the popular media in order to put present debates into perspective. As a side effect this may contribute to having more realistic expectations about the possibility of changing the public's attitudes toward science by the type of event-oriented campaigns now fashionable with science policymakers and administrators.

PATTERNS IN THE PORTRAYAL OF SCIENCE IN FILM

Some of the issues discussed in this section include which disciplines are popular in movies; the prototypical scientist; women in science; the depiction of scientists' traits; how knowledge is gained, and representations of scientific knowledge; settings of research; the dangerous discovery or invention; and utopias and dystopias of science.

Note on methodology

The following is based on the analysis of 222 films ranging over eight decades of moviemaking. The selection of films is not representative in a statistical sense but based on a search for films depicting science and scientists. Out of some 400 identified films the selection of the sample was primarily guided by availability of the films. However, an attempt was made to have a roughly equal share of examples in each decade (meaning roughly the same percentage of cases from the 400 in each decade). As can be expected there are quite a lot more recent films than older ones. The films were analyzed on the basis of a code sheet with about 120 categories. The results are based on coding by several people. Due to severe limitations of resources and time (the project was carried out in the context of a research seminar with students doing the majority of screening and coding) only in very few cases was inter-rater reliability of the codings tested. In order to keep the unavoidable impact of subjective judgments small, only those results that could be established with some confidence are presented. For the same reason we refrained from any further statistical analysis of the data since that would suggest a precision that cannot be sustained by the actual methodology used. All this implies that the percentages given cannot be seen as reliable representative figures; rather, they refer to our selection of movies only.

Fields of science—Which disciplines are popular in movies?

The relative frequency of scientific disciplines shown in movies is, first of all, a descriptive item that merely provides a background. The data reveal a picture that could at least partly be expected. "Medical research" figures most prominently, followed at some distance by the classical natural sciences: physics and chemistry. If one can assume that disciplines that have the image of being potentially life threatening or being involved in experiments with human identity receive particular attention in film scripts chemistry is clearly in a prominent position. This is especially true considering that chemistry is often involved in medical research and to some extent even in psychology. After all, Dr. Jekyll is a chemist who tries to solve a psychological problem (i.e., the separation of the "good" part of man's soul from the "bad" one). The archetypical Frankenstein in Mary Shelley's novel uses a chemical process to bring his creature to life, and it is only in James Whale's movie (*Frankenstein,* 1931) that an electrical process is used.

The fact that psychology ranks as high as chemistry and outranks biology and genetics is a small surprise. The fairly prominent role of psychology reflects the many movies featuring therapists and may be an expression of the preoccupation with behaviorism in U.S. society from the 1940s to 1960s. The wave of films dealing with genetics and the genetic manipulation of humans is probably still to come. The eugenic films that were popular until the mid-1930s have found few followers in recent times. The film *GATTACA* (1997) is an exception, as is *Boys from Brazil* (1978), which is about the cloning of humans (Kirby 2000). The disciplines of the humanities are surprisingly often the subject of film plots and, as will soon become clearer, they are associated most unambiguously with benign knowledge.

The prototypical scientist

Stories, visual stories in particular, depend on characters in action. Their messages are conveyed to the viewer by them. How are scientists portrayed in such stories? The clichés are obvious. The typical scientist in Hollywood's fiction film (more than two thirds of the films coded are U.S. productions) is white/Caucasian (96 per cent), American (49 per cent), male (82 per cent), and middle aged (40 per cent are roughly between 35 and 49 years old). The youth cult that dominates other genres has not yet penetrated to the depictions of science. Slightly less than a quarter of the movies portray scientists who are youthful (i.e., between 20 and 34 years old; 24 per cent). However, films like *Manhattan Project* (1986) or *Chain Reaction* (1996) may signal the portrayal of the preferred age cohort to come. Accordingly the vast majority of them have an inconspicuous appearance; caricatures like Jerry Lewis's *Nutty Professor* are rather rare. Finally, very little is revealed of their private lives. Almost a third of them are single, and of well over another third we are never told if they have any relationships at all.

Women in science

Science is traditionally a very male world in which women have either no place at all, or "their" place (i.e., a woman's place). As a consequence it is no surprise that less than a fifth (18 per cent) of the characters in these films are female. More importantly, women scientists are younger and more attractive than their male counterparts, and they are lower on the career ladder. In a sense, this is a quite realistic picture of science, albeit a bit out of step with recent developments.

Good, bad or... Depictions of scientists' traits

The message of a film is conveyed not only through its plot and actions, but also through its characters' traits, their motives and interests, their emotions and their deeds. In view of the notoriety of the "mad scientist" as the icon of a movie character one might expect that whenever scientists appear in film plots they tend to be descendants of Victor Frankenstein.

Instead we have compiled a slightly more complex picture that needs some explanation. On the one hand, results from a host of opinion polls show, time and again, that science *as an institution* is trusted highly by society. This is reflected in the large number of films portraying scientists as "benevolent" and "good." However, even the category of the "benevolent" scientist already includes traits of ambivalence. The benevolent scientist is naive when dealing with the interests of the powerful, means well but lives to see his or her discoveries being put to some deceitful purpose, and the like. "Ambivalent" scientists are those who are easily manipulated, idealistic but become progressively corrupted, ambitious and so lose sight of the consequences of their work, and, most importantly, grow willing to violate ethical principles for the sake of gaining new knowledge.

If one looks at the distribution of character profiles by field it is quite obvious that medical research, physics, chemistry, and psychology are the disciplines that are portrayed with the greatest ambivalence. In these fields the audience is most likely to be confronted with "mad scientists," with the Faustian scientist who oversteps ethical boundaries in order to gain forbidden knowledge and fame. Anthropology, astronomy, zoology, geology, and the humanities, on the other hand, are the fields that seem to have an unchallenged image of trust. Scientists from these fields are largely depicted as "good" and "benevolent."

This picture supports the thesis that medical research, including research on the mind as well as physical and chemical intervention in nature, is regarded with the greatest skepticism and most easily conflicts with the ethical boundaries drawn around it. In these areas the public is most likely to be confronted with the "mad scientist," the Faustian scientist who transgresses ethical boundaries in order to gain forbidden knowledge and eternal fame. Anthropologists, astronomers, zoologists, geologists, and humanist scholars, in contrast, have an image of trustworthiness. This picture supports the thesis that medical research including research on the mind as well as physical and chemical interventions into nature are regarded with the greatest skepticism and get most easily into conflict with the ethical boundaries drawn around them.

Another item in the analysis lends additional support to this interpretation: looking at how film-makers see the different ways in which knowledge is gained.

How knowledge is gained

The activity of research is usually hidden from the public eye. Precisely because the laboratory is a strange world, because the instruments used by scientists are foreign, and above all, because the methods used are obscure and powerful at the same time, the ways in which scientists gain their knowledge are of particular interest. They arouse suspicion like the methods of jugglers at county fairs who were, after all, the eighteenth-century travelling demonstrators of electricity in streets and public places (Hochadel 2003: ch. 4). Looking at the ways in which scientists gain their knowledge reveals where the lay public sees boundaries being overstepped, values violated, and crimes committed; that is, it focuses on suspicions about the scientists' doings.

The major categories in question are "experimentation on humans and animals" that represents a certain problematic type of research, and "field research and expeditions" that are associated with the adventurous sciences. The other two categories, of new knowledge being gained "through genius" or "by accident" are indications of familiar prejudices about the process of scientific discovery that are held by the lay public and supported by scientists. While "genius" may convey some residual ambivalence because of its elevated and elusive nature and because it can be associated with ethically problematic ways of gaining knowledge, it is not associated outright with danger but more with a privileged access to "opening and reading the book of nature." Accidental discovery, serendipity, suggests not a dangerous or ethically problematic approach to discovery but rather just the fruit of attentiveness and concern.

The results of our analysis show that medical research and psychology, chemistry, biology, and genetics as portrayed in film emphasize experimentation on living objects as the dominant method for gaining knowledge. Physics trails behind the other fields because it is not characterized by experimentation on humans or animals. It is also noteworthy that astronomy and the humanities are outside these concerns. Their methods of gaining knowledge (literature research or deciphering ancient knowledge) are not considered problematic, nor are they associated with genius. The status of these disciplines in the eyes of film-makers is at best one of benign marginality, at worst of complete insignificance, precisely because their methods do not collide with established societal values and ethical convictions.

Representation of scientific work

This interpretation is supported once again by looking at the patterns of representation of different types of scientific work. In the light of the preceding insights, it may be assumed that representations of the scientists' actual methods of work will only be of interest if they reveal the problematic, sometimes even criminal nature of these methods. Where these methods are not foreign to the everyday practices of the lay public representations tend to focus on the results only. Obviously, an additional factor may be if the methods in question lend themselves to visual representation in the first place. Research as practiced in the humanities is not easily depicted in visual stories, let alone interesting to the viewer.

In astronomy and the humanities the methods are not shown or discussed in detail. In fact, in all fields the results are emphasized more than the methods. Only psychology is represented equally in terms of its methods and its results. Detailed research into individual films could further clarify which methods figure prominently in the ideas of the film-makers. The dominating focus on results rather than methods probably contributes to the distance of the film medium from science and reflects the public's distance to science in general.

It is precisely for the same reason that the representation of science to the public is a problem.

Settings of research

One of the most characteristic aspects of alchemy that contrasts it with modern science is its secrecy. Likewise, the most characteristic feature of the "mad scientist" film is the secret basement laboratory, usually ornamented with gothic elements of medieval castles (35 per cent of the movies show scientists working in secret settings). Secrecy is involved even where these stylistic elements are missing (modern type basement laboratories can be found, for example, in *The Brain That Wouldn't Die* [1963] and in *The Fly* [1958]). The secret laboratory is also typically the *private* laboratory of an *individual* scientist who works at most with one assistant. It is the place in which the illegitimate experiments are carried out. This implies that dangerous research is taking place outside of public institutions such as university laboratories and government facilities (although, in fact, such institutions house their share of dangerous practices). Scientists working in their home basements are outsiders. They have isolated themselves from the critical observation of the scientific community because they feel misunderstood, often because they are obsessed by their research the questionable goals and methods of which they see justified by the success they expect to have.

A fifth of all films in the sample portray science as a secret activity carried out in private basements. Over 40 per cent of the movies that deal with chemistry are in the alchemist tradition (i.e., showing research being carried out at home). Next to chemistry no other field except medical research stands out for being associated with this characteristic. Other fields are more likely to be associated with research taking place either in the field (anthropology, zoology, biology, psychology) or at universities (humanities). Chemistry as a discipline depicted in movies has the second highest share of secrecy (behind robotics!) as a feature.

On the level of disciplines the conclusion one can draw, albeit with some simplification, is that those fields that are generally considered socially or ethically problematic are also associated with research taking place in secrecy and in places isolated from the critical eyes of scientific peers or the lay public. The unproblematic disciplines typically operate outdoors or in public settings such as universities and government laboratories (Weingart et al. 2003: 285).

The second slightly larger group of films stages research in the field. Here we may assume that this is the concession that science has to make to action film. From *Fu Manchu* (1932) to *Raiders of the Lost Ark* (1981), archaeologists and anthropologists are involved in high action adventure. Only brief scenes at the beginning or the end of the film acquaint the viewer with the scientists' mundane academic home base.

This equates to the polarization of a *public* science in which the scientist works in the context of a community of peers, and a *private* science where the scientist has chosen to leave the community or was excommunicated by it because of his or her overstepping of the boundaries into forbidden research territory. Such a boundary is a thin line, and the viewer is often unable to decide which side to take: that of the scientist, who considers his genius misunderstood or falsely judged by resentful colleagues (20 per cent), or that of his opponents. The solitary scientist (in 42 per cent of the films) who works in his home laboratory is not controlled by his peers or by public authorities. His discoveries are depicted as dangerous in more than 60 per cent of the stories. In almost half of the films (48 per cent), the "invention is kept secret from the public"; in more than a third (35 per cent) the discovery or invention "gets out of control," and in more than half of the films (58 per cent), it causes damage intentionally or unintentionally.

Dangerous discovery/invention

One of the most common stereotypes about science is the dangerous nature of the knowledge gained, of discoveries and inventions. The creation of dangerous knowledge is associated with hubris. The image of different disciplines is determined to a great extent by the kind of knowledge associated with them. In the case of chemistry only a quarter of the films in the sample show discoveries in chemistry that are not dangerous. More than half are depicted as unintentionally dangerous, the remainder being depicted as the dangerous results of ill will. The obvious next question is who is depicted as the victim of dangerous research. It turns out that in half of the movies the discoveries and inventions affect uninvolved people. In roughly a third the victim of the discovery is the scientist himself. If one includes colleagues and assistants this share goes up to about half of all films. This reflects the (alchemist) tradition of solitary research and heroic self-experimentation.

Scientific knowledge and ethical values

This ambivalence about the potentially threatening nature of scientific knowledge and technical inventions that accrue from it are reflected directly in the conflict between scientific knowledge and ethical values. In just more than half of the films (51 per cent) ethical values are challenged, undermined, or are in direct conflict with the science portrayed in the story. Again, this overall result is broken down for the different scientific disciplines, and the previous picture emerges once again: The disciplines that are ethically problematic are most frequently medical research, followed by physics, chemistry, genetics, psychology, and biology. Astronomy, anthropology, and the humanities are regarded as mostly outside of this concern. However, this has to be qualified: although physics is predominantly seen as being in conflict with ethical values (51 per cent), to a considerable extent (37 per cent) it is also seen as completely independent of these. In a surprising 29 per cent of the films, the humanities are seen as being in conflict with values. The different disciplines' "sensitivity" towards values can also be measured by the extent to which they are portrayed as having no relation to them at all. Only 10 per cent of the films about psychology and 12 per cent of those about genetics were classified in this way.

These findings lead to the question of what kind of knowledge is produced by these disciplines which then comes into conflict with established values.

Utopias and dystopias of science: Objects of fictional science

Science and scientists are relatively abstract subjects and difficult to portray in visualized stories. Fictionalization is a means to circumvent the problems of representing the world of knowledge when adapting it to the rules and constraints of visual drama. But film's overriding concern with fictional developments is also an expression of the mystique of the production of new knowledge. Science is associated with the unknown future, and it becomes the object of projections of utopias and dystopias. Science itself is often portrayed at a fictional state of development, just beyond the frontiers of contemporary research and technological achievements. In 39 per cent of the films, real scientific fields are depicted at a fictional level of development; in roughly 14 per cent, fictional fields of science are shown; less than half (47 per cent) of the movies deal with a non-fictional area of science.

If one looks at the kinds of subject matter of these fictional or semi-fictional sciences, it is apparent that the projections of the future associated with them are mostly dystopias or at least highly ambivalent utopias. Roughly a third of the movies in the sample deal with artificial, supernatural human, animal, or extraterrestrial life forms; cloning; reanimation; or immortality. If illness and cure are added to this category, the share is even larger (5 per cent more). Only a smaller segment (ca. 20 per cent) is devoted to super weapons, time travel, and other technological gadgets. The utopian or dystopian views about science are clearly dominated by concerns about the manipulation of human and animal life.

Not surprisingly, once again medical research is most often associated with fictional developments, followed by genetics, physics, psychology and chemistry.

Scientific misconduct by discipline

The misconduct of scientists without any specification of the type of misconduct may be seen as an aspect of the dubious or at least ambivalent nature of science. If science is identified with misconduct this suggests distrust. Again, certain fields are associated with questionable conduct more than others. Chemistry is among the former, together with genetics and pathology, even leading medical research, biology and the computer sciences. In literary studies and the humanities, in general, misconduct is comparatively rare.

Authenticity

Films are made to capture the imagination of the audience. Illusion is the essence of fiction film, and yet film makers mostly try hard to create plausible plots and representations, to render their products authentic in order to have impact on the public. Just a little more than a quarter of all chemistry movies are depicted as non-authentic, only about a fifth are comedies and satires. A look at *The Nutty Professor* shows that not even these are just funny.

The authenticity is obviously enhanced when gadgets and technologies are shown that look familiar to the viewer. Chemistry is often presented in conjunction with familiar instruments. As Schummer and Spector have shown some of these are iconographic for the representation of science as a whole like the chemist holding up and gazing at a flask (Schummer and Spector 2004).

Preoccupation with the past—Alchemy

There is no doubt that the legacy of alchemy had its impact on film-makers throughout the twentieth century as a selection of film titles reveals, and probably will continue to do so in the twenty-first.
Movie titles of the nineteenth and twentieth century:

The Hallucinated Alchemist (1897, USA)
The Clown and the Alchemist (1900, USA)
The Alchemist (1913, USA)
Homunculus (1916, Germany)
Der Alchimist (1918, Germany)
The Alchemist's Hourglass (1936, USA)
Alchimie (1952, France)
Une Alchimie (1966, Belgium)
Alchimisten (1968, GDR)
Alchemik (1990, Poland)
Des alchimistes/ Alchemists (1991, Canada)
Alchemy (1997, USA, TV)

The continuity of the occurrence of alchemy raises the obvious question as to what if anything has changed in the representation of chemistry in movies over the last century. Our material does not provide a definitive answer to that question, not least because the disciplinary focus did not guide the selection of films. Not surprisingly, movies dealing with science change the appearance of characters and the decorum of their research laboratories following the fads and fashions of the different genres. The creation of life by means of a cumbersome fictitious assortment of steaming and glowing chemicals that dominated the movies until as late as the 1990s is slowly being replaced by the clean microscopic techniques of molecular biology. Cloning has entered the movie scene rather late, with the exception of a very few films like *Boys from Brazil* (1978). But the basic stereotypes, the fears associated with the creation of life like the ill meaning scientist or the experiment going out of control can be found just the same way in recent productions like *Godsend* (2003), *Blueprint* (2003), and *The Sixth Day* (2000). The impression is that the underlying anxieties about new knowledge reflected in the products of popular culture are much more fundamental than the images which link them to the respective worlds familiar to their audiences.

CONCLUSIONS

Thus, the deep seated fears and expectations connected to our own lives are projected into fears and expectations about those fields of science that are concerned with the prolongation, improvement, manipulation, expansion, and termination of life. The popular fiction film, like literature, gives expression to these fears and expectations. Across all genres the creation and manipulation of life is the dominant (or at least a very prominent) category. By having detected this pattern in a sufficiently large number of movies over a considerable span of time we can appreciate much better than before that the contents of movies are, indeed, driven to a considerable extent by basic myths about the creation of new knowledge, its boundaries, and the dangers of overstepping them.

At the same time this dwarfs short-winded concerns about the public image of science and its improvement and relegates them to their proper place. No PR campaign and no slick TV-infotainment magazine even addresses the myths of dangerous knowledge, let alone has any impact on them. These myths evidently go much, much deeper than can be counterbalanced by the popular explanations of clever TV presenters or "hands-on" expositions. The detailed analysis of the representation of these myths in popular media, such as the fiction film, can contribute to further elucidating their origins and their manifestations. The distance of society from its science, and the ambivalence that characterizes the relationship, are illustrated by the fact that roughly one quarter of all scientists represented are portrayed as "unworldly," about a sixth as "eccentric" but only little more than 5 per cent as "comical." The leading genre among films about science is the horror movie. In contrast, there are hardly any comedies about science. Evidently our society does not find much to laugh about in science.

NOTES

1. This article is an edited and enlarged version of the presentation at the conference in New York on which this volume is based. Parts of it were published previously in Weingart with Muhl and Pansegrau (2003) and in Weingart (2006).
2. http://www.science-et-cite.ch/archiv/sccinema/de.aspx
3. The author learned about a project of the UCLA film school in 2000 which had this objective.
4. An analysis of TV documentary magazines today would probably reveal an increase of those focused on science. But a closer look reveals that the recent fascination with "science" is actually transporting primarily everyday knowledge. The reference to "scientific knowledge" is mostly symbolic and part of the discursive fad of the "knowledge society".
5. For the following I rely on Schummer (2004) for some detailed references to literary figures as well as on Haynes (1994).

BIBLIOGRAPHY

Back, Kurt W. (1995) "Frankenstein and Brave New World: Two cautionary myths on the boundaries of science", *History of European Ideas*, 20: 1–3, 327–332.

Crichton, Michael (1999) "Ritual abuse, hot air, and missed opportunities", *Science*, 283: 1461–1463.

Gerbner, George (1987) "Science on television: How it affects public conceptions", *Issues in Science and Technology,* Spring: 109–115.

Haynes, Roslynn D. (1994) *From Faust to Strangelove: Representations of the Scientist in Western Literature,* Baltimore, MD, and London: Johns Hopkins University Press.

Schummer, Joachim (2006) "Historical roots of the 'mad scientist': Chemists in nineteenth-century literature", *Ambix*, vol. 53, 99–127.

———— and Tami I. Spector (2007) "The visual image of chemistry", *Hyle*, 13: 3–41.

Toumey, Christopher P. (1992) "The moral character of mad scientists: A cultural critique of science", *Science, Technology and Human Values*, 17 (4): 411–437.

Turner, Graeme (1999) *Film as Social Practice,* 3rd edition, London, New York: Routledge,

Weingart, Peter (2005) *Die Wissenschaft der Öffentlichkeit*, Weilerswist: Velbrück.

———— (2006) "Chemists and their craft in fiction (film)", *HYLE—International Journal for Philosophy of Chemistry,* 12 (1): 31–44.

———— Claudia Muhl, and Petra Pansegrau (2003) "Of power maniacs and unethical genius: Science and scientists in fiction film", *Public Understanding of Science*, 12 (3): 279–288.

14 Unforgettable?

Science, Prosthetic Memory, Film

Lutz Koepnick

I

For some time now filmmakers worldwide have been preoccupied with issues of individual memory and collective recollection. We used to think of cinema as a mere witness of historical events, able to transform the flow of time into series of images and transport these images to coming generations. At best, film offered us insights about how certain ages wanted to see themselves in past, present, and future and how this seeing informed their repertoires of representation. At worst, cinema stood out as a privileged tool of emotional manipulation, a technology falsifying the real, recasting contingent histories as fate and nature, and in this way denying competing visions of the past and its future. Today, cinema no longer seems to serve as a mere servant of historical memory or an eyewitness of future pasts. Rarely do we consider it as a tool helping us to approach the past as a distant reservoir of insights and lessons; nor as a versatile technology that, by providing narratives of historical time, can help legitimize the present and incite alternative visions for the future. Instead, in our time of fast-paced image flows and networked communication, mediated images themselves—whether moving or still, whether captured on celluloid or digitally processed—assume the task of doing the work of memorialization and historical imagining for us; they become rather than merely facilitate memory in an of themselves.

The sweeping narratives, digitally improved set designs, and galvanizing surround sounds of today's mainstream cinema invite viewers to experience the past with their own bodies, as if they had actually been there. Historical memory in this context attains the status of an artificial, albeit fully operative, limb. Far from being merely second-hand, the past is designed to be worn by the body, as a sensuous memory produced by the viewer's first-hand immersion into cinematic sights and sounds. In the European context, the stunning success of so-called heritage films during the past decades adds yet another dimension to how contemporary cinema changes our relationship to both our own organic memory and the collective archives of cultural recollection (see, among others, Higson 1993: 109–129; Koepnick

2002: 47–82). Designed to present history as a lavish spectacle of positive identifications and visceral effects, heritage films fan nostalgic desires for the sheer materiality of the past. They situate the viewer as one eager to feel how it may have been to live in times preceding their own physical existences, not in order to draw lessons from the past for the future, but on the contrary to generate transhistorical empathy and thus incorporate earlier periods into the image banks of an ever-expanding present.

Unlike nineteenth-century romantics or early twentieth-century modernists, today we no longer think of memory as something deeply bound up with the past, but rather as a mode of re-presentation inherently belonging to the present—a present whose borders spread out ever-more vehemently across the formerly distinct domains of the past and the future. "The past," as Andreas Huyssen has written recently, "has become part of the present in ways simply unimaginable in earlier centuries. As a result, temporal boundaries have weakened just as the experiential dimension of space has shrunk as a result of modern means of transportation and communication" (Huyssen 2003: 1). Catalyzed by the advent of electronic communication networks and the global reach of post-Fordist consumer capitalism, this expansion of the present does not simply amount—as it was fashionable to claim in the 1980s and early 1990s—to a privileging of space over time, of geography over temporality, of topography over narrative. Instead, what we are facing today are potent changes in the very nature of the spatial and the temporal, a restructuring of human perception and experience in whose context narrative forms tend to organize events, not in the form of temporal successions and cause-and-effect chains, but across rhizomatic spaces and noncontiguous sites of representation. Expanded presents like our own might favor synchronic over diachronic dimensions, simultaneity over linear progression, information density over sequential order. They encourage the users of media technologies to cut across and rapidly switch in between disparate archives of recollection rather than find satisfaction in the single, linear, causally determined stream of classical narrative time. But in this process, neither time nor space appears to us as what it used to be. One category of experience does not simply swallow or surrender to the other, as postmodern critics seemed to suggest a while ago. What they do instead is to enter into an unprecedented dynamic of mutual interdependence which fundamentally transforms the location of knowledge, meanings, intensities, and identifications in time, and in this way, also transforms the places and procedures of individual and collective memory.

Little should it surprise us , then, that filmmakers over the course of the past decade or so have repeatedly sought to address how certain machines and scientific innovations, whether imaginary or not, may affect the function of historical and personal memory and hence expand our own or their protagonists' present. To name only a few, think of mainstream science fiction features such as Paul Verhoeven's *Total Recall* (1990), Robert Longo's *Johnny Mnemonic* (1995), Kathryn Bigelow's *Strange Days* (1995), and

Cameron Crowe's *Vanilla Sky* (2001), of gripping thrillers such as John Dahl's *Unforgettable* (1996) and Tom Tykwer's *Winter Sleepers* (1997), and of art house entries such as Wim Wenders's *Until the End of the World* (1991) and Christopher Nolan's *Memento* (2000). What we witness in films such as these are protagonists unable to remember their own lives and actions without the aid of Polaroid cameras and other mechanical recording devices. Or heroes eager to abandon their own drab existences for the sake of virtual worlds whose present shapes and past histories can be accessed with the help of powerful drugs or advanced machines of image production. Time and again, we follow the paths of tormented subjects desperately hoping to understand their own presents by exposing themselves to technologically mediated pasts. Or conversely, we encounter individuals unable to manage their own lives because they fail to recognize whether what they do remember is really their own memory or simply a product of technological simulation. Cinema today may indeed figure ever-more compellingly as the custodian and depository of individual recollection and historical memory. But contemporary filmmaking, whenever it presents alternatives to organic forms of recollection, also abounds with examples expressing substantial unease about the way in which postmodern cyber-culture reshapes temporal experience. Obsessed with prosthetic devices of remembrance and recollection, these films and their often scientifically trained heroes may serve as allegories for our need to live in extended structures of temporality, to slow down rather than accelerate the space-time compression of digital culture, and to thus open up crucial spaces for reflection and self-maintenance.

Think of the Max von Sydow character Henry Faber in Wenders's *Until the End of the World*, a scientist fanatically experimenting with the transfer of memories and perceptions in order to enable his wife to see what her body does not allow her to see due to childhood blindness. Though Faber has his son Sam (William Hurt) collect impressions and future memories across the globe with the help of a special recording device, Faber's high-tech laboratory is located seemingly beyond the circuits of global time-space compression, namely somewhere in the Australian outback. Whatever Faber does in order to advance the transfer of sight and memory, at least initially is shown as deeply embedded in and devoted to the rhythmic time-tables and ritualistic textures of the local Aborigines' lives. The end of the world here is meant to offer a window onto the world's beginning. It defines a space to recuperate lost time rather than merely obliterate the boundaries between past and present. However advanced his equipment may be, Faber's vision is to secure some kind of continuity within the accelerated streams of global temporality; it is to define a liminal space within which one can live a good, meaningful, and coherent existence.

But to think of scientists such as Henry Faber merely as symbolic figures challenging the erasure of temporality and narrative continuity in digital culture misses the point. For much more, I would like to argue in this essay, is at stake in many films today situating scientists as guardians of the boundary between past and present, between memory and history. While the quest for authentic memory, the marketable exploitation of imagined histories, or the battle against the burdens of repressed pasts might be at the center of their efforts and desires, the primary function of celluloid scientists researching the work of memory, I suggest, is to help envision powerful alternatives to how the medium of film itself shapes the viewer's perception, organizes our attention, and produces experiences of sensory immersion. In many of today's science fiction productions, to explore the vicissitudes of memory is to expand the present as much as to expand cinema itself, to probe different ways of how people may interact with machines of image production in the future. Ironically, then, contemporary cinema's obsession with memory and the past is very much one about the present and future of cinema itself. And more often than not, the figure of the scientist plays an important role by warning the viewer against the dangers and perversion of future cinemas that expose their viewers to stimulations filling the entire range of their sensory perception. In a stunning reversal of the figure of the mad scientist of earlier filmmaking, the memory researchers of contemporary science fiction often turn out to be the most persuasive advocates not only of the real, but also of traditional forms of cinematic projection and entertainment—of forms essentially tied to the idea of cinema as a well-defined frame around and window onto alternate realities rather than a hermetic space of unbound immersion.

The aim of this essay is to focus on how a number of recent science fiction films depict mnemonic interfaces and scientific experiments whose logic of immersion transcends the conventional framing devices of narrative cinema. How, I shall ask, do these films visualize scientific initiatives whose principal telos is not only to renegotiate the boundaries between past and present, but to question or even undo the very nature of cinema? If immersive technologies today emancipate the human eye from traditional laws of central perspective and invite viewers to traverse virtualized times and spaces, how can contemporary cinema represent the work of its alleged nemesis, the computer and its powerful interfaces, without giving up its own specificity? And how do science fiction films today define criteria which would allow the viewer to assess competing ways of recalling the past and envisioning alternative futures? Filmmaking might be at its best and most innovative today, is my contention, where it uses the composite and hybrid window of the silver screen to actively engage with, rather then take flight from, the ubiquity of human-computer interfaces in contemporary culture; where it opens our eyes and ears for how different interfaces may differently restructure our modes of sensory experience, perception, representation, cognition, and recollection. It is thus by discussing the two most influential paradigms of the human-computer interface, as they have been conceptualized by computer scientists in particular in the early second half of the twentieth century, that we must begin this inquiry.

II

Pressed by industry concerns and consumer demands for user-friendliness, we have come to think of the computer interface mostly as a screen or virtual space allowing for either disembodied but interactive, or vivid but passive forms of human-machine transactions. (For a more detailed version of this overview and argument, see Koepnick 2007: 15–21.) It is a common assumption that interfaces work best, that is, allow for the highest degree of immersion into alternate realities, if they either render themselves entirely invisible or at least mimic the look of older, long-familiar work or entertainment environments and thus become virtually imperceptible, the desktop metaphor here being the perhaps best known example. In a 1965 essay on the future of computer displays, MIT graduate Ivan E. Sutherland resorted to Alberti's famous Renaissance notion of painting as a window onto the real in order to describe this dream of total interface transparency: "One must look at a display screen as a window through which one beholds a virtual world. The challenge to computer graphics is to make the picture in the window look real, sound real, and the objects act real" (Sutherland 1965: 506). For Sutherland, the inventor of "Sketchpad," the first graphical user interface, computers were meant to communicate directly to their user's senses by drawing attention, not to themselves but solely to the image space they were designed to summon. In Sutherland's perspective, simulation of the real not mere illusion, sensory immersion rather than mere representation, defined the royal road to making our use of the computer's artificial intelligence most effective. In his 1970 book *Expanded Cinema*, Gene Youngblood took up Sutherland's ideas for the purpose of envisioning new osmotic relationships between viewers and cinematic images. What Youngblood called expanded cinema was to transport thoughts, emotions, and intensities without intermediary codes, signs, or processes of communication; it was meant to access the viewer's brain directly so as to create simulations of the real with no visible outside. Like Sutherland's visionary window, Youngblood's futuristic cinema was to yield experiences of total immanence and enable a seamless incorporation of the viewer into the computer's image space. No matter what they presented to the viewer, the interfaces of both Sutherland and Youngblood thus not only aimed at communicating messages without grammar and codes, but in so doing they also sought to revolutionize the structure of aesthetic experience. The ultimate computer, Youngblood concluded, "will be the sublime aesthetic device: a parapsychological instrument for the direct projection of thoughts and emotions" (Youngblood 1970: 189). Fully transparent, the perfect computer interface would restore the original sense of the aesthetic, that is to say, the sensory experience of perception. In its ability to simulate the real and immerse the viewer into alternate universes, the ideal interface would emancipate the aesthetic from nothing other but the chains of autonomous art and the bourgeois culture of reification.

Sutherland and Youngblood envisioned interfaces as surfaces, places, or spaces of contact between computers and human users whose primary purpose was either to make users believe that whatever they did to affect visual representations on screen also moved some kind of real entity, or to allow virtual worlds achieve unmediated effects on the user's perception. Both conceptions envisioned the ultimate interface as a self-effacing conduit, that by collapsing any sense of difference and alterity, enabled spontaneous, and hence, most efficient contact between human and computer. In both conceptions, users could relate to or immerse themselves into the world of machines without knowing anything about their formal operations or structures of programming. Interfaces promised to operate best whenever they stimulated acts of mystical communion between the user's and the machine's bodies and souls.

Though this model of the interface has dominantly informed the hopes and itineraries of computer scientists ever since, it was certainly not the only model developed in the early days of interface design to envisage the way in which future users should relate to and manipulate the computer's archive of images, sounds, data, and memories. Consider the pathbreaking work and visions of Douglas Engelbart, pursued with a small team of highly talented researchers at Stanford Research Institute in the early 1960s. Though essential for the later development of the mouse, the windowed user interface, and the hypertext format, Engelbart's initial research at SRI ran counter to mandates of user-friendliness and unmitigated sensory immersion. Instead, his vision for the future of personal computing was guided by concepts such as co-evolution and co-adaptation, that is, the idea that technological improvements would transform the user's capabilities to think, feel, and manage complexity; that the development of new computing systems would go hand in hand with the augmentation of new forms of human intelligence and gesture; and that hardware and software designers, by inventing new forms of artificial intelligence, would also reinvent the human user rather than merely cater to presumed needs and existing forms of cognition and sensory self-expression. As Thierry Bardini summarizes Engelbart's program,

> Engelbart's work was based on the premise that computers would be able to perform as powerful prostheses, coevolving with their users to enable new modes of creative thought, communication, and collaboration providing they could be made to manipulate the symbols that human beings manipulate. The core of this anticipated coevolution was based on the notion of bootstrapping, considered as a coadaptive learning experience in which ease of use was not among the principal design criteria. (Bardini 2000: 143)

Unlike Sutherland and Youngblood, Engelbart did not envision the computer's screen as a device empowering the untrained to manipulate data or experience sensory stimulations in more efficient ways. On the contrary, for Engelbart the user was supposed to employ future computing technologies in order to develop new ways of thinking and of exploring his or her own bodily and human identity. Whereas Sutherland's work sought to situate the computer as a popular and intuitively commensurable tool of communication, Engelbart's research was driven by Enlightenment notions of individual learning, growth, and change. Rather than absorb the viewer through self-effacing operations, the interface in Engelbart's work was considered as a representational space at which software and hardware designers were trying to construct the shape of future users, and future users were invited to test and improve these constructions like actors playing roles in a stage play. What Engelbart considered as the interface represented the designer's efforts to coevolve technology and the human subject. It was not meant to provide mystical experiences of total reciprocity between humans and machines, but on the contrary to represent a tangible third space of symbolic and physical transactions different from both the space of the user and the one of the computer. Whereas Sutherland and Youngblood sought to define computers and users as the respective others' unmediated extension, Engelbart conceived of the interface as a site where users could become other without losing a sense for what made them different from the machine. Opposed to dominant templates of user-friendliness, Engelbart envisioned the interface as an artificial limb, bridging but never closing the gap between the organic and the artificial—a performative space in which ongoing acts of negotiation provided the condition for the possibility of insight and self-transformation.

Contemporary fiction and science fiction films, in their attempt to picture the processes of memory and problematize possible tensions between organic and prosthetic modes of recollection, vacillate between Sutherland's and Engelbart's paradigmatic vision of the interface. They either display futuristic tools of remembrance as self-effacing interfaces directly accessing their users' minds and incorporating the subject into self-enclosed projections of the past. The most salient example of this conception might be seen in Bigelow's *Strange Days*, featuring a device which feeds recorded memories, experiences, and emotions directly into the user's cerebral cortex and thus allows him or her to relive other people's pasts strictly from the user's subjective point of view. Or alternatively, they present non-organic tools of recollection as tools enabling or necessitating unsettling negotiations between competing subject positions, emotions, perspectives, and temporalities, the prosthetic role of photography as a tangible, albeit highly unreliable, document of passing time in films such *Memento* and *Winter Sleepers* perhaps best illustrating this tendency. Mnemonic interfaces in all of these films promise to be powerful aids to recall the past, transfer perceptions and emotions across time, and redraw the contours of identity. What is important to note, however, is the fact that experiences of total sensory immersion, as in *Strange Days*, can yield as disastrous results for these films' protagonists as self-reflective uses of prosthetic intermediaries as in *Memento*. Neither Sutherland's nor Engelbart's vision of the interface, as recalibrated for the narratives of these films, are shown as resistant to ethical abuses, aesthetic hypertrophies, or painful misrecognitions. On the contrary, whether mnemonic interfaces in these films permit scenarios of total sensory immersion or of self-critical learning, both are shown as instruments of delusionary hubris and murderous self-inflation. And no matter whether they represent technologies preceding or potentially succeeding the medium of film itself, the moral flaws and perceptual deceptions of these mnemonic interfaces in the end seem to serve the purpose of underscoring mostly the presumed superiority of the filmic medium as a medium capturing the passing of time and connecting past and present.

In the remainder of this essay, I would like to complicate this account by focusing on three films that in their representation of prosthetic memory tend to combine, remediate, or hybridicize rather than radically oppose different conceptions of the mnemonic interface. In all three films we encounter scientists struggling via technological aids with forgotten or repressed pasts. Though these scientists do not always take recourse to computer-based technology in order to access their own or other people's memories, their quest for recollection is shown as one channeled through paradigmatic interfaces similar to the ones envisioned by Sutherland and Engelbart. However, what makes the exploration of post-organic memory in these three films interesting is the fact that all of them, instead of first playing out Sutherland's against Engelbart's version of the interface, and instead of then pitting one of the two against the mnemonic work of conventional cinema, confront their heroes and their viewers with competing versions of the interface simultaneously. This is not the place to engage in a detailed analysis of these films' narrative economy. Because the cinematic representation of memory and recollection essentially hinges on how transition techniques such as cuts, dissolves, and fades interconnect different locations in space and time, let me therefore instead solely focus on how these three films employ the formal possibilities of editing in order to represent the work of mnemonic interfaces and signify upon the experiences and dilemmas of their scientific protagonists. It is in these three films' particular editing maneuvers, I suggest, that the promises and hazards of prosthetic memory and interfaces of recollection become the clearest.

III

John Dahl's 1996 thriller *Unforgettable* follows the efforts of Dr. David Krane (Ray Liotta) to find the true culprit for the brutal murder of his wife Mary. A trained medical doctor and pathologist, Krane was suspected of having murdered his wife himself, but was freed from all charges brought against him due to some procedural errors. In his ongoing attempt to clear his reputation, Krane comes across Dr. Martha Briggs (Linda Fiorentino), employed at Washington State University and developing powerful drugs in order to transfer memory. Though Briggs's medical procedures have thus far only been tested on animals and seriously damage their brains and hearts, Krane decides to test the drug on himself. Blatantly ignoring Briggs's explicit warning against human applications, Krane steals a probe from Briggs's laboratory, appropriates a sample of his dead wife's spinal fluid from the police's medical archives, returns to the site of the murder, and injects drug and spinal fluid into his own arm in the hope of reliving the last minutes of his wife's life. As it turns out, however, repeated applications will be necessary in order to really see what Krane wants to see in order to find the person who framed him, each procedure raising the level of intensity and identification but also further damaging his own body.

Though Krane, by injecting the drug, basically takes on his wife's perception and point of view, Dahl's editing in these crucial scenes does not simply mimic the process of comprehensive sensory immersion. To be sure, similar to the editing strategies of contemporary action cinema, abrupt cuts and jerky camera movements here index the way in which Briggs's drug joggles Krane in between past and present. Discontinuous cuts and forced perspectives allegorize not simply the traumas of the past, but the physical ordeal of time travel. Moreover, repeatedly we see Krane re-experiencing the violence of the past with his own body in the present, as if being fully absorbed by what the drug reveals to him about the past. And yet, what is striking about Dahl's editing and cinematography in these scenes is the fact that his camera, in spite of showing Krane's somatic experience of full immersion, neither entirely assumes the point of view of Krane's wife nor of Krane himself. While editing and cinematography at certain moments suggest unbroken contiguity between the spaces of Krane's present and his wife's past—camera pans basically insinuate that they inhabit the same location—the visual perspective of Krane's prosthetic memory remains largely independent of any embodied viewpoint; it is not exclusively tied to what his wife in the final moments of her life was able to see. What Dahl, in other words, does in picturing the work of prosthetic memory is to sever the somatic from the visual impressions of time travel and in this way combine two different modes of interfacing past and present. Briggs's drugs open a port to other people's memory whose immediacy recalls Sutherland's and Youngblood's vision of the interface. The film's camera and editing, by contrast, frame and reframe this experience of immersion from a mediated point of view that is neither Krane's nor Mary's, but, like Engelbart's vision of the interface, opens up a third space of memory in whose horizon Krane as much as the viewer can witness in the end what exceeded the victim's own perception, namely the sight of her murder. To access and understand the past properly and effectively, in Dahl's film, is to be able to alternate between states of sensory immersion and acts of visual distanciation. One version of the interface here at once needs and produces the other, and it is only by integrating their respective functions on screen, by negotiating the somatic and the visual aspects of induced memory with cinematic means, that Dahl's *Unforgettable* in the end can present the true murderer and restore Krane's reputation.

IV

In Wenders's *Until the End of the World*, the crucial process of transferring sight and memory to Edith Farber is initially shown as a procedure of multiple framing and mediation. Essential for the success of Faber's experiment, the "medium" Claire is placed in front of a television screen so as to rewatch what she recorded at various locales weeks or months earlier. Electrodes pick up her brain waves during this process of rewatching and transport them to powerful computers, which convert Claire's input and feed visual and acoustical signals to a curious horse shoe-like device surrounding Edith's head. Wenders's cinematography emphasizes the mediated nature of this process by focusing our attention on the many video and computer monitors that populate Faber's cave laboratory and either display what Claire once recorded, is currently seeing, or what will be the kind of signals transmitted to Edith to engulf her perceptual apparatus. Furthermore, the whole scene is framed by shots showing the film's diegetic storyteller Eugene Fitzpatrick (Sam Neill) and Faber's son Sam witnessing the whole procedure on an adjacent video monitor, Eugene momentarily interrupting his typewriting of the very story we see unfold and Sam clinging to the TV set's remote control.

In the early phase of this sequence Wenders's cuts frequently back and forth between Faber's computer room, Claire's recording booth, and Eugene's improvised office, without necessarily clarifying their exact spatial relationship to each other. It is only when Claire as much as the computer is finally able to produce usable output for the transfer that Wenders de-emphasizes the role of frames within the frame and instead, first, allows Claire's images to fill the entire screen and, second, cuts to a close-up of Edith, her hands covering her sightless eyes, as she interprets the visual and acoustical signals reaching her brain like bodily perceptions. Mediation and framing at this point seemingly give way to experiences of total sensory and panoramic immersion. What is engineered with the help of advanced computing now seems to take on the impressions of the organic and unmediated, of that which exists outside and beyond the material window of the interface. And yet, that Wenders at this point rarely cuts to full-screen images showing what Edith actually sees or perceives, and that he uses lingering camera shots to focus on Edith's face as she attempts to translate prosthetic memories into linguistic expressions, encapsulates the film's unmistakable misgivings about Faber's interest in direct neurological stimulation. For in *Until the End of the World*, the benefits of memory transfer, of simulating experience through interfaces of total sensory immersion, in the end are much lower than its expenditures. It won't be long before Faber's injection of sight and memory will exhaust Edith's feeble health. Consuming other people's stories, memories, and perceptions, Edith herself will be consumed by Faber's self-effacing interfaces, her words no longer able to speak up, over, and against the power of induced impressions. Yet rather than returning him to the real, her death will cause Faber to radicalize his research by trying to record and transmit human dreams as well, a highly addictive endeavor whose self-destructive course is stopped only when Eugene, associated with the low-technology of manual typewriting and the creative energy of the poetic word, will finally step in and draw Claire away from Faber's expanded cinema of total digital attractions.

Wenders has written,

> In a few years, there will be data beamers standing next to the good old movie projectors, and in another few years, those projectors will be gone. The entire history of cinema will be available in any theater via servers, optical cables, satellites, broadband or whatever system. We want to make sure that this important transformation happens in our control, that the norms regulating the giant change do not exclude us and won't be dictated by others.... It is up to us and to the next generation of talent here in Europe to put the new digital technology into the service of all sorts of storytelling and push them into the realms nobody has opened up yet. (Wenders 2001: 38–39)

Until the End of the World envisions a future world in which digital technology, rather than enabling new stories, obliterates the conditions for the possibility of narration and storytelling altogether. Simulating memories and tapping dreams, Faber's experiments foreshadow a world which no longer knows of the old movie projector, a world which erases the frames and windows of representation, the boundaries between the imaginary and the real, in the hope of granting people access to other people's experiences, perceptions, and memories without the mediation of signs, words, and stories. Faber's hubris and delusion lie in his desire to beat artists, writers, filmmakers and storytellers at their own game, to render them superfluous. Yet we should not misconstrue the fact that at the end of the film Eugene's out-of-date mechanical typewriter prevails over Faber's dramatically expanded cinema as a mere expression of aesthetic technophobia, a nostalgic plea for old movie projectors rather than computer-generated image production, for organic rather then technically aided forms of remembrance and recollection. Instead, Wenders's typewriter reminds the viewer of the fact that we cannot own memories like objects, stick them in our pockets and transport them safely to the future. Prosthetic memory has become inescapable, whether we use cranky typewriters or cutting-edge computers to engineer the future's past. What counts, however, is whether people are able to convert archived impressions of the past into meaningful stories or whether they simply allow machines to do all the work of memory for them. Faber's hubris is that he—fetishistically, as it were—wants mechanical images to serve the purpose of recollecting and preserving the past in and of themselves. For Eugene and Wenders, by contrast, they need to be embedded in and mediated by stories in order to provide people with orientations, meanings, and a sense of identity. Cinema's future task, for Wenders, is to somehow negotiate our desire to reach out for the materiality of the past even though we will and can never succeed in this endeavor. Eugene's typewriter allegorizes the frames, windows, and interfaces of projection that at once stimulate and limit our desire for direct contact with the reality of the past. This typewriter writes against Faber's complete blurring of science and art. In the end, self-reflexive acts of storytelling are to ensure that our legitimate desire for immediate contact and immersion does not degenerate into deceptive simulations of sensory immediacy.

V

And then, there is of course the tale of Dr. Bruce Banner (Eric Bana), a brilliant scientist working together with his equally brilliant ex-girlfriend Dr. Betty Ross (Jennifer Connelly) on secret genetic regeneration projects at the University of California at Berkeley. Son of an army geneticist (Nick Nolte) who in the mid-1960s manipulated his own immune system and in so doing reconfigured the genetic code of his future offspring, Banner encounters severe anger management problems after one of his experiments goes terrible wrong and his body absorbs a normally deadly dose of gamma radiation. Banner is haunted by a traumatic past, and unable to integrate discontinuous memory flashes into a meaningful story and identity. Now, whenever he is under emotional stress, Banner turns into a powerful giant green monster. His body and emotions reprogrammed by modern science and technology, the emotionally restrained Banner transforms into a brute who comes to represent the perhaps most powerful example in film history of what Freud would have called the return of the repressed.

Ang Lee's 2003 *The Hulk* relies on state-of-the-art computer generated imagery not only to stage Banner's metamorphoses as a visceral thrill for the viewer, but also in order to find a viable cinematic style able to index and rework the story's comic book origins. As importantly, the film deploys innovative, computer-aided editing and transition techniques so as to signify the unreconciled layering of multiple realities, personality traits, and temporal orders that is at the heart of Banner's predicament. Split screens dominate the film's visual style as much as the use of multiple screens and insert shots within the screen, allowing the viewer to see one and the same action from various viewpoints at once. Time and again, we move from one scene or sequence to another, not with the help of conventional hard cuts, fade ins and fade outs, or dissolves, but pictorial layers and inserts that either grow into or out of the initial shot. For example, when Banner recalls happier times together with Betty, first we look together with Banner at a still photograph showing him and his former girl friend. Then, suddenly, this photograph becomes animated, expands across its own frame, fills up the entire screen, and thus leads Banner and the viewer effectively into the past—a past structured by yet another act of recollection. Similar techniques are used when Banner tries to find out more about the manipulation of his genetic code with the intention of better understanding the dark secrets that unsettle his present. First displayed on the screen of his computer alone, mathematical data and their graphic representations suddenly inhabit the background of the entire shot, with Banner's inquisitive face in the foreground. It is as if the computer's window here expanded drastically and incorporated our protagonist into its world of data modeling and visual simulation, as if the encoded structure of Banner's past completely absorbed or consumed the scientist's perturbed present.

Unlike Wenders, Lee in scenes like this no longer stages a scientist's struggle with and for memory as one between different conceptions of the interface, between Sutherland's quest for sensory immersion and Engelbart's hope for coevolutionary learning. Instead, his film freely borrows, recalibrates, and remediates both conceptions at once in order to explore how cinema today can respond to and complicate the much-feared waning of diachronic time in digital culture. Unable to negotiate the messy simultaneity of past and present, the scientist's troubles give Lee cause here to explore how contemporary cinema can emulate the windowed screens on our computer displays, neither in order to condemn computing as the nemesis of creative filmmaking nor to denounce cinema as an old-fashioned tool of storytelling, but on the contrary to recall historical alternatives of cinematic pleasure and narrativity largely suppressed by dominant cinema's understanding of narrative as a single, linear, causally determined stream of events.

Lev Manovich (2001) has suggested the terms of "spatial narrative" and "spatial montage" in order to theorize the stunning multiplication of pictorial spaces and visual layers in contemporary computer art as much as filmmaking. In spatial narratives, different shots and discontinuous points of view are accessible to viewers at one and the same time; in spatial montage, the classical logic of temporal replacement and succession gives way to a logic of addition, supplementariness, and co-existence. Cinematic time today becomes increasingly spatialized, contrary to the linear conception of narrative temporality in classical cinema, a matter of spatial distribution across the surface of the screen rather than sequential ordering.

In spatial montage, nothing is potentially forgotten, nothing is erased. Just as we use computers to accumulate endless texts, messages, notes, and data, and just as persons, going through life, accumulate more and more memories, with the past slowly acquiring more weight than the future, spatial montage can accumulate events and images as it progresses through its narrative. In contrast to the cinema's screen, which primarily functions as a record of perception, here the computer screen functions as a record of memory (Manovich 2001: 325).

Spatial narrative already played an important role in the Giotto's frescos and in Bosch's and Bruegel's attempts at representing various micro-narratives within one and the same painting. It became marginalized, however, with the rise of the Enlightenment and the human sciences on the one hand, and on the other the industrialization of time during the nineteenth century. Established at the cost of shutting out alternate historical possibilities of cinematic representation, the institutionalization of what Tom Gunning has famously termed the cinema of narrative integration in the first decades of the twentieth century added importantly to the truncated understanding of narrative forms in primarily temporal and sequential terms. The dawn of digital culture in the last decades of the twentieth century, according to Manovich, has galvanized a powerful comeback of spatial narratives and the possibilities of spatial montage; as a consequence, it has also allowed us to reexperience and rethink the range of modern narrativity beyond, but also within the cinema.

In Ang Lee's *The Hulk*, a scientist's struggle with both memory and his own manipulated genetic program becomes a site to reinvent narrative cinema for an age whose viewers are increasingly becoming accustomed to the dense information surfaces and heterogeneous simultaneities of ubiquitous computing. Lee's film encourages the viewers to rapidly switch between parallel frames of reference and representation similar to the way in which computer users today switch their attention from one application to another, from one window open on their screens to another. Rather than inscribe reliable points of view and nodes of identification in the film's diegesis, spatial montage here questions classical constructions of spectatorial unity and desire. The film's split and multiplied surfaces of representation no longer serve as an imaginary that sutures voyeuristic viewers into user-friendly fantasies of wholeness and unified visual control. Like the films of early film pioneers, they instead make us wonder about the nature of the medium of film itself, its very power to display the world, however fractured it might be. *The Hulk*, then, at once immerses itself into and learns from the design of different computer-human interfaces so as to reclaim the multiplicity of narrative forms, representational logics, and spectatorial pleasures that stood at the beginning of the film's century-old history. The use of new image technology here reminds us of and recuperates the fact that cinema has not always been dedicated to seamless narrative integration, that it drew its initial inspiration from various visual entertainment forms, and that its identity as a medium was and should be seen as fundamentally protean, hybrid, and inter-medial. Banner's repressed past is therefore also that of classical narrative cinema and its single streams of narrative time. In venting his anger and repressed desire, Lee's monster of science always also runs up against what classical narrative cinema since the 1910s tried to contain, anaesthetize, and reorganize in order to standardize perception and capitalize on the viewer's desire for voyeuristic pleasure. As seen through the lenses of Lee's *The Hulk*, the future of film lies always also in recognizing its past, in remembering how the pioneers of film and the cinematic avant-garde hoped to couple scientific and technological advancements to cultural and aesthetic reforms, how they sought to explore new directions at the intersection of art and scientific reason rather than reducing the medium's pluralistic and hybrid options to one dominant idiom, one essential genetic code.

VI

Science fiction has often rightly been seen and interpreted as a seismograph charting a particular present's anxiety about itself and what is normally shown as its distant future. In more recent years, by contrast, science fiction has traced the way in which electronic media and global streams of connectivity today expand our present drastically and collapse the boundaries between past and future. The genre's primary concern no longer lies with how future technologies might reshape the material architectures of human life, but how they will affect our picture of the past and ability to remember, how they might unsettle the kind of borders that formerly structured our positions in space and time. In times not so very long ago, cinematic fictions of science primarily articulated worries about how modernity's desire for and drive toward the future may redefine or express our perceptions of the present. More recently, the genre has ironically turned its attention to the question of memory and the past's future, pondering the extent to which the all-encompassing virtual spaces of contemporary consumer- and techno-culture erase the very dynamic that energized former science fiction features in the first place, namely the dialectic between different layers of temporal experience. Whenever it turns to issues of memory and prosthetic recollection, science fiction today also contemplates the possibilities of its own survival as a genre. For how can there be science fiction at all if contemporary cyber-capitalism and globalization increasingly blur all boundaries between time, space, and place, integrating future and past into the all-pervasive image banks of present-ness?

Whether literary or cinematic, conventional science fiction often used to be driven by a paradoxical technophobia. It was obsessed with and invented new technological discourses in order to warn its viewers or readers about the dangers of new technologies. In spite of their fixation on technological progress, the majority of science fiction productions remained deeply ensnared in conservative conceptions of technology as they tended to emphasize the threat of machines and invoke traditional family values as antidotes to the fear of mechanization. In Michael Ryan and Douglas Kellner's words, in earlier science fiction productions technology predominantly "represents artifice as opposed to nature, the mechanical as opposed to the spontaneous, the regulated as opposed to the free, an equalizer as opposed to a prompter of individual distinction, equality triumphant as opposed to liberty, democratic leveling as opposed to hierarchy derived from individual superiority" (Ryan and Kellner 1990: 58). Picturing scientists' dramatic struggles with memory and prosthetic forms of recollections, more recent science fiction films such as the ones discussed earlier in this essay may not always stay entirely free of the genre's earlier technophobia. What these films do accomplish, however, is to persuade the viewer that we somehow have to learn how to live with the machines and scientific forays that tend to restructure or even take over the task of memory for us today. Rather then reverting to some kind of jargon of mnemonic authenticity, these films try to probe the costs and blessings of the extent to which memory prostheses have become inevitable today. Their point is not nostalgically to deplore the prominent role of human-computer interfaces in contemporary processes of recollection, but to investigate the plurality of competing mnemonic interfaces today, and amidst this plurality to find reasonable criteria that would allow us to distinguish between the transmission of meaningful pasts on the one hand, and on the other the dissemination of deceptive illusions or merely disposable data.

The future of memory today hinges on our ability to learn how to negotiate the different kinds of interfaces that frame our perception, organize our attention, and transport us to other times and spaces. Memory can survive, not if we radically oppose the machines and interfaces that increasingly archive and transmit perception, knowledge, and experience as information today, but if we learn how to listen to its voice as it emerges from the interstices in-between different forms and frames of mediation. Our existential need for extended structures of temporality may be most effectively satisfied today if we learn how to explore the unstable intersections of competing mnemonic and scientifically advanced interfaces, their mutual framings and ongoing reframings, as the most viable portal to memory and history. Cinema today can serve this task perhaps most persuasively whenever it seeks to incorporate rather than ignore the perceptual and cognitive structures associated with ubiquitous computing, not simply in order to gratify popular expectations and viewing habits, but to negotiate alternate framings of the past and make us think about both the promises *and* pitfalls of living in an age of highly mediated and compressed simultaneity, of spatialized time as much as temporalized space.

BIBLIOGRAPHY

Bardini, Thierry (2000) *Bootstrapping: Douglas Engelbart, Coevolution, and the Origins of Personal Computing*, Stanford, CA: Stanford University Press.

Higson, Andrew (1993) "Re-presenting the national past: Nostalgia and pastiche in the heritage film", in Lester Friedman (ed.), *Fires were Started: British Cinema and Thatcherism*, Minneapolis, MN: University of Minnesota Press.

Huyssen, Andreas (2003) *Present Pasts: Urban Palimpsests and the Politics of Memory*, Stanford, CA: Stanford University Press.

Koepnick, Lutz (2002) "Reframing the past: Heritage cinema and Holocaust in the 1990s", *New German Critique* 87 (Fall 2002): 47–82.

Koepnick, Lutz (2007) *Framing Attention: Windows on Modern German Culture*, Baltimore: Johns Hopkins University Press.

Manovich, Lev (2001) *The Language of New Media*, Cambridge, MA: MIT Press.

Ryan, Michael and Douglas Kellner (1990) "Technophobia", in Annette Kuhn (ed.), *Alien Zone: Cultural Theory and Contemporary Science Fiction Cinema*, London and New York: Verso.

Sutherland, Ivan E. (1965) "The ultimate display", in Wayne A. Kalenich (ed.), *Proceedings of International Foundation on Information Processing*, vol. 2, Washington DC: Spartan Books.

Wenders, Wim (2001) "What the new technologies offer", in Sari Roman (ed.) *Digital Babylon: Hollywood, Indiewood & Dogme 95*, Hollywood: Ifilms Publishing.

Youngblood, Gene (1970) *Expanded Cinema*, New York: E. P. Dutton.

15 The Self-Referential Scientist
Narrative, Media, and Metamorphosis in Cronenberg's *The Fly*

Bruce Clarke

From the fearful cry of the human-headed fly snagged on a spider-web at the end of the 1958 movie—"Help me! Please, help me!"—to the warning against erotic entanglement with a man becoming a fly in the 1986 remake—"Be afraid! Be very afraid!"—tag lines from *The Fly* have been etched into popular mythology. *The Fly*'s cultural purchase also shows in its narrative transformations from ephemeral prose fiction to B-movie institution, its generation of continuations and variants. It first appeared as a short story by George Langelaan published in *Playboy* in June 1957 (Langelaan 1957). Within a year it was rescripted by James Clavell and made into the Twentieth-Century Fox movie directed by Kurt Neumann.[1] This was followed by *Return of the Fly* in 1959 and *Curse of the Fly* in 1965. Two decades later David Cronenberg rewrote Charles Edward Pogue's rewrite and directed his major revision of 1986, followed in 1989 by *The Fly II*.[2] In 1997 it reemerged once more in the *Simpsons* episode "Fly vs. Fly."

Langelaan's short story, Neumann's movie, and Cronenberg's remake are the versions that have effectively introduced new shapes into the social imaginary. Like *Frankenstein*, this breeding of narrative progeny has transformed a seemingly simple tale of technological horror into a collective and complex fabulation. *The Fly* is a piece of modern metamorphic mythology in the raw, a mutating narrative structure whose literal transformations, like the fantastic bodily changes under narration, are driven by the extensions of media devices. With a generation of cybernetic media culture under its belt, *The Fly*'s second narrative remediation in Cronenberg's film probes deeply into the systematic underbelly of organic-technological hybridity under a first-order cybernetic regime, drawing out the abiding nightmare of mainstream control theory, the going out of control of both organic and machinic processes.

At the same time, *The Fly* updates a perennial mythos of bodily metamorphosis. Such stories are commonly precipitated by key mistakes or misreadings—often misplaced curiosities, but also, simple wrong turns, pieces of bad luck—for which culpable or inadvertent blunders the metamorphic condition is punishment or poetic justice (see Clarke 1995: 3ff.). *The Fly*

wires its transformative complications directly into mistaken communications, blunders lodged in both the communicator and the communication system. Each version centers on a singular scientist's lone obsession
with the creation of a teleporter. The main complication results when the
inventor— André Deslambres in the 1950s versions, Seth Brundle in the
1980s, seeking to recreate either the world or himself, or both, test-pilots
the device by transmitting *himself* across space. He sends the message of
himself to himself. But unfortunately, due to an unlucky increment of noise
in the signal, the message gets garbled in transmission, and he comes out of
the receiver an insectoid monster.

In displaying the transformative power or daemonic agency of communications technology, this fable also unfolds the paradoxes of media. *The
Fly* is precisely an allegory of modern media in their equivocation between
transportation and transmission—ontologically, between matter and information. The teleporter is a paradoxical device to which organic bodies are
offered for material transportation by means of informatic transmission.
This is a topic treated in detail in a founding cybernetic document, quite
conceivably a direct source for Langelaan's tale, Norbert Wiener's *The
Human Use of Human Beings: Cybernetics and Society*:

> … We thus have two types of communication: namely, a material trans-
> port, and a transport of information alone. At present it is possible for
> a person to go from one place to another by material transportation,
> and not as a message.

> … There is no fundamental absolute line between the types of transmis-
> sion which we can use for sending a telegram from country to country
> and the types of transmission which at least are theoretically possible
> for a living organism such as a human being.

> Let us then admit that the old idea of the child, that in addition to trav-
> eling by train or airplane, one might conceivably travel by telegraph, is
> not intrinsically absurd, far as it may be from realization.…

> I have stated these things, not because I want to write a science fic-
> tion story concerning itself with the possibility of telegraphing a man,
> but because it may help us understand that the fundamental idea of
> communication is that of the transmission of messages. (Wiener 1950:
> 105–111)

In each *Fly*, the metamorphic event is the matrix for a recursive network of biological, technological, and narrative system and environment
references, unfolding the perils of unnatural and untested couplings and
interpenetrations. Informatic code and narrative text merge into analogous
communication functions: either is embedded within the other, as the mov-

ies cinematically retransmit the literary text of a story about a scientist transmitting himself through a communications device. The scientist's metamorphosis *by* a communications medium is brought about by his own prior metamorphosis *of* a communications medium. In the original prose version, the teleporter is literally a kind of matter telephone—the sending and receiving ends of the device are re-engineered telephone call-boxes. In Cronenberg's version, due to Brundle's retention of an abandoned prototype, *three* boxes or pods are available for the climactic scene of organic/mechanical merger. Moreover, and crucially for our approach, each version of *The Fly* embeds the paradoxical machinery of telematic metamorphosis within embedded narrative frames.

METAMORPHOSIS AND EMBEDDING

Narrative embeddings put self-referential processes into literary operation. Embedded narrations elicit the observation of narrative observation by enacting the narrating of narration. Prompting recursive operations—the continuous re-connection of differential elements—on the reader's part as well, stories within stories mimic the shifting of formal frames by which meanings telescope through levels of possible significations. Stories of metamorphosis simply raise these effects to another power: daemonic or transformative bodies, uncanny reframings of narrative identities, are further self-referential recursions of the formal structures of multiple and shifting narrative levels.

Moreover, the narrative shifting of embedded frames and the metamorphic shifting of bodies and minds are both about the medial transformation of matter and information. Under transmission, original messages do not remain the same, but are de- and re-formed by signals and noises traversing the very medium through which they become materialized. Narrative frames, as media forms, transform the subject of narration, just as physical metamorphoses of the body-as-medium transform the subject of metamorphosis. Both the character under metamorphosis and the story under reframing are transformed by the very mediums through which they are conveyed. The medium (narrative frame, body, celluloid, or transporter) is an active agent that self-reflexively and essentially transforms its ostensible content. In *The Fly* these dynamics are rendered hyperbolic through the constant reframing of a metamorphic story about a character caught up in a communications medium of his own devise.

THE SELF-REFERENTIAL SCIENTIST

In *The Fly*'s mediations of the metamorphic body, a story framed by forms of narrative self-reference is told about a self-referential scientist—one who takes himself as the object of his own investigations. The mediated dupli-cations of *The Fly* play off the initially unitary status of the scientist who suffers the metamorphosis. The traditional figure of the scientist occupies a mode of modern selfhood paradoxically purified of otherness by "objec-tivity", that is, by the supposed elimination of self-reference from his or her observations of objects and other subjects. In the *Fly*s of the 1950s, for instance, André comes forward as a brilliant benefactor meriting our regard and his wife's devotion. The disaster about to happen will be all the more catastrophic for destroying such a man. In Neumann's film, after André unveils his teleporter to Hélène and allays her incredulity, he rhap-sodizes over its world-transforming potential:

> The disintegrator-integrator will completely change life as we know it. Think what it will mean! Food, anything, even humans, will go through one of these devices. No need for cars or railways or airplanes, even spaceships. We'll just set up matter transmitting-receiving stations throughout the world, and later the universe. There'll be no need of famine. Surpluses can be sent instantaneously at almost no cost any-where. Humanity need never want or fear again!

Such heroic images embellish the significance of the *non*-self-reference of the proper scientist, who then personifies selflessness altogether, universal benevolence, the generosity of freely bestowing gifts upon humanity, the magnanimity of lifting old burdens off the race with his or her discoveries and inventions.

The standard formula for scientific propriety retailed here is the produc-tion of objectivity through the elimination of self-reference. One might see this as parallel to the standard goal of communications engineering, the elimination of noise and other distortions or errors from transmitted sig-nals. This demand for an uncorrupted signal certainly weighs on André's teleporter project as well. On the face of it, however, every version of *The Fly* replenishes the notion of objectivity as the elimination of self-reference by exposing the dire consequences when properly other-referential scien-tists take *themselves* as the objects of their own observations and as the subjects of their own apparatus.

And yet, it is an engineering truism that noise can never be entirely elim-inated. Random fluctuations and Brownian flutters inhabit the very matter out of which communications media are constructed. Noise can only be compensated for (e.g., through protocols of redundancy in the message) and held within tolerable levels. In mechanical systems, noise and friction are the feedback of worldly operations, while in autopoietic systems, self-referential operations accommodate and bind the noise of mediations, the materialities of metabolism, perception, and communication, within the boundaries of the system, and as a result, viable systems self-adapt, evolve, or grow more complex. In other words, in media systems, informatic noise is a marker of material self-reference in the midst of other-referential, supposedly dematerialized messages: noise is a self-referential effect asserting the indispensability of the material medium carrying the signal to its destination.

The Fly positions itself as a first-order cybernetic narrative precisely by its *demonization*, both of informatic noise—the fly in the ointment of perfect transmissions—and of self-reference, which it conceptualizes only as a positive (unregulated) feedback that, in the experiments of André and Seth, spins totally out of control. Each *Fly* renders the cybernetic coupling of media systems and living bodies by linking narrative embedding to matters of mechanical and biological embedding and reproduction, yet each version has a blind spot that covers over the necessary integration and mutual compensation of mechanical, living, psychic, and social systems. Thus each version of the *Fly* demonizes self-reference even while operating self-referentially: the 1950s versions do so straightforwardly, Cronenberg's with a self-reflexive wink.

The total mythos of *The Fly* is suffused with self-reference, individually in each version's embedded narrative structures, and sequentially in the self-referential re-entry of the metamorphic event into the narrative transformations that underwrites the tale's evolutionary social autopoiesis, the metamorphoses in the text of *The Fly* itself. Neumann's *Fly* preserves the embedded structure of Langelaan's textual original. In both 1950s versions, the story of André's catastrophe is narrated by his wife Hélène, in explanation of her role in his partial obliteration by industrial steam press. Langelaan's Hélène produces a written account embedded within a frame narrative produced by André's brother, while Neumann's Hélène offers an oral account realized cinematically as an extended flashback embedded within the main cinematic diegesis. Lacking the textual markers of prose fiction to indicate this shift in narrative level, the movie uses a standard pair of visual brackets, those classic wavy lines that relay narrative agency from one level to another. While Cronenberg's telling of *The Fly* lacks the primary embedding device of the 1950s versions, a story within a story, he produces multiple scenes of cinematic embedding, enough to suggest a certain reflexive irony in the narration of Seth's reflexive apocalypse.

In the 1950s versions, the doubling of narrative agencies, the up- and down-shift into and out of Hélène's account of the affair, is tidy and discrete, and this stability at the borders of the narrative frame resonates with the style of André's metamorphosis, which is also, while dire, tidy and discrete. Mistakenly teleported together, André and the fly swap heads and one arm apiece, and from that moment there are two Andrés, a fly-headed man and a man-headed fly. This doubling occurs along the two-sided border of the teleporter, a detail already articulated in Langelaan's text, when Hélène recalls: "It was only after the accident that I discovered André had duplicated all his switches inside the disintegration booth, so that he could try it out on himself" (Langelaan 1966: 24).[3] Here again, as a closed technological system embedded within already embedded narrative frames, the teleporter feeds back into and duplicates the narrative that narrates it. As Mieke Bal has written in a less technological narratological context, "A traveler in a narrative is in a sense always an allegory of the travel that narrative is" (Bal 1997: 137).

SELF-REFERENCE IN CRONENBERG'S *FLY*

In the 1950s *Fly*s the fly head and arm of the human metamorph emerge full-blown. Within days André feels his mind going, and with Hélène's help he does away with himself before the mental mutation into something posthuman is complete. In contrast, Cronenberg's *Fly* draws out the story time of both the bodily and the psychic transformations (see Pharr 1989: 37-46; Knee 1992: 20-34; Freeland 1996: 195-218; Wicke 1996: 302–315; Roth 2002: 225–241). Weeks elapse while the new, posthuman creature, the Brundlefly, reaches phenotypic and psychic expression from its conception through genetic merger in the teleporter. Cronenberg frontloads the self-transmission of the scientist to clear narrative space for the pseudo-evolution of the Brundlefly. Transposing the tale from the nuclear family to the singles scene, this version foregoes the familial pieties as well as the discrete narrative formalities of the prior versions and takes far less time to get down and dirty. At the same time, Cronenberg's *Fly* pays significant homage not just to Neumann's film but also, and less obviously, to Langelaan's original text. While occurring in systematic communication with them, the metamorphic action in his film transforms his precursors into more complex shapes.

The movie opens with shy technoscientist Seth Brundle's coaxing of intrepid science-journalist Veronica Quaife back to his laboratory pad set in an urban loft, furtively hoping it seems to seduce her with his telepods. Signaling Cronenberg's knowing transformation of the original teleporter Langelaan assembled from "telephone call-boxes", Ronnie jokes, "Oh, designer phone booths." When she refers to them this way a second time, Seth curtly corrects her: "Telepods." The updated substitution of "telepod" for "telephone" marks the amplification of "transportation" at stake in all versions of *The Fly*'s teleporter: the extension of the concept of transmission from informatic forms such as acoustic and visual vibrations to material substances such as inorganic and organic bodies.

The metamorphoses of media technology in Cronenberg's *Fly* occur not just in the overall form of the teleporter, however, but also in the particular forms and implications of other communications devices—for instance, the keyboards to be played on by pre- and post-metamorphic hands. In both of the 1950s versions, the media keyboard in question is the typewriter that André used to repair his metamorphic aphasia, the loss of his spoken voice, and communicate in writing with Hélène. The mainframe computer Neumann's film adds to Langelaan's teleporter is *not* keyboarded, but anachronistically operated with switches and dials. And although André's experiment goes out of control, his equipment, at least once it is perfected, does not. In a phrase introduced by Heinz von Foerster, the achieved teleporter starts out as a reliably predictable device—a "trivial machine":

> A trivial machine is characterized by a one-to-one relationship between its "input" (stimulus, cause) and its "output" (response, effect). The invariable relationship is "the machine." Since this relationship is determined once and for all, this is a deterministic system; and since an output once observed for a given input will be the same for the same input given later, this is also a predictable system. (von Foerster 2003: 208)

In Neumann's film, the perfected teleporter remains trivial: mistakenly given a multiple input, once it performs the initial, unexpected splicings of André and the fly, no amount of re-teleporting is able to make these monstrous combinations come undone or revert to original conditions. However, in a twist edited out by Clavell's script for Neumann's version, Langelaan's story climaxes with the shock produced when the teleporter surprises all concerned by recovering the "atoms" of the cat Dandelo it had previously failed to reintegrate, and monstrously compounding that third body into the already spliced man and fly. André types:

> When I went into the disintegrator just now, my head was only that of a fly. I now only have its eyes and mouth left. The rest has been replaced by parts of the cat's head. Poor Dandelo whose atoms had never come together. (Langelaan 1966: 35f.)

Cronenberg will recapitalize on Langelaan's version's lost suggestion of the possibility for the device's own uncanny transformations of function. One way to put this is that Langelaan's text already provided the template for treating the teleporter, in the lexicon of second-order cybernetics, as a *nontrivial* machine:

> Non-trivial machines…are quite different creatures. Their input-output relationship is not invariant…. While these machines are again deterministic systems, for all practical reasons they are unpredictable: an output once observed for a given input will most likely be not the same for the same input given later. (von Foerster 2003: 208)

This in itself marks the cybernetic metamorphosis of a mechanical system from passive servant to intelligent subject, and in particular, from object to agent of narration. Cronenberg immediately factors in these potential differentiations: early on, his version displays and contrasts two very different keyboards.

As the couple enters Seth's live-in laboratory straight from the Bartok Science Industries cocktail party, the cinematic frame centers Ronnie, seen from behind, between a telepod in the background on the left and an upright piano in the foreground on the right. Before drawing Ronnie's attention to the telepod and teleporter, Seth goes directly to the piano and plays, facetiously but with impressive facility, the first few bars of "Love Is a Many-Splendored Thing." This playfully-chosen melody will echo ironically against the gruesome turns of the love story to follow, but after this quick bit the piano never returns to the screen or the story. But this seeming throwaway moment makes the deeper suggestion that this keyboarded musical instrument, a communications device of a sort, is still a trivial machine, one that registers the instrumental mastery that Seth himself will throw away, as the repercussions of his coming jealous blunder—teleporting himself with a fly in the pod instead of with Ronnie in the room—take effect.

For the moment, however, manipulating this particular keyboard, Seth is still in full control, the musical output being entirely predictable from the fingered input. Cronenberg's film quickly contrasts the "trivial" status of the piano and its keyboard to the teleporter and its computerized control unit, which appears as a freestanding console, slightly smaller than but otherwise strongly resembling the upright piano, with an alphanumeric keyboard where the ivories are and a video display where the sheet music would go. And whereas in the 1950s film version, the manual typewriter and the mainframe computer remained mechanically separate and trivial in function, by the time of the Cronenberg film those real technologies had actually merged. In one of its smartest turns on the allegory of metamorphosis, Cronenberg's film literalizes the metaphor of the "personal computer" by bringing its computerized and keyboarded teleporter forth as the nontrivial machine par excellence—a *person* in its own right.

This newly autonomous and anthropomorphic form of the device now has a voice recognition function and the science-fictional abilities to communicate with its operator, to respond discursively to questions and to improvise solutions to problems put to it, both verbally by Seth and substantially by the objects to be teleported. Cronenberg's tale has coupled and fused communications and cognition, the typewriter keyboard and the computer as an artificially intelligent subject. As Jennifer Wicke has observed in a perceptive essay, the outcome of this technological metamorphosis is to supply Cronenberg's *Fly* with an additional, internal narrator:

> Throughout the film the computer screen comes to fill the frame to show the change in scene, or to effect the cut—generally, these super-closeups of the screen occupy all of it, so that we, the audience, are reading the movie screen as if it were translated into a gigantic com-

puter monitor.... Often the screen is telling the tale, on its own.... The computer is also narrating to us, while it narrates Seth into the Brundle-Fly. (Wicke 1996: 305)

Seth's sentient computer participates in the infusion of cinematic narrative embedding that pushes Cronenberg's version of *The Fly* to a new level of explicitness about the interplay of self-reference, media systems, and metamorphosis. Cronenberg's computerized and communicative teleporter becomes a fully recursive narrative device, operating both at first and at second degree within the cinematic diegesis. At first degree, it will enact the bodily metamorphoses of the organic beings at hand, and at second degree, from its video screen within the cinematic frame, it will self-referentially narrate those very acts. Cronenberg morphs the teleporter into the ultimate unreliable narrator, an agent of confusion taking the story completely out of Seth's control.

Before that comes to pass, however, Seth remains, as he informs Ronnie, the "systems management man"—the central agent and kingpin of this technoscientific project. In that role, to demonstrate for her the teleporter he is still working to perfect, he requests a unique personal article; she obliges by peeling off a nylon stocking. Corresponding precisely to Hélène's position in the 1950s versions as initially incredulous witness of André's teleportation of an ashtray, at first Ronnie sees the transmission of her stocking as a stunt. Cronenberg's film flips on the themes of narrative observation and mediated self-reference, first of all, as Ronnie starts to compute the significance of the event Seth has produced—when she secretly starts and then overtly displays her tape recorder. At first Seth protests as he resists his narrative reassignment from scientific subject to publicized object of documentation and scrutiny. However, his true proclivity for reflexivity shows itself in the narcissism that seems to induce him the next day to acquiesce and invite Ronnie to become the privileged observer of his experimental microcosm. Their romance emerges precisely under the sign of Ronnie's journalistic observation of his exhibitions. She props this kingpin up by supplying an observing frame beyond his own, and by supplementing the lack in *him*, so he says, that still prevents his machinic proxy, the teleporter, from working successfully to, as it were, convey the story of living beings from pod to pod.

Although, unlike Hélène, she does not narrate per se during the story, in anticipation of the eventual narration she will deliver to her media outlet once she has the story in hand, Ronnie's role as designated observer ratchets up Hélène's role as internal focalizor. She and Seth agree on a plan by which he will become the commodified subject of a book whose narrative will climax with his own teleportation, and she will consummate his metamorphosis from scientific agent to media object by producing the textual mediations between Seth's technology and society at large. In the fantasy all this implies, Seth will enter at once the technological enclosure of the telepods and the narrative enclosure of the text Ronnie will author. This is at least the promise if not the immediate enactment of narrative embedding, and a sort of textual reformation by media system. In fact, these anticipated media transformations resonate with the ways that the teleporter itself—a machine for the transformation of bodies into transmissible signals—will unexpectedly transform Seth's accomplishments, and so transform the story that actually occurs.

When we see them next, Ronnie has her video gear up and running. However, in the first experiment she is there to document, Seth's teleporter goes awry, turning a baboon inside out. In the aftermath of that failure, Ronnie's video cam produces the film's first cinematic embedding. Shifting both Seth and his experimental apparatus up another narrative level, the video image fills out the frame of the cinematic narration, embedding the despondent experimenter to the second degree. With a slight blur to keep the diegetic brackets perceptible, Seth is recorded and displayed on a screen within a screen. From this doubly mediated position, he atones for the destruction of his simian assistant.

Ronnie: The world will want to know what you're thinking.
Seth: Fuck is what I'm thinking!
Ronnie: Good....The world will want to know that.

Observed in deep self-examination over his and the teleporter's failure, Seth then confesses to Ronnie and her recording devices his painful lack of knowledge of "the flesh." This embedding and substitution of the video cam's image for the cinematic frame happens only once more in the movie, on the other side of Seth's fateful self-teleportation. This scene, then, is the first of a pair of formal-thematic brackets between which Cronenberg embeds the central scene of metamorphic fusion between man and fly.

In the 1950s versions, the character-bound narrator/focalizor Hélène is not there to witness or document the scene of André's mis-transmission of himself. In these earlier *Fly*s, subsequently communicating through typewritten notes, the metamorphic André remained either behind locked doors or veiled with a black cloth. Until she pulled his veil away, precipitating her own tragic transit from ignorance to revelation but enabling her later to narrate the shape of André's changes, they were held in suspense. Cronenberg's version, however, narrates Seth's first self-transmission directly, and significantly, by means of mechanical mediation. Compounding the teleporter's infolding of narrative self-reference, Cronenberg stages the video camera's automatic recording of Seth's self-experiment squarely within the main cinematic frame. In an amorous blunder, a drunken fit of jealous anger, he breaks his promise to give Ronnie firsthand witness of his first flight. However, unlike the scene after the baboon fiasco, which narrated Seth within Ronnie's video frame within the cinematic frame, this time the viewer does not see what Ronnie's video camera sees; rather, that cam is framed within the main diegesis, its tripod and mechanism substituted for her bodily presence as the internal focalizor of Seth's abrupt dash through the pods.

Again, in the 1950s versions, we understand that the main metamorphic catastrophe is completed in one swoop; what is delayed is its full revelation to another observer. Cronenberg's *Fly* transforms and extends the narrative suspense by delaying the repercussions of Seth's genetic fusion with the fly. He emerges from the receiving pod only slightly visibly altered, newly buff but also haggard and sweaty, like a coffee-drinker (Seth had earlier bragged to Ronnie about his professional espresso maker) who has just graduated to methamphetamine. Otherwise, for the moment the metamorphic consequences of his compound blunder remain embedded within his mutated genome, not yet enacted within the replica that has now reintegrated into one being the disintegrated originals of Seth Brundle and the housefly. Fresh from his exit pod, Seth articulates the ontological abyss his media technology has now opened up: "Am I different somehow? Is it live or is it Memorex?" But his joking distinction between a genuine original and its informatic duplication already mistakes the nature of his disnaturing. He is not a duplicate but a doppelganger, a posthuman hybrid constructed by the chance acquisition and inclusion of an alien genome. From this point on, Cronenberg's *Fly* spins out in a series of vicious recursions.

CYBERNETIC PURITY

Going immediately back to Wiener's and von Neumann's original conceptual splicing of organic evolution with mechanical development, against a longer background of transformation stories centered on all manner of crossed lineages, the wider cybernetic and metamorphic contexts of Cronenberg's *Fly* strongly overdetermine the notion of *purity* and its corruption. As it inflects Cronenberg's *Fly*, the violation-of-purity theme also marks that narrative's own impurity, so to speak, by linking it back to productive infections drawn from its precursor texts.

First of all, in his recovery of another detail of Langelaan's text excised from the Clavell/Neumann version, Cronenberg reprises a passage from the short story presenting teleportation as a thrill ride, that is to say, violent self-transportation that only goes around in circles. In Hélène's original account of André's experiments, she recalls:

> Our cocker spaniel...had been successfully transmitted half-a-dozen times and seemed to be enjoying the operation thoroughly; no sooner was she let out of the "reintegrator" than she dashed madly into the next room, scratching at the "transmitter" door to have "another go."
> (Langelaan 1966: 23)

In Cronenberg, this lust for repetition of "animal sensation" morphs into Seth/Brundlefly's protracted phase of teleporter mystique. Like someone on a bad Ecstasy binge, when his genetic adulteration with the fly first comes on "like a drug," Seth mystifies his affect, experiencing his self-transmission with the insectoid other as a purifying rush:

> I am beginning to think that the sheer process of being taken apart atom by atom and put back together again—why, it's like coffee being put through a filter! It's somehow a purifying process—it's purified me, it's cleansed me. ... Human teleportation, molecular decimation, breakdown, and reformation is inherently purging!

This phase of Seth's enthusiasm is predicated on his mistaken notion that he is still in total control of a trivial machine to which he can submit repeated input-output routines without concern for variation, malfunction, or "confusion"—a pure and simple machine purveying pure thrills. Seth the science geek gone to seed reverts to a stoned-out adolescent male, still clueless about "the flesh" yet momentarily empowered, treating female sex partners as interchangeable machine parts programmed to respond to his phallic input with identical productions of gratification and without complicating consequences. He implores Ronnie to trip out on the same thrill pill:

> I want you to go through. I want to teleport you as soon as possible. Right now! You'll feel incredible. Ronnie, I hardly need to sleep anymore and I feel wonderful. It's like a drug, but a perfectly pure and benign drug!

Ronnie's refusal to tag along on Seth's trip draws from him a truly harrowing rant, the hallucinatory extremity of which fleshes out one of the supreme ontological bummers of Western civilization, a pure vision of the gendered metaphysical dualism that dogs the Neoplatonic mode of thought in general and of metamorphic narratives allied with it in particular:

> You're afraid to dive into the plasma pool, aren't you? You're afraid to be destroyed and recreated, aren't you?... Drink deep or taste not the plasma spring! You see what I'm saying? I'm not just talking about sex and penetration. I'm talking about penetration beyond the veil of the flesh! A deep, penetrating dive into the plasma pool!

"The veil of the flesh" is a hoary metaphysical trope, a mainstay of dogmatic metamorphic allegories in which the manifest transformation of the body is precisely a "veil" cloaking a discourse concerning the disembodied essence of the immortal soul (Clarke 1995: 122–128ff.). In this vision, bodies arise not as fabulously complex living systems, but "purely," merely as momentary material embodiments of abiding immaterial, informatic or virtual forms. In this manic phase of self-inflation, Seth envisions his newly teleported incarnation as would some mythological god hovering over the mundane vale of wordly flesh, capable of decreating or recreating its own and others' bodies at will.

What updates that allegory in this context is the notion of the "plasma pool," a cyborgian notion combining "protoplasm" as a "pure" living medium with the instrumental controls of first-order cybernetics. As drawn from John von Neumann's "General and Logical Theory of Automata," the "plasma pool" relocates the notion of "purity" in the idea that "the flesh," living tissue per se, can be reduced to a simple substance, can subsist as pure medium, without form but in-formed by an algorithm in command of a mechanical process (von Neumann 1963).[4] His grammar elides the identity of the agent that performs this "deep, penetrating dive into the plasma pool," but that omission centers Seth's rant all the more on a classical patriarchal aporia: The occulted phallus that penetrates this watery spring is the veiled, deified masculine soul that endows the gift of form upon simple, otherwise inert feminine matter.

Cronenberg's *Fly* grants Seth his moment of superhuman masculine glory, of course, just to bring him crashing down, but not before the macho arm-wrestling match that maims his burly opponent and wins him Tawny, a barfly for the bed that Ronnie has left vacant. Seth revs himself up for sex one more time by sending himself through the telepods, but gets a rude awakening when Ronnie returns with the lab report confirming that the bristles growing from the wound he picked up from the prongs of a stray electronic component are not human, certifying that he is now a posthuman metamorph. In the terms of Deleuze and Guattari's *A Thousand Plateaus*, his becoming-animal is also a becoming-woman: no longer just the phallic diver, it/he is also the receptive plasma pool itself, a form emerging from a deep self-penetration of his own devising, by his own machines and the genome of a fly (Deleuze and Guattari 1987).[5]

A month after re-conception within his own machine, Seth has finally cognized the reality of his metamorphic situation. Fitting into the wider backdrop of tales of metamorphic mishap, Seth confirms his own culpability, his personal responsibility for the metamorphic blunder, in remarking: "I was not pure. The teleporter insists on inner pure. I was not pure." That is, he had programmed it to expect to deal only with one object at a time, thus forcing it to improvise when confronted with two at once. Debilitated and distressed, certain he is soon to die, Seth narrates for Ronnie the story of the teleporter's nontrivial narration, its positive role in his predicament: "The computer...got confused. There weren't supposed to be two separate genetic patterns—and it decided to, uh, splice us together. It mated us, me and the fly.... I'm the offspring of Brundle and house fly." In this extremity, Seth's cry repeats the famous last gasp of Neumann's version's human-headed fly: "Help me! Please, help me!"

After bargaining with her editor Stathis for his promise of aid, Ronnie returns to Seth's lab in order to document his critical condition on video tape. Prefiguring her own discovery very soon after that she has conceived a fetus of unknown, potentially metamorphic form, however, she finds Seth somehow reborn—no longer decrepit and depressed, walking with canes, but agile and playful, literally climbing the walls. With this turn of events, Cronenberg reopens the posthuman trajectory of the story that the 1950s versions foreclosed. This seeming recovery marks the onset of a potentially viable posthuman being, one ready to fuse his two lineages in an act of self-naming—or blasphemously, risen from the dead, self-christening. But the residue of Seth's humanity still registers in his claim of self-knowledge regarding the process he is undergoing. Embedded in the human capacity to self-reflect, to take oneself as the object of one's own cognition, the paradox of self-reference remains Seth Brundle's cross to bear:

Seth: The disease has just revealed its purpose. We don't have to worry about contagion anymore. I know what the disease wants.
Ronnie: What does the disease want?
Seth: It wants to turn me into something else…I'm becoming something that never existed before. I'm becoming…Brundlefly. Don't you think that's worth a Nobel Prize or two?

As previously mentioned, there are two scenes, like bookends on either side of his initial self-transmission, in which Brundle is cinematically narrated on a screen within a screen. The second of these scenes occurs just after the self-naming of the Brundlefly. In front of Ronnie's video cam for the benefit of Stathis, Brundlefly performs a kind of Mr. Wizard science skit on the self-referential topic, "How does Brundlefly eat?" With Stathis as the internal viewer of this revolting vignette, the former Seth and his vomit drop are now embedded to the third degree, multiply framed within the main cinematic frame by video cam display playing on the television monitor in Ronnie's apartment. As with the first scene of cinematic embedding, Cronenberg aligns the depth of narrative level with the depth of his main character's psychic and bodily abasement.

Cronenberg also plays on the formal resonance between narrative and reproductive embedding, between stories within stories and gestating fetuses within pregnant mothers. The similarity of Seth's telepods to maternal wombs has often been noticed.[6] This connection is compounded by the traditional correlation between metamorphic narratives, maternal transformations, and other family matters—in particular, the link between incest and monstrosity. The transformations of maturation, pregnancy, and delivery in all sexually-reproductive organisms map out a pervasive biological and natural subtext of metamorphic fantasies (see Clarke 1995:

113–147). Cronenberg's *Fly* couples mechanical, social, and biological systems into a fantastic fusion of informatic duplication, narrative embedding, and sexual reproduction.

Just after Stathis views her video tape of Brundlefly at lunch, Ronnie reveals that she is pregnant by it/him. With this turn of events, the metamorphic themes and narrative forms of *The Fly* intertwine and come full circle. The virtual destination of this tale's forms of embedding is the fusion of the media technology of teleportation with the maternal function of sexual reproduction. Seth in his transmission pod is already a posthuman fetus in a technological womb; Ronnie's troubling pregnancy doubles and confirms this reproductive frame. The informatic transmissions from pod to pod reenact genetic transmissions from womb to womb, in this case making instantaneously spectacular the potential for copying errors in the fantastic fusions of acquired genomes, not to mention the normal recombination of genetic contributions from the meiotic forms of parental sex cells. In allegorical parallel with Seth's technological rebirth—the con-fusion of transmission that delivers the metamorphosing Brundlefly— this narrative now embeds itself with a scene of monstrous miscarriage.

Without marking any shift in diegetic level, the cinematic narration cuts immediately from the scene of Ronnie's confession of pregnancy to one of her arrival with Stathis at a hospital. To complete by surgical abortion, so it would seem, a partial miscarriage of the fetus she had conceived with Seth after his genetic fusion with the fly, Ronnie is wheelchaired into an operating room, where she delivers from the crypto-pod of her womb a hideous slimy footlong squirming grub, either the premature fetus or full-formed infant monster sired by the Brundlefly. As she screams in terror, the scene immediately breaks, redistributing the shock of this horrific delivery back to the main narrative. Only now the cinematic narrator marks the embedded frame of that episode by cutting to Ronnie thrashing awake from a nightmare and curling into fetal position. The ontological downshift back to the main storyworld supplies a cheap and nasty narrative lurch, of a piece with the vicious or backhanded reflexivity of the text as a whole.

In addition to the earlier cinematic framings of screens within screens, then, this scene endows the film with an ultimately if not initially explicit moment of diegetic or narrative embedding, by placing an image of abortive reproduction into a dream fold of the main storyworld, and then aborting it. Ronnie's dream of witnessing herself deliver the monstrous grub is the direct counterpart of the scene of Hélène's witness of André as monstrous man-fly, and in Neumann's version, the fly-headed André's fly-eyed counter-shot of Hélène. As internal focalizors, both the wife and the girl-friend are victimized by visions of "scientific" monstrosity, spectacles the most monstrous precisely as beheld by these intimate beholders. And both scenes of visual assault are marked by rude twists put on forms of narrative embedding.

THE BRUNDLEFLY PROJECT

As a discrete short episode embedded within the main narrative, Ronnie's abortion dream is also embedded within the longer episode of the Brundlefly Project, the unfolding of which draws Cronenberg's version of the *Fly* narrative to a close. However, at the same time that the narrative presses toward the posthuman emergence of "something that never existed before," it is also retrofitted with a eugenic scenario that reinstates the theme of genetic purity in the midst of the genetic fusions. Echoing classical metamorphic romances seeking narrative resolution in the *return* of the lost human form, as the Brundlefly is increasingly expressed in the creature that Seth is becoming, the Brundlefly Project seeks to thwart or control that process, to "refine" it, limiting its posthuman consequences by further infusions of "pure human subjects." This mishmash of motives wonderfully twists out the ontological complications of a creature that is literally a two-sided form and of two minds about its future prospects. As Brundlefly pecks at the keyboard with metamorphosing hands bearing fused digits, the computer picks up the narration by reading out the following display:

> THE BRUNDLEFLY PROJECT
> PROBLEM: TO REFINE FUSION PROGRAM
> GOAL: TO DECREASE TO A MINIMUM
> THE PERCENTAGE OF FLY
> IN BRUNDLEFLY
> > SOLUTION: THE FUSION BY GENE-SPLICING
> OF BRUNDLEFLY INTO ONE OR
> MORE PURE HUMAN SUBJECTS

But at this very moment, the attempt to counter the posthuman is thwarted by a cybernetic twist on the posthuman turn in metamorphic tales, that aphasic moment when the human metamorph first tries to speak but can only, like Apuleius's Lucius, bray like an ass or, like Kafka's Gregor, chirp like a insect. Once Brundlefly absorbs the readout above, it voices a further oral command: "I want a disk—give me preliminary information...." But the computer's voice recognition program marks and narrates the precise moment when the process of posthumanization proceeds across the line of humanity into unrecognizable hybridity:

> > ERROR - PATTERN MISMATCH
> VOICE NOT RECOGNIZED
> VOICE NOT RECOGNIZED
> VOICE NOT RECOGNIZED...

In H. G. Wells's *The Time Machine*, accompanied by his devolved posthuman companion Weena, the Time Traveler comes across a ruined museum wherein the world of that future had forgotten a past containing his former present. Cronenberg famously places into his contemporary *Fly* the Brundlefly Museum of Natural History, wherein are collected the bodily relics of its/his human past. When Ronnie returns the next time, to tell the being she still regards as Seth about her pregnancy and decision to terminate it, before she can deliver that message Brundlefly drives her off with a warning regarding its/his irreparable bifurcation into a two-sided being, oscillating between and thus "representing" both male human and male insect subject positions:

Seth: Have you ever heard of insect politics? Neither have I. Insects don't have politics.... We can't trust the insect. I'd like to become the first...insect politician....
Ronnie: I don't know what you're trying to say.
Seth: I'm saying, I'm an insect who dreamt he was a man and loved it. But now the dream is over, and the insect is awake.

The contrast between Brundlefly's rewording of the sense of being an "insect politician" and that dialogue's probable source in Taoist scripture offers a striking comparison between the cybernetic present and the mythological past, Western rationality and Eastern religion.

> Once upon a time, I, Chuang Tzu, dreamt I was a butterfly, fluttering hither and thither, to all intents and purposes a butterfly. I was conscious only of following my fancies as a butterfly, and was unconscious of my individuality as a man. Suddenly I awaked, and there I lay, myself again. Now I do not know whether I was then a man dreaming I was a butterfly, or whether I am now a butterfly dreaming I am a man. Between a man and a butterfly there is necessarily a barrier. The transition is called metempsychosis.[7]

The classical Taoist sage sees the species barrier between man and butterfly as transcendable only after death with the transference of the soul from one to another body—a motive for the narration of metamorphosis going back to its archaic mythological sources. But given that barrier, Chuang Tzu's focus is on the undecideability of the oscillation, that is, the paradox presented by the dream life that lies embedded within the mind with a reality of its own putting one's waking reality into question. In contrast, Brundlefly is the product of an immediate living/mechanical transcendence of the species barrier; that is, it/he is a modern media metamorph and not a mythical serial metempsychotic. In the fly mode of *its* subject-oscillations, moreover, it denies the undecideability of the question. As the narrator of Kafka's *Metamorphosis* had put it regarding Gregor's uncanny awakening: "It was no dream." So it/he gives Ronnie fair warning that man and fly will cohabit this metamorph only until the awakened fly takes full control: "I'll hurt you if you stay."

And despite Ronnie's taking the hint and running away to procure an abortion for real, it is not the fly but the Brundle in Brundlefly who bursts into the operating room and carries Ronnie back to the lab. Their baby "might be all that's left of the real me," without which the Brundlefly Project will fall short of its "purest" goal, "to decrease to a minimum the percentage of fly in Brundlefly." Technoscientific to the bitter end, Brundlefly remains, as Seth had been, the object of its/his own project in posthuman self-fashioning. As Bruno Latour writes in *Aramis or The Love of Technology*, when technological designs begin, "there is no distinction between projects and objects.... Here we're in the realm of signs, language, texts"; Brundlefly's attempt to refine its/his own posthuman constitution fictively reflects the technological innovator's actual quest to "translate" or "negotiate with" both the human and the nonhuman components of the design: "the innovator has to count on assemblages of things that often have the same uncertain nature as groups of people.... You have to get a whole list of things interested *in the project*" (Latour 1996: 24, 57).

In the climactic scene that follows, Brundlefly negotiates with humans who are decidedly not "interested in the project." Ronnie refuses to abort her quest for an abortion, and Stathis shows up at the lab with a skeet-shooting rifle. It will also turn out that the "assemblages of things" involved, the teleporter and its rewired pods, will "have the same uncertain nature." This terminal episode further entrenches the traditional themes of metamorphic embedding at both the bodily and narrative levels within the cybernetic realm of machinic couplings with organic forms. As a monstrous organic being morphed by the teleporter, accordingly the teleporter has now been re-morphed by the Brundlefly. Initially conceived as a linear transportation device moving goods from pod 1 to pod 2, by splicing in the previously decommissioned "clunky" prototype pod, Brundlefly reconfigures the teleporter precisely as it had demonstrated itself to be in its own instance, a nontrivial, recursive, or recombinant gene-splicing device, now retrofitted for the "creative" merging of the contents of pod 1 and pod 2 into a composite arriving at pod 3. With Stathis once again as internal focalizor, the computer narrates:

GENE-SPLICING METHODOLOGY
HARDWARE:
    TELEPOD 1: TRANSMITTER OF SUBJECT A
    TELEPOD 2: TRANSMITTER OF SUBJECT B
    TELEPOD 3: RECEIVER OF GENETICALLY-FUSED
        A-B COMBINATION SUBJECT

After putting Stathis out of commission with vomit drop, Brundlefly accedes to Ronnie's plea to spare his life in order to reopen a final round of negotiations with her: "Help me be human…more human than I am alone." The consummation of the Brundlefly Project marks the systemic closure of the human itself in its inhospitality to the nonhuman, its doggedly in-turning self-involvement. But by now, insect politics have demented Brundlefly's traditional humanistic family values: "I go there, you go there, we come apart and come together—there: you, me, and the baby…together…. We'll be the ultimate family." One can only expect from the climactic three-way fusion of Brundlefly, Ronnie, and their fetus a monstrosity worthy of its fetishistic precursor, Langelaan's climactic three-way merger of André, the fly, and the cat Dandelo.

But we are spared that particular vision in favor of a truly cybernetic consummation. Instead of Hélène's pulling off the veil of cloth to reveal André as a composite man/fly metamorph, here Ronnie's resistance to Brundlefly's final negotiation precipitates the final collapse of its two-sided physical form, as the "space bug" or monster fly, the imago or adult stage of the metamorphic Brundlefly, hatches entirely out of the "veil of the flesh," the tattered chrysalis of Seth Brundle's human body (on the "space bug," see Kirkman 2006). It is a moment both visually horrific and conceptually spectacular, in that it renders to view the phenotypic resolution of the *organic* process set into motion by the inadvertent genome-splicing of Brundle and fly. Thus, while terrifying, it also produces a satisfaction in the narrative completion of a metamorphic plot. However, Cronenberg is not done yet.

In other, affirmative, stories of posthuman metamorphosis, such as Damon Knight's *Beyond the Barrier* or Octavia Butler's *Imago*, the arc of the plot does end here, with the achieved transcendence of the human. A being emerges into which the human has been absorbed, within which it is rendered, like the mitochondrion that was once a free-living bacterium, one of several components of a higher-order living system. Contemporary work in biological systems has strongly endorsed the view that natural evolution is driven not merely by random mutation but more importantly by the integration of separate genomes into viable consortiums. In this regard, *The Fly* actually has it half right: separate genomes can and do join forces, but when this happens, they are not just randomly scrambled together, as implied by the computer's report that Seth and the fly have been "fused at the molecular-genetic level." Rather, in the accretion of genomes within a host organism, their given or pre-evolved structures are left intact but operationally coupled together to produce a higher-order transformation, a natural metamorphosis, of the living system at hand. Lynn Margulis, the biologist at the forefront of this crucial refinement in evolutionary thought, puts it like this: "Analogous to improvements in computer technology, instead of starting from scratch to make all new modules again, the symbiosis idea is an interfacing of preexisting modules. Mergers result in the emergence of new and more complex beings" (cited in Brockman 1995: 134).[8]

From this vantage we can see again the way that Cronenberg's *Fly* teeters on the conceptual divide between first- and second-order cybernetics, or again, between the demonization and the productive unfolding of paradoxical recursions. For in the final resolution of this recursive narrative arc, as the composite teleporter feeds the output of subsystems back into its input, both the organic and the mechanical systems at hand are not just merged into a consortium of semi-autonomous modules, but scrambled together into a horrific hash. This outcome is neither literally nor fictively necessary, but dictated only insofar as in the unfolding of *Fly* mythology up to this point, the posthuman remains ultimately foreclosed and the monster must die. Thus, we are given a choice between two equally exquisite nonviable options: and instead of the "ultimate family" of Brundlefly's nostalgic organic desire, we get the terminal fusion of the organic space bug with the mechanical telepod itself: "fusion of Brundlefly and telepod successful."

This narrative closes with the consummation of Brundlefly's death spiral within its/his own cybernetic loop: the Seth-sentience left within the scrambled creature makes its/his final self-referential gesture, crawling out of the receiver pod to aim the skeet-rifle in Ronnie's hands where its/his brains would go and effectively imploring her to render its/his mercy killing. But this conclusion also marks the family-romantic nostalgia that folds Cronenberg's *Fly* back upon Neumann's version of the conclusion, as the surviving mother and son go off into the sunset with Vincent Price as the father-surrogate. This *Fly* consumes itself by its own death drive to resolve its productive detour in the restoration of its prior narrative conditions. Thus we get the message that posthuman metamorphosis is inevitably a death trip, as the "sex appeal of the inorganic" trumps, for this round of the mythic cycle at least, the living systemic imperative of autopoietic continuation (Foster 1996). But the repetition compulsion works both ways, and one anticipates that media conditions within the social subsystems responsible for *The Fly* in the first place will, at some future moment, revive this *Fly* once again.

ACKNOWLEDGEMENTS

I would like to thank Bernd Hüppauf and Peter Weingart for their support, and Colin Milburn for his insightful comments on an earlier draft of this essay.

NOTES

1. *The Fly*, dir. Kurt Neumann, Twentieth-Century Fox (1958). I treat this film extensively in Clarke (2002).

2. *The Fly*, dir. David Cronenberg, Twentieth-Century Fox (1986). For information on many of its textual and production details, see Greg Kirkman's article, *The Annotated"Fly"*(1986) at http://www.littleredman.co.uk/fly-films/articles/annotated_fly.htm.

3 All citations are from this volume. The story is also available in Ackerman and Stine (1994).

4. "The constructing automaton is supposed to be placed in a reservoir in which all elementary compornents are floating, and it will effect its construction in that milieu" (316).

5. See "1730: Becoming-Intense, Becoming-Animal, Becoming Imperceptible...." (Deleuze and Guattari 1987: 232–309). The chapter begins with a reading of *Willard*, dir. Daniel Mann (1972). Rosi Braidotti gives Deleuzian readings of becoming-woman and becoming-machine in Neumann and Cronenberg's *Fly* (Braidotti 2002).

6. Helen W. Robbins has written well on the theme of womb envy in this film and Cronenberg's following production, *Dead Ringers* (Robbins 1993).

7. Accessed at http://www.humanistictexts.org/chuang.htm: adapted from Herbert A. Giles (1926). A shorter version of this passage is offered as an example of paradox in Patrick Hughes and George Brecht (1979: 35).

8. The interview with Margulis is available at: http://www.edge.org/documents/ThirdCulture/n-Ch.7.html. For an in-depth presentation of symbiogenesis, see Lynn Margulis and Dorion Sagan (2002).

BIBLIOGRAPHY

Ackerman, Forrest J. and Jean Stine (ed.) (1994) *Reel Future*, New York: Barnes & Noble.

Bal, Mieke (1997) *Narratology: Introduction to the Theory of Narrative*, 2nd ed., Toronto: University of Toronto Press.

Braidotti, Rosi (2002) *Metamorphoses, Towards a Materialist Theory of Becoming*, Cambridge, UK: Polity Press.

Brockman, John (ed.) (1995) *The Third Culture: Beyond the Scientific Revolution*, New York: Touchstone.

Clarke, Bruce (1995) *Allegories of Writing: The Subject of Metamorphosis* Albany, NY: State University of New York Press.

—— (2002). "Mediating the fly: Posthuman metamorphosis in the 1950s", *Configurations*, 10: 169–191.

Deleuze, Gilles and Felix Guattari (1987) *A Thousand Plateaus: Capitalism And Schizophrenia*, trans. and Foreword by Brian Massumi, Minneapolis, MN: University of Minnesota Press.

von Foerster, Heinz (2003) *Understanding Understanding: Essays on Cybernetics and Cognition*, New York, Berlin: Springer.

Foster, Thomas (1996) "The sex appeal of the inorganic: Posthuman narratives and the construction of desire", in Robert Newman (ed.) *Centuries Ends, Narrative Means*, Stanford, CA: Stanford University Press: 276–301.

Freeland, Cynthia A. (1996) "Feminist frameworks for horror films", in David Bordwell and Noël Carroll (eds.), *Post-Theory: Reconstructing Film Studies*, Madison, WI: University of Wisconsin Press.

Giles, Herbert A. (trans.) (1926) *Chuang Tzu: Mystic, Moralist, and Social Reformer*, London: Bernard Quaritch.

Hughes, Patrick and George Brecht (1979) *Vicious Circles and Infinity: An Anthology of Paradoxes*, New York: Penguin.

Kirkman, Greg (2006) *The Annotated Fly,* at http://annotatedfly1986.blogspot. com.

Knee, Adam (1992) "The metamorphosis of *The Fly*", *Wide Angle,* 14(1): 20–34.

Langelaan, George (1957) "The fly", *Playboy* (June 1957): 17–18, 22, 36, 38, 46, 64–68.

—— (1966) "The fly," in *The Playboy Book of Science Fiction and Fantasy,* Chicago: Playboy Press.

Latour, Bruno (1996) *Aramis or The Love of Technology,* trans. Catherine Porter, Cambridge, MA: Harvard University Press.

Margulis, Lynn and Dorion Sagan (2002) *Acquiring Genomes: A Theory of the Origins of Species,* New York: Basic Books.

von Neumann, John (1963) "The general and logical theory of automata", in A. H. Taub (ed.), *Collected Works,* 6 vols., New York: Macmillan, vol. 5: 288–318. http://annotatedfly1986.blogspot.

Pharr, Mary Ferguson (1989) "From pathos to tragedy: The two versions of *The Fly*", *Journal of the Fantastic in the Arts,* 2 (1): 37–46.

Robbins, Helen W. (1993) "'More human than I am alone': Womb envy in David Cronenberg's *The Fly* and *Dead Ringers*", in Steven Cohan and Ira Rae Hark (eds.), *Screening the Male: Exploring Masculinities in Hollywood Cinema,* New York: Routledge: 134–147.

Roth, Marty (2002) "Twice two: *The Fly* and *Invasion of the Body Snatchers*", in Jennifer Forrest and Leonard Koos (eds.), *Dead Ringers: The Remake in Theory and Practice,* Albany, NY: State University of New York Press: 225–241.

Wicke, Jennifer (1996) "Fin-de-siècle and the technological sublime", in Robert Newman (ed.), *Centuries' Ends, Narrative Means,* Stanford, CA: Stanford University Press: 302–315.

Wiener, Norbert (1950) *The Human Use of Human Beings: Cybernetics and Society,* Boston: Houghton Mifflin.

Contributors

Morana Alač is Assistant Professor of Communication and Science Studies at the University of California, San Diego. Her work concerns ways in which scientists study cognition in environments heavily sustained by advanced technologies. By looking at everyday work of science, Alač pays particular attention to the interface between the body and technology. University of California at San Diego, 9500 Gilman Drive 0503, La Jolla, CA 92093-0503, USA. alac@ucsd.edu http://hci.ucsd.edu/morana/.

Charlotte Bigg, PhD, is an historian of science at the Max Planck Institute for the History of Science in Berlin. She works on the uses of optical instrumentation in the physical sciences and in science popularization in the nineteenth and twentieth centuries. Her recent publications include "The Panorama, or La Nature à Coup d'Oeil," in E. Fiorentini (ed.), *Observing Nature—Representing Experience. The Osmotic Dynamics of Romanticism 1800–1850* (Reimer Verlag 2007); "L'Optique de Précision et la Première Guerre Mondiale," *Revue Suisse d'Histoire*, 55 (2005).Max Planck Institute for the History of Science, Boltzmannstrasse 22, D-14195 Berlin, Germany. bigg@mpiwg-berlin.mpg.de http://www.mpiwg-berlin.mpg.de/de/mitarbeiter/members/bigg

Lars Bluma, PhD, lecturer in the department of the history of technology and environment (Ruhr-Universität Bochum) since 1999. Received his doctorate in 2004 for a study of Norbert Wiener and the origins of cybernetics. Ruhr-Universität Bochum, Historisches Institut, Technik-und Umweltgeschichte, D-44780 Bochum, Germany. lars.bluma@ruhr-uni-bochum.de http://www.ruhr-uni-bochum.de/tug/bluma.html

Lisa Cartwright is Professor of Communication at the University of California at San Diego, where she is also appointed in the programs in Science Studies and Critical Gender Studies. Her books include *Screening the Body: Tracing Medicine's Visual Culture* (Minnesota 1995), *Practices of Looking: An Introduction to Visual Culture* (co-authored with Marita Sturken, Oxford 2002), and *Moral Spectatorship*, a book on technologies of voice, agency, and affect in postwar representations of child subjects, currently in press with Duke University. Department of Communication, University of California at San Diego, 9500 Gilman Drive 0503, La Jolla, CA 92093-0503, USA. lisac@ucsd.edu http://sciencestudies.ucsd.edu/Faculty/lcartwright.html.

Bruce Clarke is Professor of Literature and Science in the Department of English at Texas Tech University. His most recent book publications are *Energy Forms: Allegory and Science in the Era of Classical Thermodynamics* (Michigan, 2001), and co-edited with Linda D. Henderson, *From Energy to Information: Representation in Science and Technology, Art, and Literature* (Stanford, 2002). His essay in this volume is part of *Posthuman Metamorphosis: Narrative and Systems,*forthcoming from Fordham University Press. Department of English, Texas Tech University, Lubbock, TX 79409-3091, USA. bruce.clarke@ttu.edu http://www.faculty.english.ttu.edu/clarke/.

Eva Flicker, Ao. Universitäts-Professor Mag. Dr., sociologist. Since her habilitation at the University of Vienna in sociology of media and communication in 2006 she has been professor for sociology in the department for sociology. Her fields of research and publications comprise sociology of media, film and communication, visual sociology, sociological gender studies, sociology of organizations and groups. She is active member of the faculty conference for the Faculty of Social Sciences at the University of Vienna.Institute for Sociology, University of Vienna, Rooseveltplatz 2, A-1090 Vienna, Austria. eva.flicker@univie.ac.at http://www.univie.ac.at/soziologie.

Bernd Hüppauf, Professor Dr., studied German literature, philosophy, and history at the universities of Würzburg, Göttingen, and Tübingen. He received his PhD from the University of Tübingen in 1970. He has held positions at the Universities of Tübingen and Regensburg; University of New South Wales, Sydney (Australia). Since 1993 he has been Professor, Department of German at New York University. He was Director of Deutsches Haus at NYU until 2003. His research areas are: literature and culture of the Weimar Republic, literature and photography, literature and philosophical anthropology, aesthetics of skepticism. New York University, 19 University Place, level 3 New York, NY 10003, USA, and Hallerstrasse 27, D-10587 Berlin, Germany. bh4@nyu.edu http://www.huppauf.de.

Lutz Koepnick is professor of German, Film and Media Studies at Washington University in St. Louis. He is the author of *Framing Attention: Windows on Modern German Culture* (Johns Hopkins University Press,

2007), *The Dark Mirror: German Cinema between Hitler and Hollywood* (University of California Press, 2002), *Walter Benjamin and the Aesthetics of Power* (University of Nebraska Press, 1999), and of *Nothungs Modernität: Wagners Ring und die Poesie der Politik im neunzehnten Jahrhundert* (Wilhelm Fink Verlag, 1994). Koepnick is the co-author of *[Grid < > Matrix] / Screens Arts and New Media Aesthetics 1* (Mildred Lane Kemper Art Museum, 2006). He is also the co-editor of *Sound Matters: Essays on the Acoustics of German Culture* (Berghahn Books, 2004), of *Caught by Politics: Hitler Exiles and American Visual Culture* (Palgrave Macmillan, 2007) and of *The Cosmopolitan Screen: German Cinema and the Global Imaginary, 1945 to the Present* (University of Michigan Press, 2007). Department of Germanic Languages and Literatures, Washington University, St. Louis, MO 63130-4899 USA. koepnick@wustl.edu http://www.artsci.wustl.edu/~/koep.

Gabriele Leidloff works with video, film, photography, and image generating techniques. She conceptualized and launched the project *l o g - i n / l o c k e d o u t—A Forum of Art and Neuroscience*—under the patronage of UNESCO. Galleries, Museums, and universities in Europe and the USA have exhibited her installations. Essays on her art have been published in books such as *Video, ergo sum*, *Video cult/ures*, and *Theater der Natur und Kunst*, as well as in catalogues and magazines. Gabriele Leidloff lives in Berlin, Rolandufer 18, D-10179 Berlin, Germany. leidloff@t-online.de http://www.locked-in.com.

Dieter Mersch, Professor Dr., Study of mathematics and philosophy in Cologne, and Bochum, dissertation and habilitation at Technical University Darmstadt. 2000–2003: Prof. for Philosophy of Arts and Aesthetics at School of Arts in Kiel, since 2004: Professor for Media Studies and Director of the Department for "Media and Arts" at the University of Potsdam. Main publications: *Umberto Eco zur Einführung* (Hamburg 1993), *Was sich zeigt. Materialität, Präsenz, Ereignis* (München 2002), *Ereignis und Aura. Untersuchungen zur einer Ästhetik des Performativen* (Frankfurt/M 2002), editor, *Die Medien der Künste. Beiträge zur Theorie des Darstellens* (München 2003), (editor with Jens Kertscher): *Performativität und Praxis*(Paderborn 2003); *Medientheorien zur Einführung* (Hamburg 2006). University of Potsdam, Am Neuen Palais, Haus 1, D-14415 Potsdam, Germany. dmersch@rz.uni-potsdam.de http://www.dieter-mersch.de.

Colin Milburn is Assistant Professor of English and a member of the Science & Technology Studies Program at the University of California, Davis. His research focuses on the cultural relations between science, literature, and media technologies. His forthcoming book about the onrushing era of nanotechnology is entitled *Nanovision: Engineering the Future* (Duke

University Press). Department of English, University of California, Davis, One Shields Avenue, Davis, CA 95616, USA cnmilburn@ucdavis.edu. http://wwwenglish.ucdavis.edu/faculty/milburn/.

W.J.T. Mitchell is the Donnelley Distinguished Service Professor of English and Art History at the University of Chicago, and the editor of *Critical Inquiry*. His books include *Iconology* (1986), *Picture Theory* (1994), *The Last Dinosaur Book* (1998), and *What Do Pictures Want?* (2005), which recently won the James Russell Lowell Prize of the Modern Language Association for a book on language and literature. He has edited numerous collections of essays, including *Art and the Public Sphere* (1994), *Landscape and Power* (2nd edition, 2003), *Edward Said: Continuing the Conversation* (2004), and *The Late Derrida* (2007). He is currently completing a book entitled *Cloning Terror: The War of Images, 9-11 to Abu Ghraib*. University of Chicago, 1050 East 59th St., Chicago, IL 60637, USA. wjtm@uchicago.edu http://humanities.uchicago.edu/faculty/mitchell/.

Sybilla Nikolow, Dr., is an historian of science at the Institute for Science and Technology Studies at Bielefeld University. She works on the history of popularization and visualization in science, especially on the relationship between science and the public in museum exhibitions in the twentieth century. English publications (selection): "'Gesellschaft und Wirtschaft.' An encyclopedia in Otto Neurath's pictorial statistics from 1930," in W. Boyd Rayward (ed.), *European Modernism and the Information Society*, London, in print; "Planning, democratization and popularization with ISOTYPE, ca. 1945. A Study of Otto Neurath's pictorial statistics on the example of Bilston, England," in Friedrich Stadler (ed.), *Induction and Deduction in the Sciences*, Vienna Circle Yearbook 11 (2003), Dordrecht 2004: 299–329; "Displaying the invisible: *Volkskrankheiten* on exhibition in Imperial Germany," *Studies in History and Philosophy of Biological and Biomedical Sciences*, 31:511–530 (together with Christine Brecht) (2000). Institute for Science and Technology Studies (IWT), University of Bielefeld, PO Box 10 01 31, D-3501 Bielefeld, Germany. nikolow@iwt.uni-bielefeld.de http://www.uni-bielefeld.de/iwt/nikolow.

Petra Pansegrau has a master's degree in linguistics, literature, and media studies and a doctorate in linguistics. Her research interests are Public Understanding of Science, Metaphors in Science Journalism, Discourse Analysis, Mass Media analysis, and Universities in a Public Sphere. Currently she is managing the Master's Programme *Interdisciplinary Media Studies* at Bielefeld University. English publications (selection): Weingart, P., A. Engels, and P. Pansegrau (2000): "Risks of communication: discourses on climate change in science, politics, and the mass media," *Public Understanding of Science*, 9: 261–283; Weingart, P. and P. Pansegrau (1999):

"Reputation in science and prominence in the media—The Goldhagen debate," *Public Understanding of Science*, 8: 1–16.Institute for Science and Technology Studies (IWT), University of Bielefeld, PO Box 10 01 31, Universitätsstrasse 25, D-33615 Bielefeld, Germany. petra@iwt.uni-bielefeld.de http://www.uni-bielefeld.de/iwt/personen/pansegrau.

Joachim Schummer is Heisenberg Research Fellow at the University of Darmstadt. He graduated both in chemistry and philosophy and received his PhD and Habilitation in philosophy from the University of Karlsruhe. He has had visiting positions at the University of South Carolina and the Australian National University. His research interests focus on the history, philosophy, sociology, and ethics of science and technology. His recent book publications include *Discovering the Nanoscale* (2004, 2nd ed. 2005), *Nanotechnology Challenges* (2006), and *Nanotechnologien im Kontext* (2006). Department of Philosophy, University of Darmstadt, Schloss, D-64283 Darmstadt, Germany, js@hyle.org http://www.joachimschummer.net/.

Wolf Singer, Prof. Dr. Dr. h.c., born 1943, studied medicine in Munich and Paris, obtained his MD from the Ludwig-Maximilians-University in Munich, and his PhD from the Technical University in Munich. Since 1981 he has been Director of the Max Planck Institute for Brain Research in Frankfurt, and in 2005 he founded the Frankfurt Institute for Advanced Studies (FIAS). His research is focused on the neuronal substrate of higher cognitive functions, and especially on the "binding problem." Most functions of the brain are based on parallel computations in widely distributed neuronal networks. How these distributed subprocesses are coordinated and bound together in order to give rise to coherent percepts and eventually conscious awareness, is a central question of current research. Max Planck Institute for Brain Research, Department of Neurophysiology, Deutschordenstr. 46, D-60528 Frankfurt am Main, Germany. singer@mpih-frankfurt.mpg.de http://www.mpih-frankfurt.mpg.de/global/Np/Staff/singer_d.htm.

Tami Spector is a professor of organic chemistry at the University of San Francisco. Trained as a physical organic chemist her scientific work has focused on fluorocarbons, the transformations of strained ring organics, and the molecular dynamics and free energy calculations of biomolecular systems. She also has a strong interest in aesthetics and chemistry and has published and presented work on *The Molecular Aesthetics of Disease, John Dalton and the Aesthetics of Molecular Representation*, and *The Visual Image of Chemistry*. Chemistry Department, University of San Francisco, 2130 Fulton Street, San Francisco, CA 94117-1080, USA, spector@usfca.edu http://artsci.usfca.edu/servlet/ShowEmployee?empID=188&deptID=2.

Peter Weingart, Professor Dr., holds a chair in sociology (sociology of science and science policy) at Bielefeld University, Germany. He is director of the Institute for Science and Technology Studies (IWT), former director of the Institute of Interdisciplinary Research (ZiF) and member of the Berlin-Brandenburg Academy of Sciences. He has published numerous articles and books in the sociology of science, among them *Die Stunde der Wahrheit?* (2001), *Die Wissenschaft der Öffentlichkeit* (2005), and *Metaphors and the Dynamics of Knowledge* (with S. Maasen,2000). Institute for Science and Technology Studies (IWT), University of Bielefeld, PO Box 10 01 31, D-33501 Bielefeld, Germany. weingart@uni-bielefeld.de http://www.uni-bielefeld.de/iwt/pw/.

Index